RECLAIMING BROWNFIELDS

Global Urban Studies

Series Editor:
Laura A. Reese, Michigan State University, USA

Providing cutting edge interdisciplinary research on spatial, political, cultural and economic processes and issues in urban areas across the US and the world, volumes in this series examine the global processes that impact and unite urban areas. The organizing theme of the book series is the reality that behavior within and between cities and urban regions must be understood in a larger domestic and international context. An explicitly comparative approach to understanding urban issues and problems allows scholars and students to consider and analyze new ways in which urban areas across different societies and within the same society interact with each other and address a common set of challenges or issues.

Books in the series cover topics which are common to urban areas globally, yet illustrate the similarities and differences in conditions, approaches, and solutions across the world, such as environment/brownfields, sustainability, health, economic development, culture, governance and national security. In short, the *Global Urban Studies* book series takes an interdisciplinary approach to emergent urban issues using a global or comparative perspective.

Reclaiming Brownfields
A Comparative Analysis of
Adaptive Reuse of Contaminated Properties

Edited by

RICHARD C. HULA
Michigan State University, USA

LAURA A. REESE
Michigan State University, USA

CYNTHIA JACKSON-ELMOORE
Michigan State University, USA

Routledge
Taylor & Francis Group

LONDON AND NEW YORK

First published 2012 by Ashgate Publishing

2 Park Square, Milton Park, Abingdon, Oxon OX14 4RN
711 Third Avenue, New York, NY 10017, USA

Routledge is an imprint of the Taylor & Francis Group, an informa business

First issued in paperback 2016

British Library Cataloguing in Publication Data
Reclaiming brownfields : a comparative analysis of adaptive
    reuse on contaminated properties. -- (Global urban studies)
    1. Brownfields--Case studies. 2. Hazardous waste site
    remediation--Case studies.
    I. Series II. Hula, Richard C., 1947- III. Reese, Laura A.
    (Laura Ann), 1958- IV. Jackson-Elmoore, Cynthia, 1965-
    363.7'3966-dc23

Library of Congress Cataloging-in-Publication Data
Reclaiming brownfields : a comparative analysis of adaptive reuse of contaminated properties / [edited] by Richard C. Hula, Laura A. Reese, and Cynthia Jackson-Elmoore.
    p. cm. -- (Global urban studies)
    Includes bibliographical references and index.
    ISBN 978-1-4094-4958-4 (hbk.)
    1. Urban renewal. 2. Urban renewal--Case studies. 3. Brownfields. 4. Land use, Urban.
5. Environmental policy. I. Hula, Richard C., 1947- II. Reese, Laura A. (Laura Ann), 1958-
III. Jackson-Elmoore, Cynthia, 1965-
    HT170.R425 2012
    307.3'416--dc23

                                                                        2012007260

ISBN 978-1-4094-4958-4 (hbk)
ISBN 978-1-138-26706-0 (pbk)

Contents

PART I: POLICY

PART II: IMPLEMENTATION AND EVALUATION

List of Figures

List of Tables

List of Contributors

Rob Alexander is Visiting Assistant Professor in the STS/Public Policy Department at the Rochester Institute of Technology in Rochester, NY, where he conducts research on the individual and organizational capacities required for local governments and their members to implement long-term, sustainability-themed public policies. Rob teaches courses in environmental policy, public administration, and place-based interdisciplinary environmental problem solving.

Laurel Berman (B.S., M.S. and PhD Environmental and Occupational Health Sciences) is the Brownfields Coordinator for the Agency for Toxic Substances and Disease Registry. She is adjunct faculty at the University of Illinois at Chicago School of Public Health, Environmental and Occupational Health Sciences Division.

Philip Catney is a Lecturer in British Politics at Keele University in the West Midlands, England. His teaching and research focus on urban regeneration and environmental policy, in addition to British politics, governance, and public policy theory. Prior to coming to Keele in 2007, Dr. Catney held research associate positions in both the departments of Politics and of Town and Regional Planning at the University of Sheffield, where he also managed the Sustainable Brownfield Regeneration: Integrated Management (SUBR:IM) consortium. He earned a PhD in Politics from the University of Sheffield in 2005.

Tiziana Cianflone is an Economist at the Institute for Environmental Protection and Research (ISPRA) in Rome, Italy, where her work focuses on the economic valuation of environmental damage, market-based instruments for environmental policy, and institutional and economic tools to redevelop brownfields. Dr. Cianflone also consults for private research centers and serves as an expert in heritage conservation and environmental economics and policy. She received her doctorate from the University of Rome III, where she has taught microeconomics and regional and urban economics.

Christopher A. De Sousa (B.A., MSc.Pl., PhD) is Associate Professor and Director of the School of Urban and Regional Planning at Ryerson University. Prior to joining Ryerson, he was the Chair of Urban Planning at the University of Wisconsin-Milwaukee, as well as a member of the Geography and Urban Studies faculty. Professor De Sousa's research activities focus on various aspects of brownfield redevelopment, urban environmental management, and sustainability reporting in Canada and the United States.

Fabian Dosch has a degree in Geography Studies with a geophysical orientation and a PhD in Natural Sciences in Germany. His current position is Senior Project Manager for land use monitoring and policies, regional and local adaptation policies on climate change; Federal Institute for Research on Building, Urban Affairs and Spatial Development, a superior Federal authority commissioned to give policy advice for the Federal Ministry of Transport, Building and Urban Affairs (BMVBS). He is mainly engaged in the development of sustainable land use strategies, project coordination and advice.

Mark L. Gillem (PhD, AIA, AICP) is Associate Professor of Architecture, University of Oregon. Gillem addresses sustainability, social responsibility, and historic preservation through his teaching, research, and professional practice. He is the author of *America Town: Building the Outposts of Empire* (University of Minnesota Press, 2007), which received the Environmental Design Research Association's Book Award in 2008. His planning and urban design work has received four national planning awards from the American Planning Association's Federal Planning Division for projects in Aviano, Italy, Iwakuni, Japan, Tacoma, Washington, and Kunsan, South Korea. Other honors include a Design Excellence Award from the U.S. Air Force and the Crocker Award for Teaching Excellence at the Air Force Institute of Technology. His diverse portfolio also includes a renovation honored by the Berkeley Architectural Heritage Association (California) and pro bono designs for projects in New Guinea, Guatemala, and Indonesia. Gillem serves as the Director of the International Association for the Study of Traditional Environments and Director of the University of Oregon's Urban Design Lab. He holds a PhD in Architecture and a Master's in Architecture from the University of California, Berkeley and a Bachelor's in Architecture with Highest Distinction from the University of Kansas. He is a licensed architect and certified planner.

Detlef Grimski has a degree in Civil Engineering. His current position is Senior Project Manager for land management and brownfield redevelopment; Federal Environment Agency Germany, a superior Federal authority commissioned to give policy advice for the Federal Ministry for the Environment, Nature Protection and Nuclear Safety. He is mainly engaged in the development of sustainable land management and brownfield redevelopment strategies in the context of environmental planning and sustainability strategies.

Justin B. Hollander is an Assistant Professor in the Department of Urban and Environmental Policy and Planning at Tufts University. His research and teaching is in the area of land use and urban redevelopment, with a focus on the changing physical form of cities in post-industrial North America. His first book *Polluted and Dangerous: America's Worst Abandoned Properties and What Can Be Done About Them*, was published in 2009 by the University of Vermont Press.

Han Hongyun is a Professor at Zhejiang University, China. Dr. Han has over 20 years teaching and research experience. She has taught courses in Natural Resources and Environment Economics, Econometrics, Intermediate Macroeconomics, and Public Administration. Her areas of interest and expertise include Economics, Resources and Environment Management, Agricultural Economy and Rural Development, and Public Administration. As a researcher who is focusing on natural resource and environmental management, she has lead nine related projects and published 40 papers related to the topics of land and water resource management.

Richard C. Hula is Professor and Chair of the Department of Political Science at Michigan State University. He previously taught at the University of Texas-Dallas and the University of Maryland. He has served as president of the Policy Studies Association (1997–1998) and was selected as a Distinguished Scholar Teacher by the University of Maryland. Hula's research and teaching interests focus broadly on environmental policy and urban politics. Current projects include changing state/local environmental policy, the impact of faith-based organizations on public service delivery, and the political consequences of state level interventions into local policy arenas. Books published include *Market-Based Public Policy* (New York: Macmillian, 1988), *The Color of School Reform* with Jeffrey Henig, Marion Orr and Desiree Pediscleaux (Princeton: Princeton University Press, 1999) and *Nonprofits in Urban America* with Cynthia Jackson-Elmoore (Westport, CT: Quorum Press, 2000). *The Color of School Reform* received a best book of the year award from the urban politics section of the American Political Science Association.

Cynthia Jackson-Elmoore is Dean of the Honors College at Michigan State University (MSU), and a Professor with faculty appointments in Social Work and Political Science and an affiliation with the Global Urban Studies Program. Her research and teaching interests and publications focus broadly on urban politics, public policy processes, and public services. Dr. Jackson-Elmoore previously served as Acting Assistant Dean of the Urban Affairs Programs, Director of the Urban Studies Graduate Program in the College of Social Science, and Co-Director of the Program in Urban Politics and Policy at Michigan State University. Dr. Jackson-Elmoore previously taught courses in urban politics, public policy processes and analysis, and social welfare policy and services; conducted evaluations on community health care reform and was project manager for a multi-state study;

and was a liaison to a national organization (Influencing State Policy) dedicated to engaging students in state politics. She has served on numerous undergraduate honors thesis, master's thesis and doctoral dissertation committees; and supervised several undergraduate and graduate research assistants. Jackson-Elmoore received a Lilly Endowed Teaching Fellowship in 1999 focused on innovation, scholarship and expertise in teaching. In 2003–2004, she was a fellow in the Academic Leadership Program sponsored by the Committee on Institutional Cooperation (CIC); and in 2008–2009 she was a fellow in the Executive Leadership Academy Program.

Robert A. Jones, PhD is an Associate Professor of Urban and Regional Planning in the Department of Geography and Geology at Eastern Michigan University. He holds advanced degrees in architecture and in urban studies and specializes in urban revitalization, land development regulation, and urban design.

Herbert Klapperich has a degree in Civil Engineering and a PhD in Earthquake Engineering. From 1984 to 1985 he was a lecturer at MIT, Cambridge. Currently he is a Consulting Engineer and University Professor at the Geotechnical Institute, Technical University Freiberg, Germany. Klapperich is on the Board of Directors CiF e. V. (Interdisciplinary Center of Excellence for Brownfields), Freiberg, Berlin, Aachen, Germany.

Courtney E. Knapp is an Economic Development and Housing Planner with the Northern Middlesex Council of Governments, a regional planning agency in Lowell, Massachusetts. She holds master's degrees in Urban and Environmental Policy and Planning from Tufts University and Gender/Cultural Studies from Simmons College, and has worked as a researcher for the Lincoln Institute of Land Policy, Project for Public Spaces, and several community development corporations in the Boston area. Her research interests include the intersection of land use planning and environmental justice, linking value capture tools to urban redevelopment projects, and community-based brownfield redevelopment.

Zhao Liange is a Professor at the College of Economics, Zhejiang Gongshang University. Dr. Zhao has over 20 years teaching and research experience. He has taught courses in Industrial Economics, Intermediate Microeconomics, and Development Economics. His areas of interest and expertise include Economics, Resources and Environment Management, and Agricultural Economy and Rural Development. As a researcher who is focusing on economics, he has led five related projects and published 30 papers on the topic of land and water resource economics.

Terri Linder (B.S., R.S.) is an Environmental and Disease Control Specialist with the City of Milwaukee Health Department. She has worked in public health for over 20 years, currently in environmental health including industrial hygiene, air and water quality, hazardous materials response and risk assessment.

Elizabeth A. Lowham is an Assistant Professor of Political Science and Director of the Masters in Public Administration Program at California Polytechnic State University, San Luis Obispo. Her research interests include leadership in collaborative policy making and environmental politics and policy. She graduated from the University of Colorado, Boulder in 2007. She has published work on brownfields and terrorism in *The Journal of Forestry* and *Defense and Security Analysis*.

David Misky (B.S., M.S.) is the Assistant Executive Director of the Redevelopment Authority of the City of Milwaukee. He has over 14 years of experience managing brownfields and has led the City of Milwaukee brownfield team in redeveloping over 100 properties retaining or creating over 3,000 jobs.

Laura A. Reese is Professor of Political Science and Director of the Global Urban Studies Program (GUSP) at Michigan State University. Dr. Reese was previously on the Urban Planning faculty at Wayne State University. She has a PhD in Political Science and a Masters of Public Administration both from Wayne State University. Her main research and teaching areas are in urban politics and public policy, economic development, and local governance and management in both Canada and the US. She has conducted large scale evaluations for the Economic Development Administration and of sub-state economic development programs including Tax Increment Finance Authorities and Industrial Tax Abatements. Reese has written or edited nine books and over 90 articles and book chapters in these areas as well as public personnel administration focusing on the implementation of sexual harassment policy. She is the Editor-in-Chief of the *Journal of Urban Affairs* and conducts training for local and state government officials on economic development incentives.

Petra Rydvalova (PhD) was born in Liberec (the Czech Republic). She graduated from Engineering and Textile College in Liberec in 1991 (currently the Technical University of Liberec) with a specialization in Economics and Management of Consumer Goods Industry. She is at present (2009/2010) Assistant Professor in the Department of Business Administration of the Faculty of Economics at the Technical University of Liberec, Czech Republic. Her main areas of education and research are: Regional disparities in the Czech Republic; Brownfield sites redevelopment; and SME's (Small and medium sized enterprises) business environment. Rydvalova has completed several internships abroad. In her scientific and innovative activities she has become involved in many projects of the Ministry for Regional Development, the Ministry of Education and Grant Agency of the Czech Republic. She also coordinated some projects at the TUL; from 2004–2006 she coordinated research projects of the Sixth Framework Programme of the European Union ENVITEX: Innovative environment of textile industry in accessing candidate countries; 2006–2008 she coordinated the innovative project financed from the European structural funds (ESF) "Regional Centre for Education in the Area of Technological Transfer." From 2009 to 2012 she is coordinating two projects of ESF: "Technological and Economic Competencies for the European Research Area" (TE-ERA) and "Development of Communication Skills in Science with the Use of Model Pilot Project of NANO" (MUNRO).

Jill A. Schreifer works with Teach for America in Washington, DC. She holds a Bachelor of Science in Engineering from Duke University and a Masters in Business Administration from Emory University.

William Welsh, PhD is an Associate Professor in the Department of Geography and Geology at Eastern Michigan University. He specializes in land use/land cover analysis, remote sensing, and the application of geospatial technologies to planning and environmental issues.

Kris Wernstedt is an Associate Professor in the School of Public and International Affairs at Virginia Tech. His work covers a variety of issues in environmental planning and policy, with emphases on climate change and variability, contaminated properties, water resources, and US regulatory and legislative reform. He received his PhD and Masters of Regional Planning from Cornell University and a MS in Water Resources Management from the University of Wisconsin. Prior to arriving at Virginia Tech in 2006, he spent 15 years at Resources for the Future in Washington, DC.

Katie Williams (BA (Hons), DipUP, DipUD, MSc (Env Des), PhD MRTPI) is an Urban Theorist, Planner and Urban Designer. She is Director of the Centre for Sustainable Planning and Environments (SPE) at the University of the West of England, Bristol. Professor Williams specializes in sustainable urban environments and is known for her work on sustainable neighborhood design (in relation to sustainable behaviors and climate change adaptation); urban form (compact cities) and land reuse (brownfield development). She has undertaken evidence-based critiques of many key urban policies such as sustainable communities and the urban renaissance. Professor Williams has over £4 million in research grants from UK research councils, government agencies and industry, and has undertaken consultancy work for central and local governments, and regional and national development agencies. She is currently undertaking EPSRC-funded research on adapting UK suburbs for climate change as part of the "Living with Environmental Change" Program. Professor Williams has held visiting lectureships in the USA, Thailand, Peru, and The Netherlands. She has authored over 100 academic papers and reports and edited three books on sustainable urbanism. She regularly presents her work at conferences in the UK and internationally. Professor Williams also holds a number of advisory positions: she is a member of a Strategic Advisory Team for the Engineering and Physical Sciences Research Council; a member of the International Advisory Board for The Stockholm Centre for Sustainable Communications; a panel member of the Swedish and Dutch Environmental Research Programmes; and a member of the Board of the International Urban Planning and Environment Association.

Andrea E. Yang is an Attorney with the law firm of Strauss & Troy in Cincinnati, Ohio with her practice focusing on real estate, zoning and land use, local government, and administrative law. She is a graduate of the University of Cincinnati School of Law, and has earned a Master's in Regional Planning from Cornell University and AICP certification. Ms. Yang has also published "Historical Criminal Punishments, Punitive Aims and Un'-Civil' Post-Custody Sanctions on Sex Offenders: Reviving the Ex Post Facto Clause" in the *University of Cincinnati Law Review* 75(3), Spring 2007.

Miroslav Zizka (PhD) was born in Liberec, the Czech Republic. He graduated from the Technical University of Liberec in 1997 with a specialization in Business Administration. In 2002 he successfully completed his doctoral studies in the field of organization and enterprise management. In 2007 he became an Associate Professor of Business Administration and Management. At present he also works as a Vice Dean for Science and Research of the Faculty of Economics, Technical University of Liberec (TUL). He is also a Visiting Professor at the University of Skoda Auto in Mlada Boleslav. Since 2003 he has been an Executive Editor of a scientific journal *E+M Economics and Management*. His publications include more than 90 monographs, study materials, and articles in journals, proceedings and reviews. His main research focuses on regional policy, economics and operation management. Between 2004 and 2009 he completed several internships at German universities (Technical University of Applied Sciences Wildau, University of Applied Sciences Ansbach). He is a member of the Scientific Board of the Faculty of Economics of TUL and a member of the Accreditation Committee of the Ministry of Education, Youth and Sports of the Czech Republic.

Acknowledgments

There are a number of people whose assistance has been invaluable in the development and completion of this project. First, we are indebted to the professionals and researchers who have contributed to this volume. Their work has endeavored to address the intractable issue of environmental degradation and alternative land uses which we believe will ultimately lead to a more sustainable urban environment. We would also like to thank graduate students Rebecca Bromley-Trujillo and Maranda Holtsclaw for their skilled assistance in literature review and in editing endless versions of the chapters. Anna Graham served as our expert and long-suffering editor for the project ensuring that everything was uniform and literate. Our colleague, Professor Kenneth Williams provided assistance with game theoretical models.

Dedication to:

- Catherine for support and patience. To Lauren and Aaron whose fieldwork contributed to my understanding of brownfield redevelopment, RCH.
- My "sisters" who help keep me sane: Trenda Baldridge, Cindy Bell, Sue Durkee, Nancy Herman, Cynthia Jackson-Elmoore, and Kellee Remer, LAR.
- Barbara O. Jackson, CJE.

PART I
Policy

One need not be an urban policy scholar to recognize that the environmental legacy of past industrial and agricultural development can simultaneously post serious threats to human health and impede reuse of contaminated land. The urban landscape in particular, is littered with sites contaminated with a variety of toxins produced by past use. Not surprisingly both public and private sector actors are often reluctant to make significant investments in properties that pose potential human health issues, and may demand complex and very expensive cleanups. And, environmental contamination is global in scope since pollution, development and land-use patterns are seldom constrained by national borders.

Environmental contamination is common in cities in both the developed and developing world. Even those cities most successful in adapting to a changing global economy continue to face significant challenges, including land area in which redevelopment is severely limited by environmental contamination from earlier usages. Environmental contamination raises several policy concerns. One, is the public health implications of contamination, particularly in relation to the confluence of equity, geographic location, poverty, and race (i.e. environmental justice). Another concern is the need to redevelop contaminated sites to reduce health risks and enhance neighborhood economic development. Efforts to encourage the redevelopment of such sites have become an important theme in both environmental and urban policy as many nations have implemented new programs to redevelop land parcels with contamination. Efforts to reintegrate contaminated land into the productive economy involve not only political and economic issues, but raise fundamental questions of environmental science. Examples range from the efficacy of remediation and containment strategies, to the proper identification of allowable contamination levels.

A review of existing literature suggests a number of important questions that serve as the focus of this book

Question 1: What do policies look like? How are programs organized? Does it matter?

There are fundamental differences in how public authorities organize efforts to reclaim contaminated properties. For example, such efforts in the United States have historically been centralized at the federal level, and have operated on what students of administration refer to as a command and control decision making process. Indeed, such systems of environmental regulation exist throughout the world. Recently, however, this system has been challenged in the United States by very aggressive state level efforts emphasizing public-private partnerships and voluntary actions. Typically the role of the federal Environmental Protection Agency is nearly irrelevant in many redevelopment efforts. Intergovernmental systems differ and the US provides only one model. The chapters in this volume illustrate how urban brownfield challenges are treated in other countries and include explicit cross-national comparisons.

Question 2: What strategies are used to promote redevelopment?

Efforts to return contaminated land to productive economic use typically rely on significant private investment. Public funding is often reserved for site cleanup efforts when no responsibility party can be identified, and subsidies to reduce development costs below those of alternative sites are required. The particular form of development subsidies varies a great deal ranging from infrastructure construction, tax relief or direct payment. The number and complexity of these mechanisms makes generalizations difficult. Considering policy options on a global level extends the array of both existing and considered policy options.

Question 3: What is the role of community preferences?

Cleanup and redevelopment efforts are often a critical concern of the communities in which the properties are located. However, there tends to be little formal requirement for community or neighborhood involvement in redevelopment planning. Most communities, nevertheless, attempt to create a forum for some neighborhood level participation. Usually this participation provides for neighborhood review rather than initiation of projects. Like health care and public health, redevelopment planning is a complex, technical issue that often requires more than lay knowledge to adequately grasp the nuances of policy decisions. As such, the extent to which meaningful citizen involvement can be sustained and community preferences incorporated into decision-making processes is likely to vary by project, community, and national context. This raises the question of the appropriate policy role of citizens most directly impacted by brownfield projects. To what extent should local communities be involved in local redevelopment efforts, and how is the desired participation level best achieved?

These questions are addressed in chapters written from perspectives around the globe. The book itself is organized into three parts. Part I focuses on extant public policy related to brownfields through a global lens. The chapters illustrate policies, intergovernmental arrangements, relations between the public and private sectors, and provide historical background on these issues in a variety of national contexts. The Chapters in Part II provide detailed examples of the implementation of brownfield policies specifically addressing questions of community preferences and participation. They also extend the discussion of the organizational arrangements through which policies are implemented and evaluated. The final section presents case studies of specific brownfield efforts cutting across the three questions just noted. The set of US cases show different organizational arrangements and policy solutions in action, specifically questioning how outcomes might differ for competing interests and communities.

Overall, the book provides academics and policy-makers with a global perspective on an increasingly critical health and development issue by providing answers to some of the important questions related to the development, implementation, and evaluation of brownfield remediation and redevelopment policy.

Policy Variation and Diffusion

As with most substantive polices, efforts to identify, clean, and redevelop contaminated properties have taken a number of forms across different jurisdictions. There are a great number of factors that would lead one to expect such a wide variance. Policy scholars have identified a myriad of factors impacting policy outcomes including context, institutional structure, public and elite preferences, culture and history. Such factors are recognized and discussed in several chapters. In an analysis of the impact of a variety of forces on US state programs, Hula identifies the severity of contamination problems as the strongest predictor of whether a state has a robust brownfield program. Rydualova and colleagues also emphasize the importance of context for the Czech Republic. They note the key historical role of a centralized planned economy and the contemporary impact of multinational actors in Czech domestic environmental policy. In addition to such traditional endogenous factors, the existence of various models and policy types serve as independent variables that can impact the design of policy. Interestingly such pressures are likely to make policy more similar as jurisdictions observe (and perhaps learn from) each other. The possible importance of such proximity is demonstrated in Hula's review of the distribution of brownfield innovation in the United States. This analysis identifies a relatively strong geographic clustering of states active in brownfield redevelopment. Although the specific chapters in this section take idiosyncratic perspectives on brownfield policy, all of the authors provide some general support for the diffusion hypothesis; at least in the United States and Europe, there has been a significant policy convergence in the past decade. The following section on policy implementation and evaluation also highlights the diffusion trend.

Policy Characteristics

A significant effort in this section is devoted to describing the parameters of brownfield policy in a variety of settings. It is hardly surprising that there is a good deal of variation in how brownfield policy is defined and implemented across political jurisdictions. This is true not only at a national level, but also at the state level in the United States. Note, however, that while program characteristics certainly differ, broad patterns can be identified. Lowham documents such patterns in US state policies by applying cluster analysis to attributes of a number of state level programs, and in this way identifies a limited number of recurring program structures. Overall, program types can be described on several independent dimensions.

Definition

Policy scholars often argue that one must identify the problem definition incorporated into policy to understand the policy process (Jones and Baumgartner 2005; Portz 1996; Rochefort and Cobb 1994). Of course, problem definition is seldom straightforward. Typically problem definitions are complex, incomplete, and implicit. One important indicator of the complexity of public policy surrounding brownfield cleanup and redevelopment is the significant variation in how brownfields are defined in different settings. Although the term brownfield is widely used, the specific meaning attached to the term depends very much on the context in which it used. Perhaps the most common definition cited is that championed by the United States Environmental Protection Agency. The EPA defines brownfields as:

> real property, the expansion, redevelopment, or reuse of which may be complicated by the presence or potential presence of a hazardous substance, pollutant, or contaminant (Kaiser 1998).

The United States EPA definition (or slight variants of it) has been cited by a number of authors in this volume. Alternate definitions do, however, exist. For example, CABERNET (Concerted Action on Brownfield and Economic Regeneration Network), a European expert network active in brownfield redevelopment defines brownfields as sites which (CABERNET 2010):[1]

---

1   The European Union (EU) is taking an emerging role in environmental matters. To date, the EU has taken a relatively modest role in brownfield redevelopment. It does, however, play a role as a central coordinating and information-sharing body. It supports a number of information networks including NICOLE (the Network for Industrially Contaminated Land in Europe), CLARINET (the Contaminated Land Rehabilitation Network), ANCORE (Academic Network for Contaminated Land Research in Europe), and CABERNET (Concentrated Action on Brownfields and Economic Regeneration).

- have been affected by former uses of the site or surrounding land,
- are derelict or underused,
- are mainly in fully or partly developed urban areas,
- require intervention to bring them back to beneficial use,
- may have real or perceived contamination problems.

The apparent consistency across these definitions is deceiving. On closer examination the seeming consensus on what constitutes a brownfield disappears. Within the United States, individual states have adopted quite different legal definitions of a brownfield. In some states, for example, properties with obsolete structures that may retard redevelopment are considered brownfields, whether or not the site suffers from environmental contamination.[2] Levels of permissible contamination also vary over time and across state boundaries. Indeed, a number of state "reforms" have been based on adjusting permissible toxic levels for a variety of substances. The effect of such a redefinition is, of course, to reduce the number of brownfields without any change in what is actually to be found in the ground or water.

Variation in what constitutes a brownfield becomes even greater when one moves to a cross-national view of programs. Perhaps most important is the absence of the implicit assumption contained in US policy that while brownfields can have substantial contamination, the most toxic sites are in a completely separate category. The federal government through the authority provided in the Comprehensive Environmental Response, Compensation, and Liability Act of 1980 (CERCLA) has responsibility for these sites.[3] Thus in the United States the term brownfield is generally not applied to sites that raise the most serious threats to human health. This is not the case in much of the world where the term is applied to a much broader class of sites, including those most heavily contaminated and most dangerous.[4]

Technical issues also contribute to debates over the appropriate definition of a brownfield. There are, for example, no universally recognized criteria for what constitutes contamination. All of the chapters in this section discuss the importance of having target sites reach some minimum or acceptable level of contamination. It is the responsibility of individual jurisdictions to operationalize what constitutes an "acceptable" level of toxins in the environment, however. Differences are almost inevitable given a lack of a clear consensus within the scientific community

2   The state of Michigan provides an example of a legal definition of brownfield that includes criteria other than site contamination. See Welsh and Jones in this volume.

3   For an overview of the development of environmental policy in the United States see Hula (2001).

4   Note, however, there is no single operational definition used in Europe. Different Europeans use a wide variety of definitions, some of which (like individual American states) do not require on-site contamination. See Oliver et al. 2005.

as to what sort of standards are necessary to protect human health.[5] Thus, a site that is formally a brownfield in one jurisdiction may not be a brownfield at all in another jurisdiction. As a result, efforts to compare program outcomes in the two jurisdictions would be very difficult (Donati, Rossi and Brebbia 2004; Whelan 2004).

Program Goals

Public policy is by its very nature a complex business. Government decisions are inevitably driven by multiple goals and priorities. Not surprisingly government programs almost always reflect this complexity. All of the chapters in this section illustrate the fact that brownfield programs have multiple goals. To be sure, much of the initial impetus for a brownfield policy was a desire to minimize potential threats to human health from ground and water contamination. There is, however, broad consensus that as programs have matured, other goals have become very important. The growing emphasis on economic development as a program goal is a particularly clear and recurring theme. Indeed, researchers argue that economic redevelopment has actually eclipsed the protection of human health as the dominant goal in most programs and has come to define the parameters of the discourse surrounding brownfield policy.

The shift from an environmental program to one that emphasizes economic issues is explained by several factors. There is, for example, uncertainty about actual health risks. The magnitude or even the existence of contamination is often unknown at any given site. Even when the source and magnitude of contamination are known, the degree of human risk is often hotly debated given that the science underlying risk analysis is far from exact. Complicating factors include the fact that contamination often has differential impacts on specific populations (children, for example, are often more susceptible to environmental toxins), and multiple toxins often coexist in a given site, giving rise to unpredictable interaction effects. This ambiguity around actual levels of risk allows for debate and possible reorienting of the program to include new and "complementary" goals.

The goal of cleaning contaminated sites is often coupled with a second environmental goal: saving undeveloped land. Typically, undeveloped sites, or greenfields, are widely viewed as a valuable and diminishing resource. Land use advocates point with alarm to a dramatic reduction in land use density. The net effect of this demographic change is to rapidly increase the consumption of land for development even where there is very modest population pressure. Some observers see in brownfields an alternative to the continued widespread development of greenfields. While land preservation is a common symbolic goal of brownfield programs, the extent to which the programs actually impact

---

5   For a general discussion of the interaction of science and policy making see Collingridge (1986), Jasanoff (1990), Pielke (2004), and Sarewitz (2004: 5294).

development patterns varies. Catney et al. compare programs in Italy, the United States, and the UK. Only the British have made the explicit political decision to force new development on brownfield sites through a combination of cleanup programs and taxes that make brownfield investment "rational" for market actors. In the United States there has also been an explicit effort to tie investment to redevelopment. However, in the US case there has not been a similar effort to redirect overall development patterns. In Italy, the link between brownfield and overall development policy is less developed since site redevelopment usually follows and is often independent of the decision to clean the site.

In addition to saving undeveloped land, a number of political and policy leaders see brownfields as a particular development opportunity for depressed urban centers, given that a large number of brownfields are found in such communities. Some commentators have gone as far as arguing that brownfields represented a key potential path to the revitalization of older industrial cities. One prominent Midwestern governor put it this way:

> The cornerstone of any urban revitalization strategy must be an aggressive brownfield redevelopment program. We have made brownfields attractive by reforming the cleanup laws and offering tax credits and low interest loans to our communities. More than anything, our success comes from making brownfield redevelopment a top economic and environmental priority in the state of Michigan (Consumers Renaissance Redevelopment Corporation 1998).

As in the case of linking brownfield redevelopment to the protection of undeveloped land, the use of brownfield redevelopment to promote community revitalization is often more symbolic than real (Hula and Bromley-Trujillo 2010). This is hardly surprising given the relatively low economic value of many contaminated properties. This is not to say, however, that some sites do not have significant redevelopment potential. For example, in many cities industrial development was tied to close proximity to water. These sites are now often very attractive for private redevelopment.[6]

Implementation Strategies

Although a separate section of this volume is devoted to a discussion of implementation strategies, it is difficult to discuss policy design in isolation from implementation strategies. Interestingly, some common program characteristics emerge even when programs are compared internationally. These include:

---

6   See Welsh and Jones (this volume) for a discussion of the special appeal of waterfront property in brownfield redevelopment.

(1) Having a development focus

As has been noted, most brownfield programs place a strong emphasis on identifying the end-use of the property. Catney et al. note that Italy provides one exception to this pattern, in that efforts to identify and clean sites remain relatively separate from issues of use.

(2) Enforcing minimal environmental protection

Consistent with the focus on site development, brownfield programs generally de-emphasize the role of environmental health. In fact, some state level programs in the United States stress the importance of identifying sites that have the potential to be contaminated, but in fact are not. These sites are then free to be redeveloped without any additional public or private investment. Sites with modest levels of contamination are sometimes also made available for redevelopment without any actual onsite cleaning. This can be accomplished by a variety of strategies including increasing permissible or officially "safe" levels of contamination, imposing differential standards that allow higher contamination for industrial and commercial uses, and implementing a variety of institutional controls and regulations that aim to limit exposure to on-site contamination.

(3) Liability protection

A common principle in environmental law is that those responsible for on-site contamination are liable for the costs incurred in cleaning up that contamination. This broad principle is in place in most jurisdictions. However, the definition of what constitutes a responsible party has been significantly narrowed in many jurisdictions. For example, purchasing a contaminated property often no longer automatically creates a liability for cleanup costs.

   In spite of these substantive similarities in tools and strategies used, there are significant differences in the sort of institutions assigned the responsibility to implement brownfield programs. Not surprisingly these differences reflect variations in overall institutional structure and political context. Examples range from locally based programs in the United States (Lowham) to the Czech Republic, which operates within a complex web of national and international organizations (Rydvalova et al.).

Program Financing

A comparative review of brownfield programs reveals a number of alternative financing mechanisms. These are best summarized on two dimensions: the relative use of public versus private investment and the relative mix of public funds from local and central government agencies. While this distinction is conceptually straightforward, real-world examples turn out to be much more complex. For example, state level programs in the United States emphasize voluntary participation and private sector investment.[7] The British also rely heavily on private investment. Each set of programs, however, is based on quite different mechanisms to generate this private sector investment. In the United States, financial incentives such as tax breaks and direct subsidies promote private sector engagement in brownfield programs.[8] In contrast, the British have fundamentally restricted the development market. They have consciously made development on greenfields so difficult and expensive that the bulk of private development has now been channeled to brownfields as a matter of "rational" economic decision making. In contrast to the British and American case, other countries rely more heavily on public funding to support brownfield programs. Not surprisingly such programs often carry more public oversight on program implementation and ultimately final land use.

No matter what the predominant source of funding may be for brownfield programs, the discussion in this section shows that funding for a specific brownfield project is typically derived from multiple sources. This suggests that at the project level, a major effort must be dedicated to identifying the multiple funding sources necessary to make the project viable. This in turn emphasizes the importance of entrepreneurial skills in a wide variety of policy settings.[9] And, as noted more fully in the implementation section, reliance on the market and/or insufficient public funding can inhibit remediation and redevelopment of brownfields, particularly in poor urban areas or more remote rural ones.

Outcomes and Assessment

Relatively absent from any of the chapters in the section is a discussion of program assessment or evaluation. There is an intuitive consensus that brownfield redevelopment has broad political support, and does in fact help redevelop contaminated properties. Assessment efforts, however, are, at best, anecdotal. Political authorities often point to the successful redevelopment of a specific site as evidence of program impact, but seldom is there any systematic

---

7 It is important to remember, however, that the federal government retains through CERCLA responsibility for the nation's most dangerous contaminated sites. State level programs are targeted to less dangerous sites.

8 Many of the state "voluntary" programs have a coercive element as well. If responsible parties chose not to enter such volunteer programs they risk being assigned to more punitive state or federal programs. The threat of federal action seems be of particular concern to many responsible parties.

9 For a discussion of the role of entrepreneurship see Mintrom (2000) and Schneider (1995).

evidence provided.[10] Lowham provides some clues as to why such evaluations are difficult and rare. She notes that developers who are, of course, key actors in US redevelopment projects, share a broad consensus about what program attributes are most important for supporting successful projects. These data are especially interesting given the fact that the factors identified by the developers seem not to be associated with overall program success. Again, evaluation issues are addressed more specifically by chapters in the next section.

Moving Beyond Description: Understanding the Policy Process

Two features of the descriptive analysis presented in this section pose interesting theoretical issues. The most prominent is the emergence and even dominance of economic development as a central policy goal in most brownfield programs. In all of the programs discussed here there has been a clear movement from environmental protection to economic development as a central program goal. This convergence directs attention to the process by which issues are defined or framed by political elites, and how such frames drive program design and overall impact. A second, and perhaps related issue is the process by which elements of innovative policy solutions diffuse across jurisdictions.

The Dynamics of Issue Framing

The transformation of what seemed to be an environmental issue to one of economic development (or "development managerialism" in the terms of Catney et al.) is a classic example of what policy scholars term "issue framing." The definition of issue framing is usually seen to be relatively straightforward. For example, Chong and Druckman (2007: XX) define framing as a "process by which people develop a particular conceptualization of an issue or reorient their thinking about an issue." Nelson et al. (1997) note that framing is distinct from persuasion. Persuasion involves the use of new information to produce belief change, while framing utilizes known information within an individual or audience. Brownfield programs appear to present an interesting case of elite issue framing. This would seem to be the case in at least two distinct ways. As has been noted, brownfield redevelopment programs blend elements of environmental protection and economic development that had previously been viewed as independent public initiatives. Equally important, although perhaps not as obvious, was a de-emphasis of a number of traditional attributes associated with earlier policy. This is particularly true in the case of environmental policy where issues of site remediation and protection of public health were given diminished priority (Hula 2002, 2001).

---

10   Obviously the absence of such data is hardly unusual. In fact this is typical of the policy world.

The literature on policy framing asserts that different frames can produce variations in policy, and simultaneously reshape public opinion. Within the environmental policy literature, for example, a tension between conservation and economic goals is often assumed (Hoffman and Ventresca 1999). Yet, most examples of brownfield programs discussed here illustrate the merging of these two goals.

Andersson and Bateman (2000) highlight the practical importance of issue framing in a study that shows that environmental groups are able to influence business organizations to take positive action on the environment if they engage in strategic issue framing. By framing environmental issues to highlight certain aspects of the problem, environmental groups can sometimes lead businesses to perceive the negative consequences of not acting. Rein and Schon (1993) argue that how an issue is framed also impacts the substance of policy. Actors create stories and metaphors to describe a policy problem, which in turn leads to different alternatives for solving the problem. They show that differing frames lead to various policy outcomes. Stone (2002) argues that groups utilize symbols, facts, and metaphors to describe policy problems in order to gain the attention of large numbers of people. This method of framing allows groups to gain members to rally around a cause. For example, welfare under the Reagan administration was framed as a "safety net" for people, which led to an emphasis on preventing "welfare abuse." They also note that different groups sometimes bring different frames to an issue. In this case, some coordination between groups may be necessary to avoid a political stalemate (Schon and Rein 1994).

Indeed, political debate can sometimes be understood as a debate over how best to frame an issue. Similarly Capek (1993) demonstrates the importance of issue framing in his discussion of how a group of homeowners was able to organize a federal buyout and relocation away from a Superfund site. He claims their success was due to the fact that the community was able to frame the issue around environmental justice. Taylor (2000) finds similar evidence for the success of the environmental justice frame. Howarth (1996), in an examination of climate policy, argues that depending on how one frames climate change, there may or may not be a case for the need to reduce greenhouse gas emissions. Note that differing policy frames may be complementary, and do not necessarily lead to political conflict. This certainly seems to be the case for the broad coalitions that come together to support brownfield programs. Agriculture interests and representatives of cities often brought different issue frames to the program, but were nevertheless in agreement as to desired policy.

Some scholars have examined policy framing within institutions. Lenschow and Zito (1998) examine how actors create policy frames that are embedded in community institutions. To better understand the impact of environmental policy frames, the authors examine waste management and agricultural policy. They identify three alternative policy frames including: conditional, classic, and sustainability. The conditional frame defines environmental regulations as equalizing the market. The classic frame defines environmental regulation as an action that will reduce health and environmental risks. The sustainability frame argues that environmental and economic goals are complementary and actually depend on each other. Findings indicate that within the European context the conditional frame has become institutionalized, and it is difficult to change.

Policy Diffusion

The chapters in this section identify a significant convergence in how goals have been framed in brownfield policy, although it should be stressed that there remains significant differences with respect to program mechanics. This observed convergence suggests that some sort of diffusion process may be occurring. To be sure, this is largely an implied rather an explicit argument of most chapters in this section. It is explicitly discussed only in the Hula chapter, and even then the discussion is limited to possible diffusion in state level programs in the United States. Hula argues that overall brownfield activity has a strong regional component. Specifically, states in the Midwest are most likely to have active brownfield programs.[11]

The interesting theoretical question, of course, is what mechanisms might be driving the observed policy convergence. Hula proposes at least a partial institutional explanation based on an empirical association between states with an active brownfield program and location within the EPA's Region 5 office. Earlier work indicates that the regional offices had a good deal of autonomy, and quite distinct cultures. Such differences might explain program differences at the state and local level. The intriguing policy question, of course, is whether there is an international process of policy innovation and diffusion at work. The possibility of such a process provides an interesting challenge for future policy research. While the analysis in this volume suggests that such broad processes might be occurring, it will require more explicitly cross national research to understand whether these national policy systems are in fact interdependent.

Concluding Thoughts

Brownfield redevelopment policy presents an interesting case for both the practitioner and the scholar. On a practical policy level these programs simultaneously address a number of citizen preferences. First of all, they target widely held environmental concerns about ground and water contamination. They also attract support from those interested in economic development, urban revitalization, and the protection of undeveloped land. And, as will be shown in the next section, brownfield policy brings together a variety of other urban goals

---

11   Note that the analysis attempts to control for overall level of contamination so that this finding is not likely to be an artifact of the magnitude of environmental issues facing the state.

and interests such as housing provision, sustainability, economic opportunity, and community social capital and empowerment.

Brownfield redevelopment also provides an interesting set of theoretical challenges to the policy scholar. These include not only specific issues of program design, implementation and impact, but also several more general conceptual questions such as the three noted in the introduction to this volume specifically: (1) What do policies look like? How are programs organized? Does it matter? (2) What strategies are used to promote redevelopment. (3) What is the role of community preferences? Of particular interest are patterns of issue definition and program convergence. The chapters in this part explore policy issues that cut across these three questions.

References

Andersson, L.M. and T.S. Bateman. 2000. "Individual Environmental Initiative: Championing Natural Environmental Issues in U.S. Business Organizations." *The Academy of Management Journal* 43(4): 548–70.

CABERNET. 2010. *Brownfields Definitions across Europe* 2010, available at http://www.cabernet.org.uk/index.asp?c=1316 (accessed: July 22, 2010).

Capek, S.M. 1993. "The Environmental Justice Frame: A Conceptual Discussion and an Application." *Social Problems* 40(1): 5–24.

Chong, D. and J. Druckman. 2007. "Framing Public Opinion in Competitive Democracies." *The American Political Science Review* 101(4): 637.

Collingridge, D. and C. Reeve. 1986. *Science Speaks to Power: The Role of Experts in Policy*. New York: St. Martin's Press.

Consumers Renaissance Redevelopment Corporation. 1998. *Michigan's Brownfield Redevelopment Program: First in the Nation.* Jackson, MI: Consumers Renaissance Redevelopment Corporation.

Donati, A., C. Rossi and C.A. Brebbia. 2004. *Brownfield Sites II: Assessment, Rehabilitation, and Development*. WIT Transactions on Ecology and the Environment. Southampton, UK; Boston: WIT Press.

Hall, T.E. and D.D. White. 2008. "Representing Recovery: Science and Local Control in the Framing of US Pacific Northwest Salmon Policy." *Human Ecology Review* 15(1): 32–45.

Hoffman, A.J. and M.J. Ventresca. 1999. "The Institutional Framing of Policy Debates – Economics versus the Environment." *American Behavioral Scientist* 42(8): 1368–92.

Howarth, R.B. and P.A. Monahan. 1996. "Economics, Ethics, and Climate Policy: Framing the Debate." *Global and Planetary Change* 11: 187–99.

Hula, R.C. 2001. "Changing Agendas in Toxic Waste Policy: The Emergence of Economic Development as a Policy Goal." *Economic Development Quarterly* 15(2): 181–99.

——. 2002. "There is Gold in Those Brownfields … Maybe." In *Urban Policy Choices*, edited by D. Thorton and C. Weissert. East Lansing, MI: Michigan State University Press.

Hula, R.C. and R. Bromley-Trujillo. 2010. "Cleaning up the Mess: Redevelopment of Urban Brownfields." *Economic Development Quarterly* 24(3): 261–87.

Jasanoff, S. 1990. *The Fifth Branch: Science Advisers as Policymakers*. Cambridge: Harvard University Press.

Jones, B.D. and F.R. Baumgartner. 2005. *The Politics of Attention: How Government Prioritizes Problems*. Chicago: University of Chicago Press.

Kaiser, S.-E. 1998. "Brownfields National Partnership: The Federal Role in Brownfield Redevelopment." *Public Works Management and Policy* 2(3): 196–201.

Lenschow, A. and A. Zito. 1998. "Blurring or Shifting of Policy Frames?: Institutionalization of the Economic-Environmental Policy Linkage in the European Community." *Governance* 11(4): 415–41.

Mazey, S. and J. Richardson. 1997. "Policy Framing: Interest Groups and the Lead up to 1996 Inter-Governmental Conference." *West European Politics* 20(3): 111–23.

Mintrom, M. 2000. *Policy Entrepreneurs and School Choice*. American Governance and Public Policy Series. Washington, DC: Georgetown University Press.

Nelson, T.E., Z.M. Oxley and R.A. Clawson. 1997. "Toward a Psychology of Framing Effects." *Political Behavior* 19(3): 221–46.

Oliver, L., U. Ferber, D. Grimski, K. Millar and P. Nathanail. 2005. "The Scale and Nature of European Brownfields." In *CABERNET 2005 – International Conference on Managing Urban*. Belfast, Northern Ireland, UK: LQM Ltd, Nottingham, UK.

Pielke, R.A. 2004. "When Scientists Politicize Science: Making Sense of Controversy over the Skeptical Environmentalist." *Environmental Science and Policy* 7: 405–17.

Portz, J. 1996. "Problem Definitions and Policy Agendas: Shaping the Educational Agenda in Boston." *Policy Studies Journal* 24(3): 371–86.

Rein, M. and D. Schon. 1993. *The Argumentative Turn in Policy Analysis and Planning*, edited by F. Fischer and J. Forester. Durham: Duke University Press.

Rochefort, D.A. and R.W. Cobb. 1994. *The Politics of Problem Definition: Shaping the Policy Agenda*. Studies in Government and Public Policy. Lawrence, Kan.: University Press of Kansas.

Sarewitz, D. 2004. "How Science Makes Environmental Controversies Worse." *Environmental Science & Policy* 7(5): 385–403.

Schneider, M., P.E. Teske and M. Mintrom. 1995. *Public Entrepreneurs: Agents for Change in American Government*. Princeton, N.J.: Princeton University Press.

Schon, D.A. and M. Rein. 1994. *Frame Reflection: Toward the Resolution of Intractable Policy Controversies*. New York: Basic Books.

Steensland, B. 2008. "Why Do Policy Frames Change? Actor-Idea Coevolution in Debates Over Welfare Reform." *Social Forces* 86(3): 1027–54.

Stone, D. 2002. *Policy Paradox: The Art of Political Decision Making*. New York: Noton and Company.

Taylor, D.E. 2000. "The Rise of the Environmental Justice Paradigm: Injustice Framing and the Social Construction of Environmental Discourses." *American Behavioral Scientist* 43(4): 508–80.

Whelan, G. 2004. *Brownfield Sites I: Multimedia Modeling and Assessment*. Southampton, UK; Boston: WIT Press.

Zabestoski, S., K. Agnello, F. Mignano and F. Darroch. 2004. "Issue Framing and Citizen Apathy toward Local Environmental Contamination." *Sociological Forum* 19(2): 255–83.

Chapter 1

Incentives for Collaboration: State-level Brownfield Remediation and Redevelopment Programs

Elizabeth A. Lowham

There is little doubt that the role of governments in policy formation and implementation is changing in a number of policy arenas (Pierre and Peters 2000). Government is increasingly attempting to serve as a facilitator rather than a leader. One important example of this transformation is state-level programs for the remediation and redevelopment of brownfields. While researchers know that states combine common policy features in their state level programs (ICF Consulting and The E.P. Systems Group, Inc. 1999), they know relatively little about how such combinations of features may facilitate or deter collaboration and action on brownfield projects (Reisch and Bearden 2003; GAO 2004).

This chapter investigates the policies that states develop to manage, remediate, and redevelop contaminated sites within their borders. Such policies require multiple public and private agencies to work together to remediate and redevelop sites, in part because no actor has the legal authority *and* fiscal resources *and* connections to local communities to reinvigorate sites successfully alone. Using cluster analysis on features of state programs (financial incentives, liability provisions, public participation, long-term stewardship, economic impacts, and a proxy variable for program age), results presented here indicate that there are discernible and meaningful patterns of incentives that states utilize in their voluntary cleanup programs.

Such brownfields are an ideal policy area in which to investigate incentives for collaboration because, in the past 25 years, the cleanup of contaminated sites has transformed from largely top-down, federally-led structure to a context where the cleanup and redevelopment of many sites is accomplished through networks of multiple actors from various public and private institutions. While brownfield remediation and redevelopment occur on individual pieces of properties, such local site activities occur in the context of state-level programs that outline the incentives and parameters for participation. These state-level programs are the focus of this research; they provide the context for specific remediation and redevelopment actions. The patterns of state program incentives shape who participates, how they participate, and to what extent they participate in the remediation and redevelopment of contaminated properties.

Brownfields

Commercial and industrial growth since World War II has left its imprint on the United States. During the socio-economic changes following the war, pollution increased roughly ten times faster than the gross national product (Andrews 1999). At the same time, the number of commercial and industrial processes utilizing complex and hazardous chemicals increased (Urban Institute et al. 1997). As commercial and industrial firms sought to increase efficiency, many firms either abandoned their traditional property locations or manufacturing processes, leaving behind older buildings and remnants of processes which have contaminated the property. These remnants are known as brownfields.[1]

The legal definition of a brownfield is "real property, the expansion, redevelopment or reuse of which may be complicated by the presence or potential presence of a hazardous substance, pollutant, or contaminant" (U.S. Congress 2001: 6). In layman's terms, brownfields are sites, generally in urban areas, where actual contamination or the perception of contamination prevents or hinders redevelopment or reuse.[2]

The EPA estimates that there are upwards of 450,000, and perhaps as many as one million brownfields, in the United States (GAO 2004). Many of these sites are located in blighted urban areas which are thought to be unattractive to residential, industrial, or commercial developers (Urban Institute et al. 1997). Reisch and Bearden (2003) argue that brownfields are in many ways paradoxical sites. On the one hand, brownfield sites are generally not contaminated enough to be cleaned through enforcement-driven programs, like Superfund, but developers avoid the sites because of the fear of contamination and the liability for cleanup costs associated with them.

Yet, the remediation of brownfields within a community can often be a win-win situation. Remediation of sites removes contamination or prevents its further spread to other properties, thus reducing the risk of exposure for human and environmental health. Redevelopment of these sites can spur local economic redevelopment, taking properties which were once unprofitable tracts of land and turning them into new commercial, industrial, or residential areas. These activities can also relieve low income and minority populations' greater exposure

---

1 Certainly, a brownfield can exist on a site from which a commercial or manufacturing firm has not moved. In these cases, the firm seeks to remediate its own site and its own contamination. Such brownfields are not the focus of this research in large part because the contaminator still owns/operates the property. As a result, the legal requirements of the program are much different from voluntary remediation programs where the owner is generally not associated with the contamination.

2 Brownfields can be found in suburban or rural areas. They are, however, less common in such areas because, in a practical sense, a brownfield is a brownfield because someone wants to redevelop it. The push for redevelopment is lower in rural and many suburban areas where non-contaminated land is relatively close.

to contamination, thus addressing concerns over social justice. Further, these societal benefits are intended to be done in a manner that generates profits for those individuals who undertake activities at these sites.

Development of Brownfields Policies

The development of brownfields policy is, in many ways, a bottom-up story, initiated by a small set of environmentally progressive states in the late 1980s. In 1978, the discovery of Love Canal prompted a national search for other abandoned hazardous waste sites.[3] As communities and governments began investigating, they discovered hundreds of abandoned hazardous waste sites around the country, spurring Congress to pass the Comprehensive Environmental Response, Compensation, and Liability Act (CERCLA) in 1980.

The passage of CERCLA recognized that state and local governments lacked both the resources and capacity to deal adequately with such contaminated sites and, further, that then-existing federal laws regarding disaster relief could not address such sites (Reisch and Bearden 1997, 2003). Congress intended CERCLA to respond quickly to the worst hazardous waste sites around the country by allowing EPA to (1) compel responsible parties to remediate a site or (2) remediate a site itself utilizing what became known as Superfund, then to seek compensation from responsible parties.[4] CERCLA creates a situation where EPA, a federal level agency, takes the leadership role in cleanup, though the role of state governments has increased over time (Ramseur and Reisch 2006).

CERCLA contains a very stringent set of liability provisions to ensure that responsible parties are held accountable for remediation costs. CERCLA uses strict, joint and several, and retroactive liability provisions which taken together means that anyone involved in a site at any point in the site's history, as owner, manager, transporter, or sender of hazardous materials, could be held liable for the entire cleanup cost of the site (Ramseur and Reisch 2006) as well as for damages to publicly held natural resources (Reisch and Bearden 2003).[5] The CERCLA

---

3   Love Canal was the first and only time when a pollution incident received federal emergency relief (Reisch and Bearden 2003).

4   Until 1995, Superfund was maintained through excise taxes on the chemical and petroleum industries as well as income taxes on corporations (Reisch and Bearden 1997). Since 1995, Congress has been paying for cleanup costs from general revenue funds (Ramseur and Reisch 2006).

5   Strict liability implies that the government need only prove involvement with the site; proof that a particular party strictly caused any part of the pollution is unnecessary (Reisch and Bearden 1997). Joint and several liability means that any responsible party may be held liable for the entire cleanup costs for the site, regardless of the proportion of damage a particular party may have caused (Reisch and Bearden 1997). Finally, retroactive liability means that parties are liable for costs even if the contamination or pollution occurred before Congress passed CERCLA (Reisch and Bearden 1997).

liability provisions are so stringent that, in a few cases, lenders who have foreclosed on properties have been held liable for cleanup costs (Urban Institute et al. 1997). These liability provisions created strong incentives on the part of "responsible parties" to vigorously fight any assignment of liability by the government, in part because they could be held liable for the entire cost of cleanup, even if their contribution to the contamination had been minimal. As a result the remediation of CERCLA sites proceeded quite slowly, if at all (GAO 2004).[6]

The slow speed of CERCLA remediations was especially concerning to states as they discovered numerous sites within their borders with lower levels of contamination. Given the number of severely contaminated sites across the United States, it was highly unlikely these sites would ever be eligible for CERCLA cleanup activities. As a result states were forced to craft their own cleanup strategies. Most states created their own Superfund programs which mimicked the liability structure of CERCLA or relied on CERCLA to remediate even these lower level sites. Remediating a site through either state or federal level Superfund programs, raised the same liability issues faced by federal authorities. In addition these programs often "tainted" the property with a public perception that even though remediated, the sites still carried a danger to public health. This perception clearly interfered with the later redevelopment of the site.

As a result of the perceived limitations of CERCLA and CERCLA-like programs, a small group of states began developing their own voluntary cleanup programs. Such programs allowed a private developer, generally an owner who is not associated with the contamination, to submit a voluntary cleanup proposal. Once this plan was implemented, the private developer would be free to redevelop the site according to an approved plan. These programs allowed states to bypass their enforcement-driven cleanup programs and work cooperatively with other participants (Reisch and Bearden 2003). In most cases, if a site was properly remediated under a state plan, the state would release the owner from future state liability associated with it.[7] Many of these initial state programs also obtained a Memorandum of Agreement (MOA) with EPA. Such MOAs allowed site owners to be released from most federal liability, with a few caveats, as long as the site went through an approved state program. Thus, in general, the incentives for parties to participate in brownfields remediation and redevelopment programs were much different from those under CERCLA. Under CERCLA, the process of remediation was contentious, slow and costly. State voluntary cleanup programs,

---

6   There is some variation in the importance of the strict liability provisions of CERCLA in preventing site redevelopment. An Urban Institute, Northeast-Midwest Institute study (1997), for example, finds that while strict liability provisions were a concern of developers, those concerns were never the factor that would make or break the deal.

7   Generally, such releases from liability are only provided to parties who are not associated with the release of contamination. Thus, most of the remediation and redevelopment occurring through these programs happened with the transfer of property.

on the other hand, offered financial incentives, liability protection, and the promise of profit for participation in remediation and redevelopment activities.

As a response to the development of these state voluntary cleanup programs, EPA created a host of programs related to brownfields during the early to mid-1990s (e.g., the Brownfields Economic Redevelopment Initiative, the Brownfields Action Agenda, and the Brownfields National Partnership). For the most part, these federal programs offset state and local government remediation costs and provided incentives for further brownfield remediation and redevelopment.[8]

In 2002, Congress passed the Small Business Liability Relief and Brownfields Revitalization Act (Brownfields Act). The Brownfields Act clarified some remaining liability issues such as funding to support state voluntary cleanup programs.[9] The Brownfields Act outlined four broad requirements that state programs must meet to obtain federal funding: (1) a timely survey and inventory of brownfield sites within the state, (2) oversight and enforcement authorities to ensure the protection of human and environmental health, (3) meaningful public participation (including public access to documents and prior notice of action),

---

8   In 1993, EPA created the Brownfields Economic Redevelopment Initiative (BERI). The purpose of BERI was to provide seed money to bring participants together to remediate sites that were less contaminated than Superfund sites, but not to fund the cleanup activities (Reisch and Bearden 2003).

EPA started the Brownfields Action Agenda in 1995. This initiative essentially created a set of funds that states or individual sites could access to help alleviate the costs of remediation or to provide funds to create interest in redeveloping a site. Whereas BERI funds could only be utilized to bring participants together, Action Agenda funds could be utilized to defray actual remediation costs. The Action Agenda also allowed the EPA administrator to delist sites from the Superfund program. De-listing dramatically increased the number of sites which private interests could and would consider redeveloping.

In 1997, Vice-President Gore announced the creation of the Brownfields National Partnership (Reisch and Bearden 2003). This partnership represented the commitment of over fifteen federal agencies, mainly EPA, HUD, and the Economic Development Administration (Reisch and Bearden 2003). Other federal agencies participating in the National Brownfields Partnership include the Department of Agriculture, Department of Defense, the Department of Education, the General Services Administration, the Department of Health and Human Services, the Department of Treasury, the Department of Commerce, the Department of Labor, the Department of Veterans Affairs, the Department of Transportation, and the Department of the Interior. Together, these agencies directed over $300 million from their budgets to the brownfields issue (U.S. Environmental Protection Agency 1997). The purpose of BERI, the Action Agenda and the Brownfields National Partnership was to provide money to serve as "catalysts to bring together other resources in the communities to provide the environmental cleanup component of redevelopment efforts" (Reisch and Bearden 2003: 88).

9   The Brownfields Act clarified that owners of properties contiguous to contaminated properties could not be liable if the pollution migrated and that owners who purchased land known to be contaminated, for the purposes of redevelopment, could not be held liable.

and (4) a mechanism for the approval of remediation plans and certification that the cleanup is completed.[10]

In summary, current state and federal brownfields policies stress collaboration, not command. Such state cleanup programs are clear departures from the top-down, liability-based mechanisms in CERCLA. Further, there is recognition on the part of states that remediation and redevelopment would likely require multiple agencies at multiple levels of government working in conjunction with private interests and citizens to redevelop sites.

Research on Brownfields Programs

Existing research on brownfields programs focuses on two issues: factors which participants in the brownfields process consider important and statistical predictors of successful site remediation and redevelopment. Interestingly, there is little overlap between participants' beliefs and statistical predictors. This review highlights those aspects of state programs—including liability provisions, financial incentives, and community support—that are generally considered to be most important in successful brownfield remediation and redevelopment.

States have a good deal of flexibility to design programs, including liability provisions, cleanup requirements, and funding mechanisms. In a nationwide survey, Lange and McNeil (2004a) asked stakeholders to identify which factors contributed to successful remediation and redevelopment. Survey findings indicated specific site/redevelopment features are important factors in successful site redevelopment, including time, cost, and consistency with community master plan, land use, and the number of jobs created.

Research on predictors of successful redevelopment, however, has largely found that those policy features on which states have flexibility largely determine successful redevelopment, not the site/redevelopment-specific factors. In their research comparing successful and "not-so-successful" brownfield redevelopment projects, Lange and McNeil (2004b) find that specific site descriptors (including urban/rural, previous site usage, acreage, level of contamination, etc.) do not predict successful site redevelopment. Further, contradicting previous practitioner surveys (e.g., Lange and McNeil 2004a), they find that job costs per acre, time and cost of remediation, cost of redevelopment, or consistency with master plan are not significantly related to successful outcomes (Lange and McNeil 2004b).

---

10   In other words, states and sites are released from future federal action upon the completion of cleanup. There are exceptions to this release. EPA can step in if (1) states request federal assistance, (2) pollution migrates across state borders or onto federal property, (3) pollution presents an imminent danger to public or environmental health, or (4) new information initially unavailable to the states makes more remediation necessary (U.S. Congress 2001).

Lange and McNeil (2004b) conclude that in successful redevelopment efforts, participants develop creative and multiple funding mechanisms, seek policies to limit liability, work with communities to garner their support for projects, and encourage investment in existing infrastructure around brownfield sites. A separate survey finds that mayors believe that the top three impediments to successful redevelopment are the lack of funds for cleanup, concerns over liability, and environmental issues (including the need for environmental assessments and cleanup standards) (U.S. Conference of Mayors 2006). These three factors have ranked at the top of the list of mayor's concerns for the past five surveys (U.S. Conference of Mayors 2006). Additionally, research on state brownfield programs indicates that regulatory relief in the form of liability protections and cleanup requirements is important to prospective developers and thus to the initiation of site remediation and redevelopment (ICF Consulting and The E.P. Systems Group, Inc. 1999; Alberini et al. 2005).[11]

This literature points to several important facets of brownfield remediation and redevelopment programs, including liability structures, financial incentives, and community support. Researchers know that states combine some form of regulatory flexibility with financial incentives and liability relief when designing their voluntary cleanup programs (ICF Consulting and The E.P. Systems Group, Inc. 1999). Despite the knowledge that these aspects of brownfields policies are crucial for successful redevelopment, relatively little is known about the variation in brownfield redevelopment programs across states. That is, there is very little knowledge about how states piece together their own redevelopment programs and incentive structures from the myriad of redevelopment aspects available to them (Reisch and Bearden 2003; GAO 2004). The remainder of this chapter describes the existence of discernible patterns of features of redevelopment programs across states based largely on what previous research has identified as important factors for redevelopment: liability provisions, financial incentives and community support. Beginning with a description of the measurement and methodology applied to uncover patterns of state level programs, this chapter concludes with a discussion of the results and implications of different program designs. Such patterns in program design have implications for both the performance and politics of redevelopment programs.

---

11   Alberini et al. (2005) find that the attractiveness of financial incentives or subsidies for remediation and redevelopment projects varies by the developer's experience with contaminated properties. That is, developers without previous experience with contaminated properties tend to value financial incentives less than developers with previous experience with contaminated lands.

Data Collection, Measurement, and Methodology

The dataset on brownfield remediation programs used in this study distinguishes features of state cleanup and redevelopment programs for all 50 states as identified in the EPA's *State Brownfields and Voluntary Response Programs* report (SRA International 2006) and the Northeast-Midwest Institute's *Brownfield Voluntary Cleanup Program Impacts* (Bartsch, Anderson and Dorfman 1999). Using the descriptions of state programs and the summaries of state impacts, each program was coded using a series of variables and indices measuring financial incentives, liability provisions, public participation, long-term stewardship, economic impacts, and a proxy variable for program age. While these variables are descriptive of programmatic structures, they also provide incentives for and limitations on participation in state brownfield programs. To reduce the risk of overweighting certain facets of brownfield remediation programs, each variable or index exists on a 0–1 scale.

Measurement

Financial Incentives

Financial incentives are measured in two ways. First, an index of financial incentives is created by summing a series of dichotomous measures of whether or not a state offers grants, bonds, loans, tax abatements and/or tax credits as methods of offsetting the costs of cleanup and redevelopment. These financial incentives encourage participation from various private parties. Without these incentive programs, developers and businesses may not seek to remediate and redevelop brownfields because of a perceived lack of governmental support for the project. Financial incentives also provide an avenue for state-level agencies, particularly those that deal with funding programs, to participate in brownfields programs. A second financial indicator is based on whether a state offers environmental insurance. Insurance differs substantively from the traditional incentives in that it allows developers to insure themselves against the cost of remediation. The existence of environmental insurance may indicate that the state plays a somewhat different role in the remediation process because the state itself cannot afford regularly to require cleanup beyond that for which the developer is insured.

Liability Provisions

Two measures of liability incentives are presented. The first, utilizes an index of CERCLA-type liability provisions that measures how closely a state's brownfield liability provisions match those used by CERCLA. This index delineates the state's role in determining responsibility for contamination of brownfields.

Stronger liability provisions (that is, more CERCLA-like) may actually deter some actors from participating in projects. This index also captures elements of the relationship between the state and private enterprise seeking to redevelop a site. Given past experience with strict, joint and several, and retroactive liability, if states utilize only CERCLA type liability provisions in their brownfields program, the relationship between private enterprise or site owners and the state could be quite contentious.

States, however, can also utilize forms of liability other than those outlined in CERCLA. In many ways, these forms of liability represent increased state flexibility. Alternative liability provisions provide state protection to some parties, thus increasing the incentive to remediate and redevelop. Forty-three states provide some sort of liability protection. Examples of alternative liability mechanisms include: proportional liability, covenants not to sue, and restricted use agreements. Because of the wide variety of alternative liability rules, the variable used to measure it simply indicates whether or not a state has some alternative liability provision in place.

Public Participation

Public participation is measured in two ways. First is an index of more traditional forums for public participation in environmental policy such as use of the public record, public notice, public comment and public hearings. This index highlights the ways in which the public can participate in brownfields remediation. The higher a state's score on this index, the greater the potential for public participation; thus, the greater potential role for the public.

A small number of states (four) also offer direct grants to citizen groups. Such grants differ substantively from other forms of citizen participation by catalyzing investment in remediation and redevelopment activities. Such grants allow citizens to drive the remediation and redevelopment of particular sites. In such situations, the leadership role of these citizen groups may be of a different nature and level than in states where citizen groups cannot receive grants. Citizens in these states have a substantially different set of incentives for participation in site projects.

Long-term Stewardship

Two indicators of long-term stewardship are used. The first is an index of different types of long-term stewardship for brownfield sites including whether or not the state tracks sites through cleanup and redevelopment projects, oversees sites through the process, and monitors remediated sites. The stewardship index highlights the long-term and on-going role of state environmental and health agencies in the brownfield site after the remediation and redevelopment is complete.

A second indicator is based on whether or not states allow for reopeners. This is rare, only a small number of states (three) allow reopeners. Reopeners permit a state regulatory agency to require additional cleanup to a site after the remediation and redevelopment plan has been approved. A reopener is actually a disincentive to participate because it allows the state to renegotiate cleanup standards once a private party has agreed to remediate and redevelop a site.

Additional Variables

This research includes two additional indicators to characterize brownfield remediation programs. First, it includes whether or not a state has a Memorandum of Agreement (MOA) with the EPA. This variable serves as a proxy variable for the maturity of a program. After the passage of the Brownfields Act in 2002, states were no longer required to have an MOA with EPA as long as state remediation programs met basic requirements. Thus, states which have an MOA with EPA were generally fully operational and mature before 2002. These differences in program maturity may lead to differences in incentive structures and program operations.

This research also considers whether or not a state is currently tracking economic impacts related to the cleanup and redevelopment of sites. Such impacts may include the number of jobs, residential units, tax revenues, and businesses created. Because of differences in size, staffing, and budgeting of state programs, the type (if any) of economic impacts a state may track varies widely (Bartsch, Anderson and Dorfman 1999). Thus, this variable only indicates whether or not a state is tracking some economic impact of redevelopment. Since so few states track economic impacts of brownfield remediation and redevelopment, this provides some indication of the development of program goals.

Methodology

A series of cluster analyses on all 50 states using the variables discussed above is used to identify types of state programs. Cluster analysis identifies groups (clusters) of states that have similar characteristics (scores on variables). It is different from most types of factor analysis, which reveal patterns across variables, because cluster analysis keeps the case whole, revealing patterns across cases. Cluster analysis, thus, allows the researcher to investigate patterns rigorously and systematically while still maintaining the context of a case.[12]

Cluster analysis detects naturally occurring groups within a dataset based on the information available. Relationships begin to emerge without the researcher imposing meaning or divisions on the data in an *a priori* manner. Such classifications can provide useful information on possible policy interventions and improvements or simply categorizations of programs based on variables for more in-depth analysis.

---

12   Cluster analysis is similar to particular types of pattern recognition. In general, however, pattern recognition is used to investigate patterns across known groups while cluster analysis is used to explore and classify different cases (Garnham 1976).

The choice of the clustering algorithm can create crucial differences in how state programs group together. In order to reduce the probability of a grouping based on the idiosyncratic characteristics of any given algorithm, this research conducted a series of cluster analyses (complete linkage, average linkage, and weighted average linkage) using the Euclidean distance as a measure of similarity.[13] If states grouped together in two of three algorithms, they created a cluster (see cluster output in Figure 1.1).[14] Utilizing two of three algorithms increases the robustness of the clusters and hence ensures that patterns and similarities are not artifacts of the method. For the most part, states clustered together consistently across the three data runs.

---

13   In this case, cluster algorithms combine state programs based on their similarity. In the average linkage algorithm, for example, the first step of the cluster analysis is to find a proximity matrix based on the selected distance measure, in this case, Euclidean distance. After finding the states with the smallest Euclidean distance (most alike), those most similar states are merged into one unit and the proximity matrix is recalculated based on the average Euclidean distance of the merged unit. The analysis program then searches the new proximity matrix for the next most similar state or states, averages their Euclidean distances, recalculates the matrix, and begins searching again. These steps continue until, in the last step, the entire sample becomes one cluster. The complete linkage and weighted average linkage algorithms follow the same basic steps using different methods of recalculating the proximity matrix. The complete linkage algorithm uses the greatest distance between any two individuals in the sample as a way of determining clusters. Rather than recalculating using the average Euclidean distance, the complete linkage (furthest neighbor) algorithm selects the value between the merged pairs that has the largest Euclidean distance (thus, the most dissimilar value). The weighted average algorithm recalculates the proximity matrix by averaging, but taking the number of objects in each cluster into account as a weight. Weighted averaging is thought to be useful when the researcher suspects that the number of individuals in a cluster may be uneven.

14   To ensure robustness of the results from choice of similarity measure, this research also used the same three algorithms with the Squared Euclidean measure of similarity. Using the same two-out-of-three robustness measure, the clusters are identical.

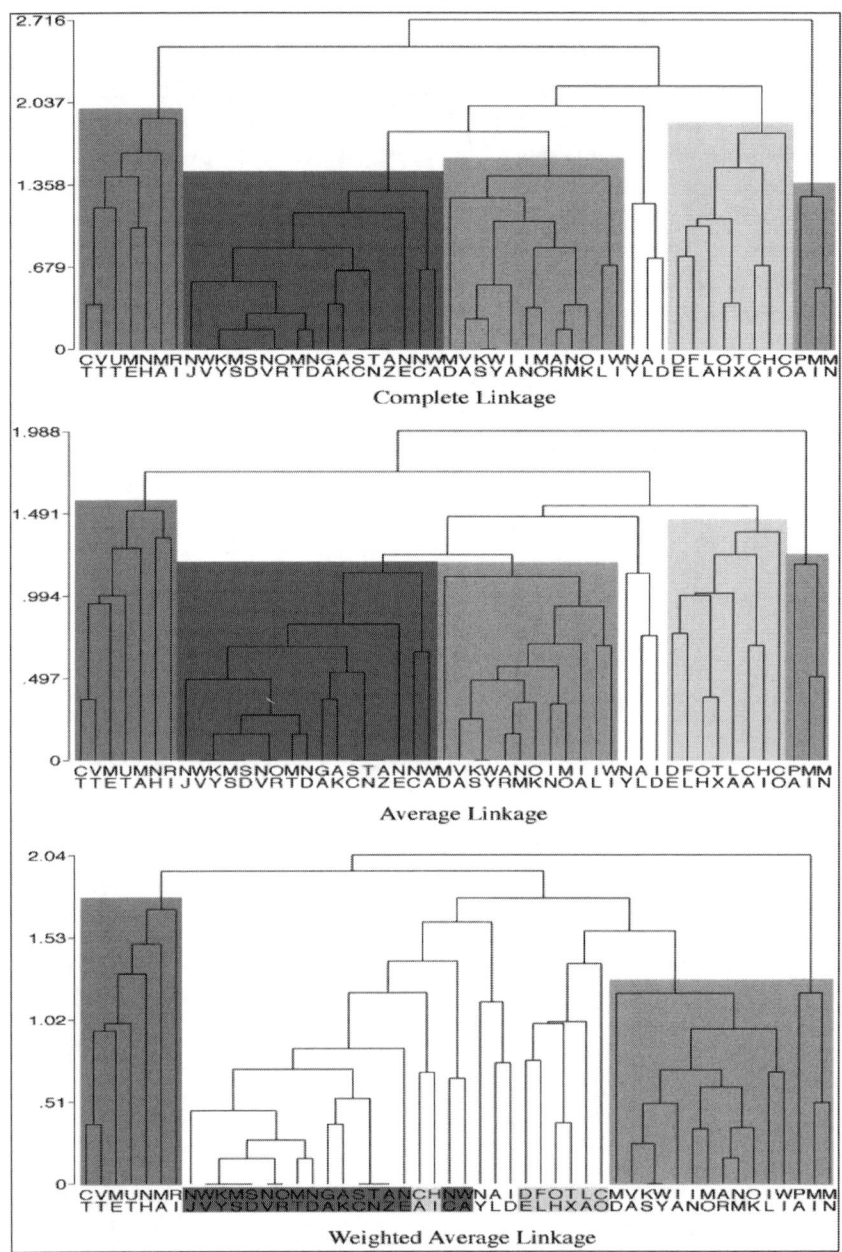

Figure 1.1   Cluster analysis using three algorithms

Table 1.1   Ideal types. Cluster averages across each variable and index used in cluster algorithms

	High Liability, Participation, Incentives, and Stewardship Programs	High Participation, No CERCLA Liability Programs	High Incentive, Low Participation Programs	CERCLA Liability, Low Stewardship Programs	Mid-Range with EPA MOA Programs
Liability					
CERCLA Index	0.89	0	0.084	0.67	0.11
Other Liability Provisions	1	1	1	0	1
Public Participation					
Participation Index	1	0.87	0.66	0.71	0.85
Citizen Group Grants	1	0	0	0	0
Financial Incentives					
Incentives Index	0.5	0.30	0.46	0.41	0.36
Environmental Insurance	0	0	0.13	0.14	0
Long-term Stewardship					
Stewardship Index	0.67	0.53	0.54	0.14	0.50
Re-openers	0.33	0.12	0	0	0
Tracking Economic Impacts	1	0	1	0.14	0
MOA with EPA	1	0	0.75	0.29	1
States within Cluster	3	16	8	7	12
Best Specimen*	MN (0.97)	MT, ND (0.98)	TX (0.93)	CT (0.94)	NM (0.99)

Note: * Value in parenthesis is the Pearson's Correlation between individual state and program profile.

Table 1.2   Breakdown of indices by cluster*

	HLPIS	HPNCL	HILP	CLLS	Mid
Liability Provisions					
Joint and Several	1	0	0.13	0.86	0.17
Strict	0.67	0	0.13	0.86	0.08
Retroactive	1	0	0	0.29	0.08
Other Liability Statute	1	1	1	0	1
Public Participation					
Public Record	1	1	1	1	1
Public Notice	1	0.88	0.63	0.71	0.83
Public Comment	1	0.82	0.5	0.57	0.83
Public Hearings	1	0.71	0.5	0.57	0.75
Citizen Group Grants	1	0	0	0	0
Financial Incentives					
Grants	1	0.71	0.63	0.43	0.58
Bonds	0	0	0	0.29	0
Loans	1	0.47	1	0.71	0.5
Tax Abatement	0	0.29	0.25	0.29	0.5
Tax Credit	1	0.29	0.63	0.57	0.5
Environmental Insurance	0	0	0.13	0.14	0.25
Long-Term Stewardship					
Track Sites	0.33	0.24	0.38	0.14	0.17
Oversight	1	0.88	0.75	0.14	0.92
Monitoring	0.67	0.47	0.5	0.14	0.25
Re-openers	0.33	0.12	0	0	0
Tracking Economic Impacts	1	0	1	0.14	0
MOA with EPA	1	0	0.75	0.29	1

Note: *Breakdown of indices by cluster. Cells can be interpreted as the percentage of states within a cluster that has the particular characteristic.

Results

With identifiable clusters of state programs, a cluster profile was created by averaging the scores of states within a given cluster across the ten indices/variables (Table 1.1) as well as the individual variables used to create the indices (Table 1.2). These cluster profiles can be thought of as ideal descriptions of the state programs within the cluster.

This study found a wide range of state voluntary cleanup programs, descriptions of which appear in more detail below.

None of the clusters are negatively correlated (Table 1.3). Given that states generally combine some form of financial incentives, liability provisions, avenues for public participation, and stipulation for long-term stewardship, these positive correlations are not surprising. However, there is substantial and substantive variation in program design and structure as evidenced by the wide range of correlations among cluster profiles (ranging from 0.009 to 0.69). Given the variability in the policy structure of these programs, each ideal type described below represents a different set of (dis)incentives for participation.

Table 1.3   Pearson's Correlation Coefficients among cluster profiles

	HLPIS	**HPNCL**	**HILP**	**CLLS**
HPNCL	0.26			
HILP	0.52	0.51		
CLLS	0.22	0.12	0.009	
Mid-Range	0.47	0.69	0.65	0.22

High Liability, Participation, Incentives, and Stewardship (HLPIS) Programs

Three states (MN, MI, and PA) have high liability, participation, and financial incentives programs. In many ways, these are hybrid programs because while they are progressive on many important facets of brownfield legislation, including stewardship, public participation and funding mechanisms, these states score highest on the CERCLA liability index (0.89).[15] Each of these states also makes use of other forms of liability. This cluster also scores highest of all clusters in terms of participation (1), financial incentives (0.5), and long-term stewardship indices (0.67). They are also endeavoring to track some form of economic benefit for the state from brownfield redevelopment (1). All three of them give grants to citizen groups and strongly encourage public participation. Of the four states in the sample that utilize all five forms of identified public participation, three of them are in this cluster.[16]

These three states are in some sense an outlying cluster, implying that, as a complete cluster, their programs are more like each other than any other state or cluster in the sample. In programmatic terms, these three programs are highly unique formulations of voluntary cleanup programs. The best specimen for this cluster is Minnesota.[17]

---

15   Numbers in parentheses indicate the cluster's average score on the index or variable.

16   The fourth state to make use of citizen group grants and the other four traditional means of public participation is New York. New York is in a cluster peripheral to the High Participation, No CERCLA Liability cluster.

17   Best specimens are found by correlating individual state programs with the cluster profile. The state with the largest Pearson's Correlation Coefficient is the best specimen.

High Participation, No CERCLA Liability (HPNCL) Programs[18]

Seventeen states (NJ, WV, KY, MS, SD, NV, OR, MT, ND, GA, AK, SC, TN, AZ, NE, NC, WA) have voluntary cleanup programs that score highly in terms of public participation and do not utilize any form of CERCLA liability provisions. This cluster is progressive in terms of public participation (0.87), though no members offer citizen group grants, and in stewardship (0.56). This cluster scores the lowest on the fiscal incentive index (0.30). All states in this cluster make use of other forms of liability protection (1). None of these states are currently tracking economic impacts or have MOAs with EPA. Montana and North Dakota are the best specimens of this cluster.

High Incentives, Low Participation (HILP) Programs

Eight states (CA, HI, CO, FL, DE, OH, TX, LA) have programs that score highly on the financial incentives index and low on the public participation index. These states offer a bevy of financial incentives (0.46), mostly in the form of loans but provide the fewest avenues for public participation of any cluster (0.66). All of these states are currently tracking economic impacts of redevelopment and cleanup.

These states all use forms of liability protection other than CERLCA (only Delaware makes use of any form of CERCLA liability). This cluster scores lowest on the public participation index, although all states within this cluster still encourage public participation through public record. No states in this cluster give grants to citizen groups. The best specimen for this cluster is Texas.

(CLLS) Programs

Seven states (CT, VT, MA, ME, UT, RI, NH) have programs that generally utilize some form of CERCLA liability and no other forms of liability provisions. These states also score lowest on the long-term stewardship index (0.14) and only three states in this cluster make use of some type of stewardship provision. This set of programs are the only programs in the nation that offer bonds as a method of

---

18   There is a peripheral cluster of three states (NY, AL, ID) that clusters together in all three data runs. However, the cluster that it attaches itself to changes depending on the algorithm. This peripheral cluster has the same characteristics as the HPNCL cluster or of a combination of the HPNCL and the Mid Range cluster, but in a weakened form. In other words, this peripheral cluster does not represent a different set of programs, merely a weaker version of the core to which it is attached. This is, however, the only cluster of states to offer environmental insurance to participants in their brownfield redevelopment programs.

offsetting remediation and redevelopment costs. This is the only cluster in the sample that offers no other forms of liability besides those utilized in CERCLA.[19]

This cluster also has the second lowest score on the public participation index (0.71) and no state in this cluster offers grants to citizen groups. Interestingly, this cluster, with the exception of one state, is geographically contiguous. All states in this cluster, except Utah, belong to EPA Region 1. Rhode Island is currently the only state in this cluster that is tracking economic impacts of remediation and redevelopment. Connecticut is the best specimen for this cluster.

Mid-range Programs

Twelve states (AR, NM, OK, KS, WY, VA, IN, MO, IA, IL, WI, MD) offer brownfield remediation programs that contain characteristics from all other clusters. These programs score in the middle of all clusters on all indices and variables. No states in this cluster are currently tracking economic impacts of remediation and redevelopment.

In terms of liability, all states in this cluster offer other forms of liability and only two states utilize some type of CERCLA liability. No state in this cluster offers grants to citizen groups, but nearly all require public notice and comment, while two-thirds also require public hearings.

The states in this cluster offer a range of financial incentives to help with the costs of remediation and redevelopment. None of the states offer environmental insurance, but most offer loans as a way of financing remediation activities. The best specimen for this cluster is New Mexico.

Implications and Discussion

One of the ways of framing the forgoing results is that each cluster represents a different incentive structure for participation in brownfields programs. For example, in the CLLS cluster, states do not offer much incentive to participate through flexible liability structures, thus potentially limiting voluntary participation in brownfields projects. The CLLS cluster also has a fairly low public participation score, indicating that there is little opportunity for the citizens to participate in remediation or redevelopment aspects of projects. Additionally, once a site has gone through a voluntary program in these states, there appears to be very little involvement from state agencies to oversee changes or potential issues. It appears, then, that these programs may function more similarly to Superfund (i.e. relying on litigation and state- mandated action) more than a collaborative, interorganizational policy process.

---

19   Utah is the only state which does not offer some sort of liability provision for its state response program.

The HILP cluster, on the other hand, offers several types of financial incentives and flexible liability provisions, but does not offer many avenues for public participation. Across the board, these types of programs offer the fewest opportunities for public participation, regardless of the type of participation. The role for the private sector, however, may be greatly enhanced because of the opportunities for financial aid and/or reimbursement for projects. Thus, while these states may represent an inter-organizational collaboration, the role of citizens is potentially diminished in favor of private enterprise.

As a final point of contrast, the HLIP type programs are hybrid programs, combining traditional CERCLA elements with more progressive environmental policy elements. For example, the programs in this cluster make use of liability strategies associated with CERCLA. All programs make use of joint and several and retroactive liability; all but one program in this cluster also utilizes strict liability. Thus, like the CLLS cluster and CERCLA, these programs could have limited voluntary participation. However, all the states in this cluster also make use of more flexible liability statutes and require all the traditional forms of public participation. These states all offer grants to citizen groups to encourage participation. The implication here is that these states are progressive in terms of their relationship with citizens on project sites. These states also seem to understand the economic and environmental aims of brownfield redevelopment; they offer the most opportunities for funding and track economic impacts of site activities, but they also maintain a prominent role for the state in oversight after remediation. Thus, these programs have several structural elements that are consistent with complex, interorganizational policy making while maintaining a strong role for the state.

The next empirical step is to link these different clusters to measures of outcomes and participation. This move, however, is remarkably difficult to make because most states do not track outcome data, environmental, economic, or otherwise on their brownfields programs (Bartsch, Anderson and Dorfman 1999). The EPA itself has been chastised recently for not measuring any sort of environmental outcome of the Brownfields program (GAO 2004). Additionally, there is a tremendous amount of variation in projects within state borders that makes generalizations difficult to draw. Thus, on any particular site, the individuals involved in the remediation and redevelopment activities play a prominent role in determining whether or not the project is successful.

What is clear is that there are discernible and substantive patterns in the incentives for participation in state voluntary cleanup programs. State programs vary by the number of avenues for public participation, types of financial support, liability protections, and provisions for long-term stewardship. In addition to impacts for participation, these incentives may have long-term impacts for outcomes and the results of state brownfields programs.

References

Alberini, A., A. Longo, S. Tonin, F. Trombetta and M. Turvani. 2005. "The Role of Liability, Regulation and Economic Incentives in Brownfield Remediation and Redevelopment: Evidence from Surveys from Developers." *Regional Science and Urban Economics* 35: 327–51.

Andrews, R.L. 1999. *Managing the Environment, Managing Ourselves: A History of American Environmental Policy*. New Haven: Yale University Press.

Bartsch, C., C. Anderson and B. Dorfman. 1999. *Brownfield Voluntary Cleanup Program Impacts: Reuse Benefits, State by State*. Washington: Northeast-Midwest Institute.

Garnham, D. 1976. "Power Parity and Lethal International Violence, 1969–1973." *The Journal of Conflict Resolution* 20(3): 379–94.

ICF Consulting and The E.P. Systems Group, Inc. 1999. *Assessment of State Initiatives to Promote Redevelopment of Brownfields*. Report prepared for U.S. Department of Housing and Urban Development, Office of Policy Development and Research (HC #5966, Task Order 13). Washington, DC.

Lange, D. and S. McNeil. 2004a. "Clean It and They Will Come? Defining Successful Brownfield Development." *Journal of Urban Planning and Development* 130(2): 101–8.

——. 2004b. "Brownfield Development: Tools for Stewardship." *Journal of Urban Planning and Development* 130(2): 109–16.

Pierre, J. and B.G. Peters. 2000. *Governance, Politics and the State*. New York: St. Martin's Press.

Ramseur, J.L. and M. Reisch. 2006. *Superfund: Overview and Selected Issues*. Washington, DC: Congressional Research Service.

Reisch, M. and D.M. Bearden. 1997. *Superfund Fact Book*. Washington, DC: Congressional Research Service.

——. 2003. *Superfund and the Brownfields Issue*. New York: Novinka Books.

SRA International, Inc. 2006. *State Brownfields and Voluntary Response Programs: An Update from the States* (Appendix B). Report prepared for U.S. Environmental Protection Agency, Office of Solid Waste and Emergency Response. Washington, DC.

U.S. Conference of Mayors. 2003. *Recycling America's Land: A National Report on Brownfields Redevelopment*. Volume IV. Washington, DC.

U.S. Congress. House. 2001. *Small Business Liability Relief and Brownfields Revitalization Act*. 107th Congress, 1st sess.: HR 2869.

U.S. Environmental Protection Agency. Office of Solid Wasted and Emergency Response. 1997. *EPA: Brownfields National Partnership Action Agenda*. Quick Reference Fact Sheet. Washington: Office of Outreach and Special Projects.

U.S. Government Accountability Office. 2004. *Brownfield Redevelopment: Stakeholders Report that EPA's Program Helps to Redevelop Sites, but Additional Measures could Complement Agency Efforts.* Washington, DC: Government Printing Office.

Urban Institute, Northeast-Midwest Institute, University of Louisville, University of Northern Kentucky. 1997. *The Effects of Environmental Hazards and Regulation on Urban Redevelopment.* Washington, DC: Report Submitted to U.S. Department of Urban Development and U.S. Environmental Protection Agency (UI Project No.; 06542-003-00).

Chapter 2

Changing Agendas in State Environmental Policy: Development versus Cleanup in Brownfield Programs

Richard C. Hula

Brownfield redevelopment has become an important theme in environmental policy as both federal and state governments rush to implement new programs to redevelop land parcels that are "abandoned, idled, or underused industrial and commercial facilities where expansion or redevelopment is complicated by a real or perceived environment contamination."[1] Collectively brownfield programs challenge several fundamental assumptions driving American toxic waste policy since the 1960s. First, there has been a major shift in authority to the states. A second important change has been the movement from a punitive regulatory framework to one stressing cooperation, self-interest and incentives. Perhaps the most interesting is a shift in the fundamental justification for toxic cleanups. While public health remains an important goal, economic development now plays a central role in the design and implementation of site cleanups. A number of states now factor in the probability that a site will actually be placed back into productive economic use, as well as the potential threat to public health, when making decisions about where to invest state cleanup funds.

This chapter explores the diffusion of these state level brownfield programs. As an introduction, the first section provides a brief description of the policy context in which these new programs have developed. The second section outlines a framework to describe key elements of emerging state programs. The third section reviews past research comparing state level environmental policy. An effort is made to sketch a more adequate theoretical basis for interstate comparisons in the fourth section. The final section of the chapter tests a set of hypothesis about the diffusion of innovative state brownfield programs.

---

1   This is the definition of brownfields commonly cited by the federal Environmental Protection Agency (EPA). See Kaiser (1998).

The Policy Context[2]

For many years the federal government was the dominant force in public efforts to clean contaminated properties. This authority was asserted in the 1980 Comprehensive Response, Compensation, and Liability Act (CERCLA), and its 1986 reauthorization, the Superfund Amendments and Reauthorization Act (SARA). Key to understanding CERCLA is its strong commitment to central (federal) decision making, restoring sites to a "natural" condition and imposing cleanup costs on those responsible for the site pollution.[3]

While CERCLA reserved the nation's worst toxic sites for federal action, thousands of potentially contaminated sites that were not sufficiently dangerous to qualify for federal action were left to the states. In general, state policy makers opted for a legislative framework similar to CERCLA. States implemented command and control programs in which most key decisions about specific cleanups were made by state authorities. State laws usually imposed a liability similar to that imposed by CERCLA. For example, 41 states have strict liability, 36 have several and joint liability to allocate responsibility for costs among responsible parties. Forty-three states impose retroactive liability. Thirty-two states have authority to recover for damages to natural resources.[4]

For almost two decades, CERCLA and various "state superfund" programs defined the parameters of national toxic waste policy. Increasingly, however, this policy network has been challenged from a number of different sources including the EPA itself, Congress and state authorities. The result has been important changes in policy including levels and type of public engagement, the overall role of federal authorities in toxic waste policy, and even the justification for public action.

Many state authorities see the federal preeminence that was built into CERCLA as no longer appropriate, arguing that it is time to shift major responsibility for toxic cleanups to the states. This argument is not simply based on a critical view of CERCLA, but on a positive assessment of existing state capacity. This general view is supported by annual reports issued by the Environmental Law Institute that document a broad increase in the capacity of state programs to deal with toxic waste cleanups. This capacity has developed along several dimensions. Most states have a set of laws in place governing the cleaning of contaminated property, and have established funds to underwrite cleanups where no responsible party

---

2   For a more complete discussion of past cleanups policies see Hula (2001).

3   Although CERCLA liability is often described as implementing the principle of "polluter pay," it is actually a good deal more punitive. For example it extends to owners who fail to take sufficient care to discover previous contamination. Responsible parties are theoretically liable for the full cost of a site cleanup, no matter what proportion of the contamination they actually caused. See Reisch (1997) andMeltz (1998).

4   For a more complete description of the state "superfund" programs see Environmental Law Institute 1989, 1991.

can be found. A number of state programs have developed the necessary technical expertise to deal with NPL caliber sites. In fact, Copeland (1997) reports that state authorities have assumed the lead responsibility for cleanups at about 10% of active NPL sites. In addition, state authorities have largely assumed responsibility for the identification and initial assessments of potential NPL sites (United States General Accounting Office 1999).

Although state participation in CERCLA cleanups has increased, the procedures that govern such participation severely restrict the overall impact this participation has on the fundamental approach and structure of the program. Since CERCLA provides no mechanism for the assumption of authority for cleanups by state agencies, states are allowed to undertake or supervise cleanup by responsible parties only through specific contracts or cooperative agreements with the EPA.[5] Such arrangements are developed on a case by case basis. Typically, state participation demands extensive negotiation between state authorities and the EPA. Even where agreement is reached, the EPA retains the right to reenter the site if there is reason to suspect that the state has failed to adequately perform its oversight functions. Although such intervention is rare, the symbolic force of such qualifications has not escaped state actors.

As states began to question CERCLA authority, they have become much more innovative in dealing with contaminated sites under their jurisdiction. This trend began in the early 1980s when a few states began to experiment with voluntary cleanup programs. These programs emphasized a less complex administrative organization and partial liability relief from cleanup costs if private parties would take the initiative to rehabilitate their properties to the point that they could be returned to productive economic use. Usually these voluntary programs were used in tandem with mandatory CERCLA-like programs. Indeed, the existence of mandatory programs was often seen as a powerful inducement for property owners to enter a "voluntary" program.[6]

State voluntary cleanup programs have become very common across the nation. The Environmental Law Institute reports that 44 states had such programs

---

5   In this respect CERCLA is unlike a number of other environmental programs that explicitly promote state participation. For example the Solid Waste Disposal Act (RCRA) allowed for state authorized programs (Copeland 1997). As of 2010, 50 states and territories have been granted authority to implement the base, or initial, program. Many are also authorized to implement additional parts of the RCRA program such as Corrective Action and the Land Disposal Restrictions. Since the regulatory responsibilities of CERCLA and RCRA often overlap, there is often a good deal of tension between state and federal actors over site jurisdiction.

6   State authorities continue to use the threat of federal EPA action as an additional powerful inducement for parties to state joint voluntary programs. See Hula (1999a). This explains what might seem to be a curious finding of widespread support among state officials for voluntary programs at their level along with continued mandatory program at the federal level. The ability to threaten federal intervention provides a powerful incentive for state level "cooperation."

in place in 1998. The speed at which these programs have been adopted is striking. A 1995 survey reported 35 states with voluntary programs. In 1993 there were only 14 (Environmental Law Institute 1989; United States General Accounting Office 1997). Almost all of these voluntary programs offer some set of inducements for private parties to participate. It is difficult to summarize these programs, however, given the wide range of specific provisions they contain. For example, 40 states now offer some measure of liability relief to the new owners of contaminated property. However, the form and magnitude of that relief varies substantially across states. Adding to this complexity is the fact that some states do not have a single voluntary program, but rather operate several, each with separate targets, conditions and inducements.

A number of states have initiated more specialized voluntary programs targeted specifically to brownfield redevelopment. However, the distinction between targeted brownfield programs and more general voluntary cleanup programs is often arbitrary. In some states the distinction has little relevance. In general, however, two elements of brownfield programs stand out. First is a relatively greater emphasis on urban industrial and commercial sites. Second, these programs also stress the redevelopment of property rather than simply efforts to clean it. As a result brownfield programs stress the need to identify the planned use of the property after the cleanup is complete.

Measuring State Innovation in Brownfield Policy

An important issue in the existing literature on state environmental policy is the appropriate level of aggregation at which policy should be measured. Much of the initial research in this area focused on relatively narrow media based policy such as air or water quality. Such a narrow focus may give a false impression of overall environmental policy. Much of the more recent literature has relied on a broader conceptualization of environmental policy (Klyza and Sousa 2008; Vig and Kraft 2008). This is a reasonable argument to the extent that a single linear policy dimension can represent various environmental programs. However, aggressive brownfield programs are not simply "more" of the traditional environmental effort, but rather an effort of a different sort. A more program-focused analysis is justified to the extent brownfield programs represent a redefinition as well as an expansion of state environmental effort.

Innovation is often difficult to define with any degree of precision. This is certainly true for brownfield programs. The enormous variation among states in terms of concrete programs and strategies introduces a set of serious conceptual and practical difficulties in comparing state efforts. There is simply no "model plan" which can be used as a standard to compare programs. In an effort to develop a useful benchmark, three general criteria will be used here. The first is based on the set of legislative and policy initiatives that form the legal basis of state programs. A second standard will be the level of state resources (such as financial and staff support) that have been committed to the programs. A final standard will be based on the magnitude of state efforts to promote community participation in the planning and implementation of specific site projects.

Legislative and Policy Basis

Several national organizations have published summaries that attempt to capture the legislative and formal policy basis of state brownfield programs. Reports have been issued by the Environmental Law Institute (1991), the Northeast-Midwest Institute (Bartsch and Anderson 1998), and the Consumers Renaissance Development Corporation (1998).[7] Each provides a useful overview of state level programs, and documents the significant variation across state programs, both in terms of specific strategies used and the state's overall commitment to brownfield redevelopment. Often significant variation exists even when states seem to embrace a similar approach. For example, almost all states with active programs have made a commitment to provide some measure of relief from cleanup liability and economic subsidies for developers. However, the particular form these incentives take in practice can be quite different.

The report by the Consumers Renaissance Development Corporation (1998) provides the most systematic effort to assess the substantive content of state brownfield programs. Table 2.1 provides a summary of 10 specific dimensions on which CRDC ranked state programs. The CRDC report argues that the most important dimension is liability relief. Almost all state programs modify traditional CERCLA liability to provide some protection for non-responsible developers from the liability for environmental cleanups. However, state programs typically have other operational components as well. Often these include the use of more flexible standards to control site cleanups. Many states tie cleanup standards to the final use of the property. In practice this means that industrial and commercial sites are held to less demanding standards than property developed for residential use. States also reduce cleanup standards by allowing greater use of physical barriers, and institutional controls to reduce human exposure to on-site contamination. Finally, some states have simply lowered cleanup standards required of developers.

---

7   Consumers Renaissance Development Corporation was formed in 1996 to support the implementation of a brownfield program in Michigan. It was financed by the Michigan Jobs Commission, the Michigan Department of Environmental Quality, the Michigan Municipal League, the Michigan Developers Association and Consumers Energy of Michigan.

Table 2.1   CRDC key characteristics of state brownfield programs*

Standard	Description
Liability Relief	The extent to which liability protection existed for property owners not responsible for site contamination.
Cleanup Criteria	Amount and type of cleanup required by the state. Whether the state permitted the use of institutional controls on property.
Financial Incentives	The extent to which the state relieves developers of costs associated with site remediation.
Climate/Attitude of State and Local Officials	Levels of commitment by state government and local government to brownfield redevelopment. Considers whether state and/or local government provides incentives for redevelopment.
State Oversight	The extent to which state oversight was considered sufficient to be helpful, but modest enough not to be overbearing.
Agreement with USEPA	Whether the state has (or is negotiating) a Memoranda of Agreement (MOA) with U.S. EPA that permits state to take lead role (without EPA supervision) in remediation of sites in state cleanup program.
Related Policy Issues	The extent to which there are other policies and regulation in effect which had the capacity to either encourage or impede brownfield redevelopment.
Participation Requirements	The degree to which program rules made it easy or difficult for developers to participate in the brownfield redevelopment effort.
Fee Structure	The total set of costs associated with participation in the brownfield program.
Eligible Parties/ Sites	The rules that determine whether a specific party or site is eligible to participate in the program.

Note: *Derived from Consumers Renaissance Redevelopment Corporation (1998: 3–4).

Two CRDC dimensions are based on financial incentives offered to interested developers by state and local governments. One set of incentives targets the direct costs of site remediation. The second dimension captures pubic subsidies targeted to general costs associated with site redevelopment. The CRDC also ranks the level of state oversight.[8] One dimension is something of a legislative "catch-all" in which the CRDC identifies other legislation and executive efforts that positively promote brownfield redevelopment. The CRDC review also considers the capacity of the state to work independently of the federal EPA. States that have signed a memoranda of agreement (MOA) with the U.S. EPA in which the federal agency agrees not to take action on a contaminated site which is enrolled in a state cleanup program are seen to have greater autonomy, and thus capacity, than those states which have no such agreement. Finally the CRDC survey examines several structural characteristics of the brownfield program itself. These include

---

8   Quality of oversight was defined as a curvilinear function. It was assumed that some state oversight was essential for program success. However, "excessive" oversight was seen as negative program characteristic.

the relative ease of participating in the program, who is eligible to participate and the costs association with participation.

In an effort to summarize the relative effectiveness of each state program, the CRDC ranked state programs using a composite index consisting of a weighted sum of the dimension scores.[9] However, given the fact that states do not follow a single model in developing their brownfield programs, the CRDC assumption that program components can be combined into a simple linear index seems problematic. Rather than simply assume a linear structure, unweighted CRDC dimensions scores were factor analyzed to assess whether there was, in fact, a more complex underlying structure. The rotated matrix of factors generated by this analysis is reported in Table 2.2 and provides strong evidence that the CRDC assumption of a single linear policy dimension is an oversimplification. The analysis describes four rather than a single dimension. Fortunately these dimensions display a relatively clear intuitive structure. Given the attention it has received in the literature, it is not surprising that one of the components (Factor 4) centers on liability relief. Specific indicators captured by this factor include both formal state liability relief, and whether the state has an agreement with the EPA that the federal agency will not intervene in a site properly enrolled in a state program. Developers almost certainly see such an agreement with the EPA as protection from strict federal liability. Another factor (Factor 1) centers on cost issues facing developers. Indicators include available financial incentives to reduce cleanup costs, climate (which includes non-cleanup financial incentives) and breadth of eligibility. It is interesting to note that liability relief also contributes to this factor. This is consistent with the view that traditional CERCLA liability generates both a known cost (expected cleanup costs), and a less predictable liability if unexpected contamination is found. State policies to reduce overall liability actually address both issues simultaneously. A third factor (Factor 2) identified a number of regulatory issues including cleanup standards, state oversight policies, and a number of "related" issues. A final factor (Factor 3) is primarily composed of rules governing participation and program costs.

---

9   The weighting applied to the index was based on the judgment of CRDC staff as to what elements of a state program were the most important in generating redevelopment opportunities. Since a great emphasis was placed on the importance of liability relief, the final CRDC index weighted liability relief so that it accounted for 50% of the final index. A 0.1 weight applied to flexible cleanup criteria, financial incentives and overall state policy climate. State oversight, the existence of a MOA with EPA, "related issues," participation rules, fee structure and eligibilityfix format collectively account for 20% of the final index.

Table 2.2   Rotated component matrix of factor analysis of unweighted CRDC scores*

Demension	Component			
	1	2	3	4
	Developer Costs	Regulatory Issues	Participation and Program Costs	Liability Relief
Liability Protection	**.55**	.23	.005	**.55**
Cleanup Criteria	.23	**.57**	.13	.18
Financial Incentives	**.65**	.39	-.31	.01
Climate	**.80**	.15	.0006	.19
State Oversight	.002	**.87**	.009	.12
Agreement w/U.S EPA	.002	-.003	-.11	**.88**
Related Issues	.008	**.69**	-.008	-.21
Participation	.004	.005	**.85**	-.003
Fee Structure	.005	.005	**.87**	-.006
Eligibility	.75	-.005	.19	-.21

Note: Numbers in bold are included in factor. * Extraction Method: Principal Component Analysis. Rotation Method: Varimax with Kaiser Normalization. Rotation converged in 5 iterations.

Commitment of Public Resources

Students of implementation have convincingly documented that legislative action and executive pronouncements are only one element of a complex network of factors that determine public policy outcomes. One key factor is the commitment of public officials to the goals of formal policy (Ringquist 1995). This notion of commitment is itself complex, ranging from a collective willingness to invest public funds to a more personal commitment by political leaders to work to see specific efforts are successful. This chapter focuses on the more straightforward collective decisions to staff and finance brownfield programs. A review of published data reveals significant interstate variation in staff assigned to programs as well as public outlays.

A reasonable objection might be taken to the ranking of state brownfield programs based on public investment given the strong commitment in all state programs to engage private development capital in the redevelopment process. It is conceivable that state programs that are most successful in attracting private capital might use less public funding than less effective programs. Even if this is the case, however, it is reasonable to suggest that innovative programs will require some additional public investment.[10]

---

10   Note that there is some evidence that the level of public expenditure is positively correlated to the number of reported sites cleaned and redeveloped (Hula 2000).

Public Participation

A number of the federal initiatives to support brownfield redevelopment have stressed the importance of engaging the impacted community in planning for site redevelopment. The logic is straightforward: a public program to redevelop brownfields should benefit the broader community subsidizing the development process. Not surprisingly, this emphasis on public participation is controversial. Developers and local officials often argue that stringent requirements for public participation can slow, and sometimes kill, specific site initiatives. It is argued that local residents may have a clear idea of what sort of facility they would like to have in their community, but often have no idea of the economic reality that might constrain such development. Advocates of community participation counter that if local residents are not engaged in the planning process, fatal opposition to the project is likely to develop in later stages. If such opposition does emerge the prospective developer is likely to face greater economic losses than those produced by early community engagement.[11] It is interesting to note that a relatively large number of states (13) do not require any sort of public participation in the formulation of cleanup/redevelopment plans. Where participation is required, it typically takes the form of publishing a public notice (36 states), providing for public comment (34 states), and holding public hearings (31). A handful of states require other forms of participation. For example, two states provide funds to impacted parties and organizations to develop the capacity to review specific cleanup/redevelopment plans (Environmental Law Institute 1998).

Identifying Correlates of State Environmental Policy

A substantial literature describing correlates of state level environmental policy exists (Lester 1990; Hays 1996). Typically an effort is made to statistically explain state level variation on a measure of environmental policy using a set of independent variables thought likely to be associated with programmatic decisions. Although this literature is relatively atheoretical, it does provide valuable clues about factors driving environmental policy decisions. Variables that have been tested include pollution levels, state resources, state bureaucratic capacity, the relative strength of environmental and industrial interest groups, political culture, and partisan control.

It is difficult to summarize the conclusions of this literature because of differing designs and operational definitions. The wide range of indicators used to measure environmental policy has already been noted. For example, a number of

---

11   Given the contentious debate that often surrounds the issue of public participation, it is noteworthy that there is actually very little empirical evidence as to what impact it actually has on redevelopment. For a more extended discussion of the debate about the appropriate role of citizen participation in brownfield redevelopment see Hula (1999b).

efforts target state allocations in relatively narrow media-specific programs such as air and water pollution. Other papers have focused on state efforts to implement federal environmental mandates such as RCRA and CERCLA. Still others have attempted to assess overall environmental policy using a variety of indices of state environmental effort. While this research is clearly complementary, significant variation is likely to occur across such measures. For example, Ringquist (1993) reports significant differences between factors that appear to drive state-level water and air pollution policy.

While recognizing the diversity of past efforts, some general patterns in results do exist. There is a positive association between environmental problems and level of state environmental regulation (Lester et al. 1983; Lester and Bowman 1989; Bacot and Dawes 1997).[12] A number of studies have found that state resources measured by income and wealth predict increased environmental effort (Hays, Esler and Hays 1996; Bacot and Dawes 1997). Legislative and administrative capacity also appears to be related to effort (Lester and Bowman 1989; Hays, Esler and Hays 1996; Bacot and Dawes 1997).[13] Other independent variables reported to have a statistical association with environmental policy include region, political culture, and the strength of industrial and environmental groups (Lester 1980; Lester et al. 1983; Lester and Bowman 1989; Bacot and Dawes 1997; Andrews 1998). Interestingly, Hays (1996) reports that industry and environmental groups both drive the state toward increased environmental regulation.

A number of scholars have sought to measure the impact of political party on environmental policy. Kamienieke (1995) has documented systematic differences between formal environmental positions adopted by the Democratic and Republican Party in their national political platforms. In general, Republicans are seen as more hostile to environmental protection efforts than Democrats. The capacity of federal policy makers to affect state level actors is also in dispute. The more common view, however, is that federal authorities have, at best, a modest capacity to compel state level action. Davis and Lester (1987) note that for the most part states did not replace federal environmental budget cuts implemented during the Reagan administration. In a study of overall state environmental efforts Hays and Esler (1996) conclude that the federal government has virtually no impact on state policy. In contrast, Lowry (1992) argues that the degree of federal influence varies significantly across issues areas. He suggests that while federal influence is constrained by competition among states, it can have a significant impact on state decisions.

---

12   For a dissenting view see Hays (1996).

13   Administrative capacity is often measured by the existence of a state authority charged with enforcing environmental regulations. However, Burby (1993) offers some evidence that such state level centralization may not always lead to the highest possible level of compliance. Davis (1999) also casts some doubt on the direct link between political resources and environmental policy.

There is some evidence that federal influence may take forms other than traditional command and control models. Kraft and Scheberle (1998) argue that recent federal initiatives to build more positive working relationships with state actors have had important impacts on environmental policy at both the federal and state level. Scheberle (1997) explicitly argues that federal impact will be greatest in a cooperative environment. Similarly Davis and Davis (1999) stress the importance of federal-state interaction and cooperation rather than direct federal dominance.

Innovation, Agenda Setting, and Brownfield Policy

Although the existing literature on state level environmental policy identifies correlates of the scope of environmental programs, it fails to address two key questions. The first centers on the issue of policy innovation. Measures of environmental policy typically focus on overall scope of state effort, the magnitude of specific media focused programs or the state response to federal mandates. Brownfield redevelopment programs, however, are better conceived as a redefinition of environmental policy than a linear increase in overall state effort. Indeed, scholars and practitioners point out that these efforts are as much economic development programs as environmental. The failure of existing correlation studies to explain much about program redefinition identifies the second weakness of the existing literature, an absence of a theoretical context. While variables included in past analyses are generally intuitively reasonable, they remain ad hoc. A more conceptual analysis of policy change is required.

There is a good deal of debate about how one ought to conceptualize the process of policy innovation in the American political system. There are two distinct themes in the political science literature. One perspective stresses the underlying stability in public policy. This stability is seen as a function of the capacity of a set of dominant interests to maintain control over the decision making process in a specific substantive area. Baumgartner and Jones (1993) argue that such policy monopolies require two critical conditions. First, policy decisions must take place in an identifiable institutional structure and this structure must limit external access to the decision making process. In addition, there must be a powerful supporting idea that provides a popular "explanation" for the dominant institutional structure. Examples of such ideas include claims of expertise, public interest, or other goals having broad support in American political culture. To the extent that an institutional structure is strongly identified with such goals, it is more likely to be able to sustain itself as a policy monopoly.

Much of the early interest group literature argued that many policy areas operated in relative isolation, and were often dominated by very narrow interests (McFarland 1991). A number of labels were attached to these areas including issue networks, iron triangles and policy monopolies. A relatively large case study literature in which scholars documented how institutional structures were receptive to only a narrow set of interests seemed to support the claim that policy monopolies were common (Baumgartner and Leech 1998). However, the key variable of time was often missing from this literature. Later studies of alleged policy monopolies showed that they were, in fact, quite fragile and subject to successful external attack (Baumgartner and Jones 1993).

A substantial body of empirical work that stresses the relative instability of policy outcomes has challenged the equilibrium paradigm. Students of agenda setting argue that so-called policy monopolies are seldom as stable or impenetrable as they might appear at any given point in time. This instability is partly a function of the complexity and decentralization of the American political process. There are simply too many opportunities for policy entrepreneurs to engage the political process for any narrow set of interests to remain immune from external assaults (Baumgartner and Jones 1993). Schattschneider (1975) argues that the political process is not simply a clash of ideas and interests. An important element of any political debate is the venue in which the debate takes place. Change in the decision venue will often change the mix of core decision-makers. This is likely to change the mix of resources available to various interests, and can ultimately change political outcomes. Note, that there may be significant disadvantages to Schattschneider's notion of expanding conflict. For example, Vizzard (1995) claims that broad pluralist conflict tends to produce policy crafted for political advantage not practicality. Successful policy needs to be grounded in the mechanics of the field. When it is not (for example, in the case of gun control) the results are unlikely to be very effective.

Cobb and Elder (1972) describe four conditions as necessary for a redefinition of a public agenda. These include: 1) broad public awareness of the issue; 2) consensus that the issue defines an important problem; 3) that it is a legitimate public issue; and 4) that corrective action by the government is possible. Kingdon (1984) argues the process is best seen as one in which political entrepreneurs and policy streams occasionally converge for a short period of time. Both Kingdon (1984) and Cobb and Elder (1972) point to the critical importance of public perception. That is, a critical element of any public debate is the public definition of the issues at stake. Success or failure in a policy debate is as much a function of being able to frame the terms of public debate, as it is convincing the public to support a specific position. Indeed, once an issue is "defined," the identification of appropriate public policies may be nearly automatic. Nothing in this argument implies that policy change is easy; it clearly is not. Nor does this view deny that strong structural biases operating in favor of certain interests exist (Elkin 1987). That said, it seems beyond doubt that policy can and does change in significant ways over time. Such change involves not only substantive political outputs, but also the mix and location of key decision-makers. Any policy change model must identify strong structural constraints to change, while identifying factors that can promote change (Ramsay 1996).

Baumgartner and Jones (1993) have attempted to reconcile these competing perspectives—a stable versus unstable policy process—through their notion of a "punctuated equilibrium." They argue that policy networks do emerge and are often stable for a time. However, this stability is subject to external disturbances, which can result in quite a different policy. The transformation of regulatory state cleanup programs focused on public health to more cooperative efforts emphasizing both health and economic development provide a fascinating example of such a process. Although the agenda setting literature is not sufficiently developed to allow one to deduce concrete hypotheses, it does provide broad direction about what factors ought to predict policy innovation.[14] Five hypotheses are proposed.

Hypothesis 1: States with the greatest resources are most likely to promote innovative brownfield programs.

The agenda setting literature is quite clear that the adoption of policy is clearly dependent on a perceived a capacity to implement it. Since new programs will inevitably require additional public resources, the level of available resources is a critical factor for policy makers. Not surprisingly some measure of state resources is typically incorporated into efforts to model state environmental policy.[15] Earlier efforts to predict state environmental policy have a variety of resource measures. One approach emphasizes basic demographic indicators such as state size, population, income and state revenue (Bacot and Dawes 1997). Hays (1996) defined state economic resources as the absence of debt rather than income. Other research has focused on the more specific issue of institutional capacity. Such capacity has been measured by indicators ranging from the level of "professionalism" in the state legislature, to the extent to which the legislature is able to structure implementation of environmental policy (Lester 1980).

Hypothesis 2: States with the greatest problems (i.e. number of contaminated sites) are most likely to promote brownfield programs.

Policy change will occur only if there exists a popular and elite perception that there is in fact some problem that needs to be addressed by public authorities. In general, we assume that the perception of environmental problems is, at least, related to objective problem levels. As noted above, most efforts to model state level environmental policy introduce some measure of environmental quality as an independent variable. The specific indicator used varies from general measures of environmental quality to more specific indicators targeted to measuring specific

---

14   It is reassuring to note that a number of the variables which are consistent with the agenda setting literature have been incorporated into past empirical studies of state environmental policy.

15   Although there is a relatively large literature that examines correlates of state environmental policy, the selection of independent variables to predict policy outcomes is done atheoretically. See, for example, Lester and Bowman (1989).

media. The indicators are usually chosen to match the level of the environmental dependent variable.

Hypothesis 3: States with innovative brownfield programs will be found in geographic proximity to each other.

Hypothesis 4: States receiving relatively greater federal brownfield assistance will have greater brownfield innovation.

Policy innovation requires the perception of a solution to a problem. Hypotheses 3 and 4 offer two alternative mechanisms by which state authorities might learn of such alternatives. Hypothesis 3 posits that state authorities learn of new programs from neighboring states. Hypothesis 4 tests the proposition that a set of federal EPA pilot programs promoting brownfield redevelopment have encouraged state level innovation.

Hypothesis 5: States that have the most competitive political systems are most likely to have innovative brownfield programs

The literature on state environmental policy typically argues that the Democratic Party is generally more supportive of public environmental programs than the Republican Party (Hula and Hemmond 2004; Vig and Kraft 2006). It is not clear, however, whether this logic fits this case. Brownfield programs do not comfortably fit with many traditional environmental programs. Supporting this view is the fact that brownfield programs are often supported by extraordinarily diverse political coalitions. Major environmental groups seldom lead such coalitions, and in some states are hardly visible at all. In many states, it is a coalition that defies traditional expectation, cutting across political party, ideology, and region. The problem set which brownfield redevelopment addresses is equally diverse. While it certainly includes site remediation, it also incorporates other important goals. For urban policy makers the dedication of contaminated sites to future industrial and commercial use transforms brownfield redevelopment into an economic development tool. To the extent that brownfield redevelopment reduces development pressure on rural and agricultural land, it serves as a greenspace preservation effort. Indeed, brownfields are seen by some as a means to achieve key elements of comprehensive land use policy without the need to actually impose direct regulation (United States Conference of Mayors 1990; Michigan State University Center for Urban Affairs 1998).

The nature of program support clouds expectations about the partisan link to brownfield redevelopment. As noted above, many state brownfield programs not only stress traditional environmental values, but also emphasize reduced government regulation, public-private partnerships, and economic development. These latter values seem more consistent with policy positions of the Republican

Party.[16] Given this convergence, it seems possible that in highly competitive states, brownfield policy will emerge as a policy capable of drawing widespread support across a bipartisan coalition. In less competitive states, parties may be willing to implement more "pure strategies" consistent with their ideological view.

Analysis

Each of the hypotheses outlined above was subjected to a simple bivariate test. Following these individual tests, a summary model is presented to test the combined effects of the five independent variables.[17]

Hypothesis 1: States with the greatest resources are most likely to promote innovative brownfield programs

Five indicators measure state resources: (1) total population—1997 estimate, (2) median household income—1997 estimate, (3) state general expenditure, (4) public welfare expenditure, and (5) total personal earnings generated by manufacturing—1995 estimates. Size (population) and wealth (median income) can be seen as relatively crude capacity measures of the states to engage in a number of policy initiatives. If such capacity is associated with innovation in general, then it is likely that innovative brownfield programs will emerge in these states. A slightly different sort of capacity measure is the overall level of state expenditures. These budget figures capture both the wealth of a state and the political will to use this wealth for state initiated programs. Public welfare expenditures measure the state's willingness to engage social policy. Total earnings generated by manufacturing not only taps state capacity, but is also associated with the importance of manufacturing in the state economy.

Table 2.3 presents a correlation matrix for each of the state resource indicators and indicators of policy innovation. In general, the correlations reported in Table 2.3 are weak, and not statistically significant. The major exception is the strong and consistent pattern between staffing levels and state resources. This is consistent with the general expectation that larger, wealthier states will have a more developed public sector. Resource measures are also positively associated with state enactment of supportive brownfield regulations. There is little association between resources and other legislative innovation or requirements

---

16   The difficulty of identifying brownfields with a particular party is revealed by the complicated congressional politics surrounding the issue. There seems to be broad support for similar brownfield legislation in both parties. However, it has been impossible to pass such legislation because of partisan differences on the reauthorization of CERCLA. See Reisch (1998) and Reisch (1999).

17   The last complete report on state brownfield initiatives was public in 1998, and reported on the status of programs in 1997.

for public participation. It is particularly interesting to note that there seems to be no correlation between the importance of manufacturing (as measured by income generated) and liability relief.

Table 2.3   State size/wealth and brownfield innovation, FY 1997 [i, ii]

	Legislative and Policy Innovation				Required Public Participation	Staffing Levels
	Direct Cost Subsidy	Supportive Regulations	Ease of Participation	Liability Relief		
Population, 1997	.04	.44**	-.21	-.14	.20	.62*
Median Household Income, 1997	.02	.17	-.18	.14	-.05	.37**
Earnings (BEA), 1995	.21	.50*	-.22	-.01	.24	.65*
State General Expenditure, 1995	.04	.42*	-.18	-.17	.19	.70*
Public Welfare General Expenditure, 1995	.04	.37*	-.18	-.19	.16	.68*

Notes: [i] Measured by Pearson correlation with n=41 states. [ii] Demographic data taken from US Census, State and Metropolitan Data Book, 1997–98. * Indicates correlation is significant at the 0.05 level (2-tailed). ** Indicates correlation is significant at .01 level (2-tailed).

Hypothesis 2: States with the greatest problems (i.e. number of contaminated sites) are most likely to promote brownfield programs

Five indicators have been identified to measure the magnitude of statewide contamination: (1) number of National Priority List (NPL) sites, (2) known and suspected state sites, (3) state sites identified as needing attention, (4) state inventory and priority list, and (5) the number of manufacturing establishments in the state. The National Priority List is comprised of the nation's most seriously contaminated sites. These sites are not directly relevant to state programs since they are reserved for federal action under CERCLA. Nevertheless, it is reasonable to argue that there should be a correlation between the number of NPL sites and other, less severely contaminated sites. The next three indicators correspond to categories used by the Environmental Law Institute to estimate non-NPL contaminated sites. Known or suspected sites represent the broadest category of potential sites tracked by state programs. Sites identified as needing attention refers to sites that have been evaluated by authorities, and have been confirmed by state authorities as needing attention. The third category of state inventory refers to a formal state priority system. The final indicator is the number of manufacturing sites in the state. It assumes that manufacturing is in general correlated with the number of existing contaminated sites.

The empirical link between contamination levels and program innovation is examined in Table 2.4. There is a strong positive association between all indicators of contamination and program staffing. There is also a general pattern of positive associations between measures of contamination and both the implementation of direct cost subsidies and regulations to encourage brownfield redevelopment. For direct cost subsidies, only the association with NPL sites is statistically significant. For regulations, the correlation between NPL sites and the number of manufacturing sites is statistically significant. However, there is no link between any measure of contamination and liability reform or requirements for public participation.

Table 2.4   Level of contamination and brownfield innovation, FY 1997[i]

	Legislative and Policy Innovation				Required Public Participation	Staffing Levels
	Direct Cost Subsidy	Regulations	Ease of Participation	Liability Relief		
Final NPL Sites (n=41)	.36*	.40*	-.13	-.05	.23	.86*
Known and Suspected State Sites (n=37)	.26	.25	-.32	-.16	.19	.75*
State Sites Identified as Needing Attention (n=39)	.26	.28	.10	-.10	.22	.76*
State Inventory or Priority List (n=32)	.23	.15	.35**	.08	-.10	.44**
Number of Establishments[ii] 1994 (n=41)	.11	.48*	-.24	-.14	.21	.66*

Notes: [i] Measured by Pearson correlation. An * indicates correlation is significant at the 0.05 level (2-tailed). [ii] Data taken from US Census, State and Metropolitan Data Book, 1997–98. ** Indicates correlation is significant at .01 level (2-tailed).

Hypothesis 3: States with innovative brownfield programs will be found in geographic proximity to each other.

To test whether geography is related to innovation, mean innovation scores were compared across the nine federal EPA regions in the country. A simple visual inspection of these scores revealed that EPA Region 5 (Illinois, Indiana, Michigan, Minnesota, Ohio, and Wisconsin) was a consistent outlier on innovation measures. To further test this "outlier status," Table 2.5 compares mean innovation scores for states in and out of EPA Region 5. A comparison of mean scores for Region 5 and all other states supports the view that Region 5 is, in fact, "different." Region 5 states, on average, show higher degrees of innovation than non-Region-5 states on all measures of innovation. These differences reach levels of statistical significance (as measured by a t-test) for direct cost subsidies, ease of participation, and liability relief.

The mechanism that drives innovation in Region 5 is not clear. A number of rival hypotheses are consistent with these empirical results. For example, the key variable might be physical proximity. It might be a shared economic and political history. Policy intervention of the EPA regional office provides a third potential explanation. Without much more detailed contextual data, the findings in Table 2.5 point to an intriguing policy puzzle that demands further attention.

Table 2.5   EPA region and brownfield innovation, FY 1997

	Legislative and Policy Innovation				Required Public Participation	Staffing Levels
	Direct Cost Subsidy*	Regulations	Ease of Participation*	Liability Relief*		
States Not in EPA Region 5	-.18 (n=35)	-.01 (n=35)	.13 (n=35)	-.13 (n=35)	2.3 (n=35)	78 (n=35)
States in EPA Region 5	1.02 (n=6)	.53 (n=6)	-.75 (n=6)	.71 (n=6)	2.7 (n=6)	122 (n=6)

Notes: * T-test for difference means significant at .05 level.

Hypothesis 4: States receiving relatively greater federal brownfield assistance will have greater brownfield innovation.

Although states have been the key actors in brownfield redevelopment, the federal government has also attempted to encourage local development efforts through a series of pilot programs. Table 2.6 presents the empirical association between federal grants and state innovation. There is a strong correlation between each measure of grant program and staffing levels. There is a strong link between favorable regulations and targeting of federal pilot programs. There is also a positive link between levels of required public participation and pilot projects, although the association is statistically significant only for the number of projects, and not the total value of the projects.

Table 2.6   Federal pilot programs and brownfield innovation, FY 1997[i,ii]

	Legislative and policy innovation				Required public Participation	Staffing levels
	Direct Cost Subsidy	Supportive Regulations	Ease of Participation	Liability Relief		
Total Number of Pilot Programs (as of 1997)	.12	.45*	-.17	-.20	.30**	.59*
Total Value of Pilot Programs (as of 1997)	-.04	.46*	-.17	-.20	.22	.61*
Total Number of Showcase Communities (1998)	-.18	.28	-.34	-.06	.24	.51*

Notes: [i] Measured by Pearson correlation with n=41 states. [ii] EPA pilot data obtained from EPA web site: http://www.epa.gov/brownfields/pilot.htm. * Indicates correlation is significant at the 0.05 level (2-tailed). ** Indicates correlation is significant at .01 level (2-tailed).

Hypothesis 5: States that have the most competitive political systems are most likely to have brownfield programs

Table 2.7 examines the association between brownfield innovation and two measures of political party competition. The first is an average deviation from "perfect competition" in five electoral indicators: presidential votes in 1992 and 1996, makeup of the state's upper and lower legislative house, and composition of the state's congressional delegation. Specifically the percentage of votes or seats held by Republicans was subtracted from 50%. The absolute value of this difference was then summed and averaged. The second index is a measure of Republican Party dominance and is the average of Republican percentages on the five indicators used in the competition index.

Table 2.7 provides little support for the hypothesis that party competition generates innovation in brownfield policy. Nor is there any evidence in Table 2.7 to support the view that one party is more likely to implement such policy. The coefficients reported in Table 2.7 are uniformly small and so have no clear pattern in sign. None reach a minimal level of statistical significance.

Table 2.7   Partisan political variables and brownfied innovation, FY 1997 [i]

	Legislative and Policy Innovation				Required Public Participation	Staffing Levels
	Direct Cost Subsidy	Regulations	Ease of Participation	Liability Relief		
Party Competition Index (n=40)	-.26	-.15	.27	.05	.08	-.15
Republican Party Dominance Index (n=40)	.02	-.17	.22	-.03	-.18	-.04

*Note*s: [i] Party competition index is the average deviation from 50% for five indicators: percentage of Republican vote 1992 and 1996, percentage of Republican seats in upper and lower state house in 1996 and the percentage of Republican seats in the state's Congressional delegation. Republican party dominance measure is computed as the average of the same five indicators: percentage of Republican vote 1992 and 1996, percentage of Republican seats in upper and lower state house in 1996 and the percentage of Republican seats in the state's Congressional delegation. Political data were taken from US Census, *State and Metropolitan Data Book, 1997–98.*

Summary Model

Table 2.8 presents the results of a set of multiple regression models that incorporate indicators from each of the hypotheses discussed above. Thus, policy innovation is predicted using an indicator of state resources (BEA Income), the degree of contamination (# NPL sites), geographic region (EPA region 5), level of party competition (party competition index) and federal investment through EPA pilot programs (# federal pilots). Indicators were selected on the basis of explanatory power displayed in the bivariate analysis.

Table 2.8   Summary models prediciting brownfield innovation, FY 1997[i]

Independent variables	Legislative and policy innovation				Required Public Participation	Staffing levels
	Direct Cost Subsidy	Regulations	Ease of Participation	Liability Relief		
	Model 1	Model 2	Model 3	Model 4	Model 5	Model 6
BEA Income	-.45	.34	.04	.003	.19	.12
Number of NPL Sites	.48*	.04	.07	.012	.16	.75
EPA Region 5	.46*	.07	-.29	.34	.16	-.04
Party Competition Index	-.19	-.05	-.11	.13	.20	-.16
Number of Federal Pilots	.08	.27	-.22	-.26	.13	.15
N	39	39	39	39	48	48
R^2	.34	.33	.13	.16	15	.79

Notes: [i] Reports standardized regression coefficients. * Indicates coefficient significant at .05 level. ** Indicates significance at .01 level.

Certainly the most striking aspect of these models is their relatively modest predictive power. This is particularly true for legislative innovation in program participation rules (Model 3) and liability relief (Model 4), and levels of mandated public participation (Model 5). None of the models explain more than 16% of the variance in the dependent variable, and none of the individual coefficients are statistically significant (at the .05 level). The remaining three models are somewhat more powerful. For example, the decision by a state to provide cost reduction subsidies (Model 1) appears to be driven by levels of contamination and being part of EPA Region 5. Although not rising to the level of statistical significance, there are other relationships worth noting. In Model 1 state resources (measured by BEA income) are positively associated with state decisions to provide direct cost subsidies. Positive regulatory climate is positively associated with levels of state resources and participation in federal pilot programs (Model 2). Staffing levels are strongly driven by reported contamination levels (Model 6).

Perhaps more interesting than specific models is whether it is possible to identify general patterns across the models. Contamination levels, measured by the number of NPL sites, predict the scale of brownfield programs as measured by state direct cost subsidy (Model 1) and program staffing levels (Model 6). There are also important regional effects. These are particularly clear in subsidy efforts (Model 1), but may also exist in participation rules (toward less permissive) (Model 5), and liability relief (Model 4). Federal intervention through brownfield pilot programs has a modest relationship with permissive regulations (Model 2), restrictive participation rules (Model 5), and a lack of liability relief (Model 4). State resources as measure by BEA income are positively associated with cost reduction efforts (Model 1) and permissive regulations (Model 2). Perhaps the most interesting finding is the very modest predictive power of resources across the elements of innovative brownfield programs.

Conclusions

State level brownfield redevelopment programs present an intriguing set of stories to the policy scholar. On a substantive level they provide a case of innovative environmental policy. In fact, these efforts involve a number of significant policy shifts, including a redefinition of environmental policy goals from a pure cleanup strategy to one that also stresses redevelopment. These programs also redefine the role and responsibility of developers in site cleanups. Brownfield programs provide a number of interesting process issues as well. For example, they signal a shift from federal to state authority. Indeed, this decentralization often extends to the local level. Although specific projects are typically supported with state funding, they are often designed and implemented by local authorities. Thus, brownfield programs offer an empirical test to common claims that radical policy decentralization will result in better-designed and implemented policy.

This chapter has considered several elements of this complex story. Two were largely descriptive. An initial effort was made to describe the policy context in which brownfield initiatives have developed. The content and focus of specific state programs was then sketched. This showed that states use alternative strategies to promote redevelopment. While similar elements may be adopted in different states, quite different policy configurations are possible. One needs to take care when speaking of state brownfield redevelopment policies as if they were all the same. They clearly are not.

Which states are most likely to implement a brownfield program? Two factors stand out. The first is the existence of contaminated (or at least potentially contaminated) property. States with high concentrations of suspect properties are much more likely to invest program resources and initiate subsidy legislation. However, as noted, contamination levels were not associated with major elements of reform efforts including regulations, rules for participation, liability relief, and rules for public participation.

Geographic proximity is a second factor related to policy innovation. There was a significant clustering of innovating states in the EPA's Midwest Region 5. There are several possible explanations for this finding. It may be a spurious correlation generated by a shared history of manufacturing that has produced an above average number of contaminated sites. Proximity itself may play a role. States that are geographically close to innovators might model innovative behavior of their neighbors. It is also possible that federal authorities have played an important role in the diffusion of brownfield program innovation. The vigorous effort to promote brownfield redevelopment by the national EPA has provided regional offices with resources to promote policy innovation in their areas. It would hardly be surprising if the impact of such efforts varied systematically across agency regions given the high degree of autonomy regional offices have been given (Scheberle 1997). Central office policy can assume quite different appearances when executed by regional authorities.[18]

One of the more interesting findings reported here is the relatively weak explanatory power of state resources. While there is no question that resources serve as an important constraint on policy, these data argue against the view that economic factors drive policy independently of the political process. Indeed, the relative weakness of these aggregate models strongly argue that we will need to explore state level decision making processes in much greater detail to understand the dynamics of policy innovation.[19]

Brownfield redevelopment has generated widespread public support. It is likely that this support is a double-edged sword. On one hand, its popularity drives adoption in many states. However, its attraction to very diverse constituencies may generate unrealistic expectations. Nevertheless, brownfield redevelopment programs have the potential to simultaneously serve as a practical strategy to recycle contaminated industrial and commercial property, and offer the policy scholar an intriguing window from which to observe policy innovation.

---

18   For an interesting example of how regional offices can take the lead in resolving policy issues facing the EPA see Smith and Silver (1998).

19   One key element in the innovation process that demands further analysis appears to be leadership within the state or local political system. See Schneider et al. (1995) and Mintrom (2000).

References

Andrews, C.J. 1998. "Public Policy and the Geography of US Environmentalism." *Social Science Quarterly* 79(1): 55–73.

Bacot, A.H. and R.A. Dawes. 1997. "State Expenditures and Policy Outcomes in Environmental Program Management." *Policy Studies Journal* 25(3): 355–70.

Bartsch, C. and C. Anderson. 1998. *Matrix of Brownfield Programs by State*. Washington, DC: Northeast-Midwest Institute.

Baumgartner, F.R. and B.D. Jones. 1993. *Agendas and Instability in American Politics*, American Politics and Political Economy Series. Chicago: University of Chicago Press.

Baumgartner, F.R. and B.L. Leech. 1998. *Basic Interests: The Importance of Groups in Politics and in Political Science*. Princeton, NJ: Princeton University Press.

Burby, R.J. and R.G. Paterson. 1993. "Improving Compliance with State Environmental Regulations." *Journal of Policy Analysis and Management* 12(4): 753–72.

Cobb, R.W. and C.D. Elder. 1972. *Participation in American Politics: The Dynamics of Agenda-Building*. Boston: Allyn and Bacon.

Consumers Renaissance Redevelopment Corporation. 1998. *National Comparative Analysis of Brownfield Redevelopment Programs*. Jackson, MI: Consumers Renaissance Redevelopment Corporation.

Copeland, C. 1997. *Superfund and the States: The State Role and Other Issues*. Washington, DC: Congressional Research Service.

Davis, C. and S.K. Davis. 1999. "State Enforcement of the Federal Hazardous Waste Program." *Polity* 31(3): 451–68.

Davis, C.E. and J.P. Lester. 1987. "Decentralizing Federal Environmental Policy: A Research Note." *Western Political Quarterly* 40(3): 555–65.

Elkin, S. 1987. *City and Regime in the American Republic*. Chicago: University of Chicago Press.

Environmental Law Institute. 1989. *An Analysis of State Superfund Programs: 50-State Study*. Washington, DC: U.S. Environmental Protection Agency Office of Emergency & Remedial Response Hazardous Site Control Division.

——. 1991. *An Analysis of State Superfund Programs: 50-State Study, 1990 Update*. Washington, DC: U.S. Environmental Protection Agency, Office of Emergency and Remedial Response Hazardous Site Control Division.

——. 1998. *An Analysis of State Superfund Programs: 50 State Study, 1997 Update*. Washington, DC: Author.

Hays, S.P., M. Esler and C.E. Hays. 1996. "Environmental Commitment among the States: Integrating Alternative Approaches to State Environmental Policy." *Publius: The Journal of Federalism* 26(2): 41–58.

Hula, R. 1999a. *The Michigan Brownfield Initiative*. Paper read at American Political Science Association, September 2–5, at Atlanta.

——. 1999b. *An Assessment of Brownfield Redevelopment Policies*. Washington, DC: PricewaterhouseCoopers Endowment for the Business of Government.

——. 2000. *Recycling Urban Land: Understanding Changing Patterns in Local Economic Development and Land-Use Policies*. Paper read at American Political Science Association, August 31–September 4, at Washington, DC.

——. 2001. "Changing Agendas in Toxic Waste Policy." *Economic Development Quarterly* 15(2): 181–99.

Hula, R. and A. Hemond. 2004. *Race and Public Perceptions of State Environmental Policy*. Paper read at the Midwest Political Science Association, April 15–18 at Chicago.

Kaiser, S.-E. 1998. "Brownfields National Partnership: The Federal Role in Brownfield Redevelopment." *Public Works Management and Policy* 2(3): 196–201.

Kamienieke, S. 1995. "Political Parties and Environmental Policy." In *Environmental Politics and Policy*, edited by J. Lester. Durham: Duke University Press.

Kingdon, J.W. 1984. *Agendas, Alternatives, and Public Policies*. Boston: Little Brown.

Klyza, C.M. and D.J. Sousa. 2008. *American Environmental Policy, 1990–2006: Beyond Gridlock*. Cambridge, Mass., MIT Press.

Kraft, M.E. and D. Scheberle. 1998. "Environmental Federalism at Decade's End: New Approaches and Strategies." *Publius: The Journal of Federalism* 28(1): 131–46.

Lester, J.P. 1980. "Partisanship and Environmental Policy: The Mediating Influence of State Organizational Structures." *Environment and Behavior* 12(1): 101–31.

Lester, J.P. and A.O'M. Bowman. 1989. "Implementing Environmental Policy in a Federal System: A Test of The Sabatier-Mazmanian Model." *Polity* 21(4): 731–53.

Lester, J.P., J.L. Franke, A.O'M. Bowman and K.W. Kramer. 1983. "Hazardous Wastes, Politics, and Public Policy: A Comparative State Analysis." *Western Political Quarterly* 36(2): 257–85.

Lester, J.P. and E.N. Lombard. 1990. "The Comparative Analysis of State Environmental Policy." *Natural Resources Journal* 30(2): 301–19.

Lowry, W.R. 1992. *The Dimensions of Federalism: State Governments and Pollution Control Policies*. Durham: Duke University Press.

McFarland, A. 1991. "Interest Groups and Political Time." *British Journal of Political Science* 17: 129–47.

Meltz, R. 1998. *Superfund Act Reauthorization: Liability Issues*. Washington, DC: Congressional Research Service.

Michigan State University Center for Urban Affairs. 1998. *Urban Summit: Proceedings*. East Lansing: Author.

Mintrom, M. 2000. *Policy Entrepreneurs and School Choice, American Governance and Public Policy Series*. Washington, DC: Georgetown University Press.

Ramsay, M. 1996. *Community, Culture, and Economic Development: The Social Roots of Local Action*. SUNY Series: Democracy in American Politics. Albany: State University of New York Press.

Reisch, M. 1998. *Superfund Reauthorization Issues in the 105th Congress*. Washington, DC: Congressional Research Service.

——. 1999. *Superfund Reauthorization Issues in the 106th Congress*. Washington, DC: Congressional Research Service.

Reisch, M. and D.M. Breardon. 1997. *Superfund Factbook*. Congressional Research Service.

Ringquist, E.J. 1993. *Environmental Protection at the State Level: Politics and Progress in Controlling Pollution*. Bureaucracies, Public Administration, and Public Policy. Armonk, N.Y.: Sharpe.

——. 1995. "Political Control and Policy Impact in EPA's Office of Water Quality." *American Journal of Political Science* 39(2): 336–63.

Schattschneider, E.E. 1975. *The Semisovereign People: A Realist's View of Democracy in America*. Hinsdale, Il.: Dryden Press.

Scheberle, D. 1997. *Federalism and Environmental Policy: Trust and the Politics of Implementation*. Washington, DC: Georgetown University Press.

Schneider, M., P.E. Teske and M. Mintrom. 1995. *Public Entrepreneurs: Agents for Change in American Government*. Princeton, NJ: Princeton University Press.

Smith, R. and D. Silver. 1998. "Getting the Job Done: Resolving State-Federal Conflicts in Superfund." *Policy Studies Journal* 26(4): 735–47.

United States Conference of Mayors. 1998. *Recycling America's Land: A National Report on Brownfields Redevelopment*. Washington, DC: United States Conference of Mayors.

United States General Accounting Office. 1997. *Superfund: State Voluntary Programs Provide Incentives to Encourage Cleanups*. Washington, DC: Author.

——. 1999. *Superfund: Progress, Problems and Outlook*. Washington, DC: Author.

Vig, N.J. and M.E. Kraft 2006. *Environmental Policy: New Directions for the Twenty-first Century*. Washington, DC: CQ Press.

Vizzard, W.J. 1995. "The Impact of Agenda Conflict on Policy Formulation and Implementation: The Case of Gun-Control." *Public Administration Review* 55(4): 341–7.

Chapter 3

Revitalizing Contaminated Land in Italy, the United Kingdom and the United States: A Comparative Perspective

Philip Catney, Tiziana Cianflone and Kris Wernstedt

Introduction

Brownfields in both the United States and Western Europe comprise previously used properties, with abandoned or underused facilities, where expansion or redevelopment faces barriers to reuse.[1] In the United States and some western European countries, real or perceived contamination is an important element of brownfields, although other parts of Western Europe do not define this as a necessary component. In both settings, however, the focus is on interventions that may be necessary for beneficial reuse of largely unwanted properties—interventions that may reduce or mitigate liabilities that current owners, purchasers, or developers may have to absorb if they attempt redevelopment—or those that address other site and neighborhood characteristics that deter investment interest.

Land contamination did not prove to be a particularly controversial issue in either the United States or Western Europe when it was first identified as a potential environmental and economic problem in the late 1970s. Other types of pollution appeared to present a more direct threat to human health and to the functioning of ecological systems. Particularly in Western Europe, so long as demand for land for development did not exceed the available greenfield supply, developers could readily avoid entanglement with contaminated sites. Moreover, on the supply side, the process of returning severely polluted sites to productive use did not—and still does not—present merely a technical problem. To the contrary, it brings with it a complex array of legal, political, and social issues—such as the assignment of responsibility for cleaning up contamination, definition of required

---

1 The United States covers roughly twice as much land area with roughly 160 million fewer residents than Western Europe. However, the number of brownfields in the two settings appears comparable, depending on how they are counted. For instance, some estimates place the number of brownfields in the United States as high as a million sites (Heberle and Wernstedt 2006), while a report from CABERNET, the Concerted Action on Brownfield and Economic Regeneration Network, indicates that a similar number of brownfields lie in Western Europe (Ferber et al. 2006).

remediation standards, inter- and intra-governmental coordination, subsidization, and redistribution of resources to disadvantaged areas—that pose difficulties that may be simpler to ignore than confront.

To varying degrees across the United States and western European countries, the challenges of addressing brownfield remediation and development can fall upon the central government, local authorities, the original polluters, or current owners, insurers, or developers (Walker 2002: 10). The ways in which these challenges are distributed and managed reflect the rationale and character of different national approaches to contaminated land. In this chapter, we examine the experience of regenerating brownfields and other previously developed or contaminated sites across three countries: the United States (US), United Kingdom (UK), and Italy. In particular, we frame our analysis around three related thematic areas.

First, national approaches to brownfield redevelopment in each of the three countries have been guided to varying degrees by some form of "development managerialism." In its purest form, this politico-administrative perspective frames brownfields in economic terms, with contamination standing out as an obstacle to economic progress and urban (re)development, instead of as an issue principally related to environmental quality or public health. Health and environmental problems are thus secondary concerns. This framing appears in the way that state actors in all three countries structure the palliative response to brownfields through existing administrative apparatuses, primarily planning entities in the UK, environmental agencies in Italy, and economic development offices in the US. However, while "development managerialism" provides the overarching discourse that structures the response to land contamination, differences in the strength of the discourse exist and the precise policy regimes that have emerged in the different countries covered depends on national and local circumstances.

Second, the particular structure of policy, governance, and land use planning in the three countries fundamentally shapes the regeneration terrain. In the UK, land use planning plays a greater integrating role in creating a setting for effective brownfield regeneration by connecting brownfields and heavily contaminated sites to the redevelopment market. In addition, the UK planning system's limitations on development on greenfield areas has increased land values to the point where developers no longer consider decontaminating such sites a major financial drain when compared to overall development costs, making them more ready to absorb remediation costs without recourse to public subsidy (see Catney et al. 2006; Dixon 2007). In Italy, by contrast, a separation has persisted between remediation and redevelopment due to the primacy that brownfields policy has given to the former. If a site meets the legal definition of contamination, redevelopment considerations typically are not broached until remediation has been completed. Oversight of remediation rests at the sub-national level, except for contaminated sites on the

National Priority List,[2] which the Environmental Ministry must manage at the central level. In the US, some aspects of both the UK and Italy models have appeared, but with stronger reliance on economic and regulatory incentives and voluntary cleanups, consistent with devolutionary trends in environmental governance since the mid-1990s (Wernstedt 2001). The pre-eminence of state and local (rather than national) authority over land use means greater variation across the country, with states serving as the primary regulatory authorities for cleanups and other environmental issues, and local governments using a range of tools, including blight elimination, comprehensive and economic development plans, state grants and loans, and tax increment funding to address land use activities on brownfields in targeted areas. In addition, the large number of sites privately remediated and redeveloped in stronger markets evinces the operation of an opportunistic, real estate investment decision model in some settings.

Third, while all three countries have sought to harness market processes to address brownfields cost-effectively, substantial variation exists across them. In part, these variations mirror the extent to which larger-scale, cost-effective sustainable development concerns have penetrated public authorities. The UK, for instance, has sought to deal with brownfields and contaminated land in a more holistic fashion than the US, adopting a comprehensive area-wide planning approach and tying site level regeneration to larger policy-driven efforts to accommodate urban growth and other changes. This has offered the potential to create more sustainable future urban forms through coordinated remediation and development across multiple properties and jurisdictions simultaneously. Italy's traditional separation of remediation from redevelopment has privileged health and environmental problems over the economic and social dimensions of sustainability, although this balance is shifting perceptibly to a greater focus on redevelopment. In this respect, Italy has adopted a weaker version of development managerialism, even as it strives to increase the effectiveness of its brownfield policy through market instruments. Moreover, while the simultaneous engagement of regions, provinces, and municipalities in Italy remains attractive in principle, it often has substantially slowed regeneration efforts in practice because of poor coordination and inefficient duplication. US efforts, although sometimes tied to larger areawide objectives, more traditionally adopt a highest-and-best use development model. In addition, the greater reliance on voluntary cleanups in the US has not lent itself as readily to a holistic, systematic approach to brownfields.

---

2 Contaminated sites included on the National Priority List have particular features that elevate their importance to the national level, such as a large amount of hazardous contaminants, high health or ecological concerns, or risk of damage to cultural and environmental heritage (ISPRA 2009).

Development Managerialism

Before pursuing our argument, it is necessary to clarify what a discourse is and how we employ it in this chapter. A discourse represents a shared way of understanding and talking about the world that is embedded in language. It provides a means for individuals and groups to articulate, interpret, and organize social complexity to produce coherent stories or accounts (Dryzek 2005: 8). Discourses are constructed from assumptions and judgments and are used to filter the range of peoples' experiences of the world. Discourse analysis starts from the assumption that that all actions, objects and practices are socially meaningful and that these meanings are shaped by the social and political struggles in specific historical periods .... At the level of everyday interaction, discourses represent specific systems of power and the social practices that produce and reproduce them (Fischer 2003: 73).

One set of more powerful actors/institutions may have the ability to force less powerful others to subscribe to their particular discourse (Dryzek 2005: 9). Discourses are distributed across institutions and a key task for analyzing them is to account for the "viewpoints and positions from which actors speak and the institutions and processes that distribute and preserve what they say" (Fischer 2003: 76). The utility of the discourse approach for policy research is that by recognizing the existence of a discourse and exposing its tenets, it can help unearth the submerged and often opaque meanings of policy regimes. Such meanings often tend to be taken for granted and thus remain unchallenged by those, particularly policy implementers, who work with them on a regular basis.

In the brownfields realm, we argue that the policy regime and institutional mechanisms that have emerged in the US and parts of Western Europe to deal with the problem of brownfield land are structured around the dominant discourse of "development managerialism" (Catney et al. 2006). This discourse, while recognizing that contamination poses health and environmental problems, leads actors to frame the issue primarily in economic terms, as an obstacle to economic progress and urban (re)development. As such it can privilege a vision of economic gains that can be captured by a relative few (both public and private actors), rather than broader gains that may be more diffuse. The emphasis of the discourse is on cost-effectiveness of remediation techniques, protecting economic interests and harnessing market-led development processes to bring back into use contaminated land instead of adopting more proactive state-led models (as the Netherlands once adopted). Where development pressures are strong, adopting such an approach can be an effective means of delivering rapid land remediation and containing the potentially high (financial) costs associated with remediating polluted sites (a key factor in the Dutch changing their approach). Used in the context of an effective system of land-use planning, "development managerialism" thus can enable the swift release of land in urban cores, helping to promote regeneration in these places while potentially helping to prevent sprawl elsewhere. However, where little scope for development exists, the potential for contamination to be addressed is severely diminished (Catney et al. 2008). In such cases—in areas generally characterized by economic deprivation—"development managerialism" offers less benefit, potentially leaving in place a hazard which without direct governmental intervention would inflict harm on affected groups.

Our concern is to examine the institutional context and practices in which the "development managerialism" discourse is stated, understood and feeds into practice (Fischer 2003: 90). Operating at the level of policy-institutional design and shaping the way that individuals and organizations operate, the framing of this priority manifests itself through the use of existing (economically-oriented) administrative apparatuses. This does not imply that development managerialism operates to the exclusion of other agendas of brownfield regeneration across the three countries. In Italy, for example, the historical strict separation of remediation and redevelopment has tended to elevate the importance of environmental concerns since these must precede any redevelopment discussions. The origins of US brownfields in the 1980 federal Comprehensive Environmental Response, Compensation and Liability Act (the Superfund law)—which imposed cleanup liability on a wide range of entities involved with a contaminated property and thereby often discouraged reinvestment in them—means that environmental and social benefits from the redevelopment of brownfields continue to play a strong role in brownfields policy and planning, along with economic considerations (Heberle and Wernstedt 2006). Yount (2003) argues that despite corporate concerns about economic liabilities driving policy, brownfields have never lost their environmental "essence" and the importance of the Environmental Protection Agency (EPA) as a funding agency reinforces this. Voluntary programs, for instance, arguably have been developed primarily to accelerate reuse, yet they still must meet environmental standards set by the EPA or by states that have signed a memorandum of agreement with the EPA. Multiple agenda clearly operate in the brownfields policy arena.

Table 3.1, for example, shows evidence from two surveys in the US, one of public officials from federally-supported brownfield pilot projects, and the other of public and private brownfield stakeholders in the State of Wisconsin (see Heberle and Wernstedt 2006 for a description of these surveys). The highest proportion of public respondents in both surveys (columns 1 and 2) indicate public health risks are a "very important reason" for redeveloping brownfields, with reducing environmental risks and removing eyesores attracting the next highest proportions of respondents. The highest proportions of private stakeholders (column 3), in contrast, indicate that increasing tax revenues and creating jobs both provide a "very important reason" for such redevelopment.

The different perspectives do not mean that the discourse of development managerialism does not dominate, but rather that its relative emphasis may shift across different players. In fact, at one level an implicit endorsement of development managerialism appears in the very definition of "brownfields" in the US on which the EPA brownfields program rests—the Small Business Liability Relief and Brownfields Revitalization Act defines brownfields as "real property, the expansion, redevelopment, or reuse of which can be complicated by the presence of a hazardous substance, pollutant or contaminant" (42 *USC* 9601). This language arguably elevates redevelopment objectives and includes environmental concerns as obstacles to this, a focus that local revenue considerations noted below amplify.

Table 3.1   Percentage of respondents selecting as "very important" reason

Reason to Redevelop Brownfield Property	National public (1)	Wisconsin public (2)	Wisconsin private (3)
Increase tax revenue	55	28	42
More efficient use of infrastructure	47	27	32
Remove eyesores	58	42	28
Create jobs	51	25	36
Reduce public health risk	69	45	30
Reduce environmental risk	57	45	32
Reduce sprawl	55	21	23
Diversify business mix	19	8	7
Promote greenspace	27	14	11
Part of area-wide redevelopment agenda	33	15	21
Number of respondents	89	155	74

Note: See Heberle and Wernstedt 2006 for a description of these surveys.

Many European countries, in contrast, have taken a broader view, both in terms of what constitutes a brownfield and with respect to the object of brownfield interventions. No official European Union definition of brownfields exists— each country defines brownfields according its unique policy goals—but the multiple country CABERNET[3] process supported by the European Commission defines brownfields as sites that have been affected by former use, are derelict or underused, may have real or perceived contamination problems, lie in developed urban areas, and require intervention to bring them back to beneficial use (Ferber et al. 2006). This enlarges the focus of urban land management beyond sites that are contaminated and stresses the need for additional actions to realize site potential.

---

3   Concerted Action on Brownfield and Economic Regeneration Network (see www. cabernet.org.uk, accessed: 17 June 2009).

Policy and Governance Framework

The differences in approaches to brownfields between the US and western Europe reflect definitional bounds on what types of land are considered to be brownfields, alternative perspectives on the goals of sustainable brownfields development, and other factors. In particular, the structure of policy, governance, and planning in the different countries has fundamentally shaped the form that brownfield regeneration takes in each setting.

Policy and Governance in the United States

The institutional backdrop to the brownfields problem in the US lies in the federal Superfund law and consequently federal agencies have played key roles in brownfields policy. The EPA both sets the broad regulatory framework for contaminated properties and, since 2002, has provided roughly $160–$170 million per year in assessment and cleanup grants and loans to state and local public and non-profit entities for brownfields work. However, under the US federalist system most of the direct, day-to-day environmental oversight of brownfield projects comes from individual state programs for contaminated sites, rather than from the national level as in the UK and Italy. Many states have a signed memorandum of agreement or understanding with the EPA that defines a very circumscribed role for EPA in contaminated sites that enter a state brownfields or voluntary cleanup program. Even for those states without such an agreement, EPA plays no direct role in regulatory decision making for the majority of sites and brownfield activities. Instead, these typically are led by a combination of state agencies and local public and private actors. As a consequence, the US experiences a greater diversity across the country of laws governing cleanup and land use than do most countries in Western Europe.

The role of local, municipal-level agencies, vis-à-vis remediation in the US, typically is limited and any environmental regulatory power they wield remains subordinate to the state and federal levels.[4] In contrast, local governments exercise planning and code oversight, making them the principal public player in the redevelopment aspect of brownfield revitalization. In some situations, these governments have funded brownfield coordinators who merge the environmental and developmental aspects of brownfields and often serve a public entrepreneurial function to find assessment and cleanup money from external sources and to match developers with available brownfield properties. Absent such a dedicated position, planners or economic development practitioners often incorporate brownfields into their mix of responsibilities.

---

4   In contrast to current practice, Fortney (2006) argues for the development of locally-run cleanup programs, contending that these could provide needed attention to the lesser contaminated sites that are overlooked in urban areas.

Furthermore, municipalities and the state typically do not take responsibility for "orphan sites"—sites where contamination is evident but original polluters cannot be traced—unless they are themselves a responsible party for the contamination or want the site for some reason. As will be discussed below, this contrasts with Italy and the UK where public agencies take responsibility for contamination as a last resort. As Meyer, Williams and Yount (1995) argue, in Europe liability is fault-based because the state is often the potentially responsible party, as a result of the activities of (previously) nationalized industries. Other potentially responsible parties are innocent unless they can be proved to be the original polluter.

Policy and Governance in the United Kingdom

The UK government—driven by many factors such as high population densities, rapid inflation of house prices, the perceived need to protect the country's rural heritage, and a desire to create more compact urban areas to combat sprawl—has supported brownfield recycling through a variety of plans, institutions and initiatives. In perhaps its most visible effort, it has set targets through national planning documents for regional and local planning authorities to locate 60% of new housing on brownfield land. In addition, the central government has given its main regeneration body, the Homes and Communities Agency (formerly English Partnerships), a lead role in identifying and supporting development activities— e.g. by entering into partnerships with private developers—that bring brownfields into productive reuse. It also has made the redevelopment of such sites a key measurement of the success of the Regional Development Agencies that it created in the late 1990s for England's regions and the Greater London area. Furthermore, the UK Treasury offers a range of tax incentives similar to the US, albeit with a lower budget and without the full benefit of still-nascent tax increment financing. Other central government players include the Department of Communities and Local Government (urban development), the Department of Environment, Food and Rural Affairs (environmental policy), and the Environment Agency (a non-departmental public body with responsibility over "special sites"[5]).

For "hardcore" sites, the rehabilitation of contaminated (non-agricultural) land has been operationalized mainly through the development planning process under the auspices of local planning authorities. The emphasis is on site-specific remediation. Liability attaches to the original polluter and/or the current owner, but remediation is required only when there is a significant or potentially significant

---

5 "Special sites" represent sites with particularly defined problems that make it unreasonable for a local authority to deal with. Factors determining this status include, for instance, sites where pollution is actively entering controlled waters, land contaminated by waste acid tars, Ministry of Defence land, and land containing explosives or chemical weapons.

threat to public health or the environment. More specifically, the current policy regime consists of two main elements.

First, only properties that meet formal criteria under Part 2A of the Environment Protection Act of 1990—where contamination poses a "significant possibility of significant harm, or pollution of controlled waters" and *where a pathway to receptors can be demonstrated to exist* (the "source-pathway-receptor" mode)— are deemed "contaminated land." Under Part 2A, local authorities are required to survey their jurisdictions for such properties. If found, the local authority must either 1) take action itself to break the source/pathway/receptor linkages and end the "contaminated land" status of the site; or 2) propose the site for listing as a "special site," if it meets the appropriate criteria, to be adopted by the Environment Agency (see Catney et al. 2006; Catney et al. 2008; Luo, Catney and Lerner 2009).

Second, remediation of all land that contains potentially hazardous contaminants but which does not satisfy the legal definition of "contaminated land" falls under the local planning system (Catney et al. 2006).[6] As a normal part of the planning and development process, local planning authorities may impose conditions on planning permissions that require site principals to ensure that the land they are developing is remediated to a standard fit for the proposed use. Intending applicants must normally carry out investigations and devise proposals for effectively dealing with contamination in accordance with any relevant Supplementary Planning Guidance issued by local authorities. Failure to do this may result in planning permission being refused, or granted subject to condition(s) requiring that a risk assessment be carried out prior to the start of development to assess remediation needs. In addition, the local authority has responsibility for remediation of "orphan sites." Where it identifies such a property, it can apply to a central government fund for support of the capital costs of land remediation.

Part 2A and the planning system constitute the core policy processes charged with providing a coherent policy framework. Yet the degree to which this has been achieved in practice is questionable. Catney et al.'s (2006) analysis of the institutional structure for dealing with contaminated land in the UK suggests that the system is not as interlocking as government documentation purports. They argue that problematic misalignments in regulatory philosophies exist within the UK system on contaminated land. While the planning system is goal-seeking, relational and systems-based, holistic and participatory, the Part 2A process is problem-solving, project-focused, specific, technical, and exclusive (Catney et al. 2006: 349). In addition, Part 2A is proactive, partly publicly funded, and a process focused on dealing with contaminated sites for which there is no immediate prospect of development. The planning system, on the other hand, is reactive (to development proposals) and achieves privately funded treatment of contamination through the exercise of public regulation (Henneberry et al. 2005).

---

6   The planning system also can deal with sites that meet the Part 2A criteria if there is a proposal to redevelop it.

Policy and Governance in Italy

Until 2008, Italy had no specific integrated national program for brownfields. Rather, the two elements of brownfields—environmental remediation of contamination and redevelopment of sites—remained separate in time and with respect to responsible agencies and objectives at the national level. Redevelopment did occur on previously developed lands that in many cases had required prior interventions—such as demolition or even cleanup—to make a site ready for beneficial reuse, but at contaminated sites, redevelopment remained distinct and unsynchronized with remediation.

On the environmental side, the management of contaminated land was regulated until 2006 by the Waste Management Act of February 1997 (Legislative Decree n. 22/97). The Ministry for the Environment introduced regulations for remediation of contaminated sites under this Act in October 1999 (Ministerial Decree n. 471/99), which defined acceptable limits for contaminant concentrations in soil and subsoil as a function of land use and in groundwater. It also provided guidelines for sample collection, preparation and analysis, as well as general criteria for cleanup project design and remedial actions.[7] To address these sites, regional authorities must approve Regional Remediation Plans—these prioritize remediation according to human health and natural resource risks and plan investments for remedying and recovering the contaminated sites according the more hazardous situations—while municipal and provincial administrations have principal responsibility for approving and certifying remediation projects and providing institutional controls.[8] In addition, while parties identified as liable for causing pollution must pay for remediation, where orphan sites exist, municipalities have the primary responsibility for funding remediation.

On the redevelopment side, oversight exists in part through building and urban controls, in particular DPR n. 380/2001 (Vanetti 2008). The central government

---

7 Legislative Decree no. 152 in 2006 changed the approach by endorsing risk assessment for quantifying contamination risk and targeting remedial action plans for the first time. This new approach applies if allowable contaminant concentrations determined by land use considerations in the Waste Act are exceeded. Its application could significantly alter the selection of future remedies.

8 Italy has 20 Regions. In the remediation process each region must: adopt directives for local governments to develop guidelines and criteria to clean up contaminated sites; finance remediation according to existing programs and support municipalities that need to clean sites if polluters are insolvent or not identified; participate in program agreements (accordi di programma) that the Ministry of Environment has entered into with Provinces and Municipalities for remediation; and approve remediation projects for sites on the National Priority List. Each Regional Plan thus operates in a functional hierarchy. Unlike differences across individual states in the US, in principle this approach is consistent across Italy, but registries of contaminated sites are not yet fully developed and are not consistent, making it difficult to compare actual practice across regions (ISPRA 2009).

possesses the regulatory power in urban policy,[9] with each municipality the principle authority and front-line actor in land use planning and implementation. Sub-national plans cover all exigencies of urban planning in terms of forecasts of future development of urban forms and include coordination between different competencies across different levels. Similar to the Regional Remediation Plans, a functional hierarchy exists across development; that is, regional and provincial plans guide the general form and location of development at the broad scale of a territory, while general urban plans and programs make this guidance operational on the ground. This hierarchical approach provides a useful framework for planning brownfield redevelopment since it recognizes brownfields as previously developed land within the plan. Unfortunately, the approach omits sufficiently detailed information on socio-economic activities surrounding the contaminated site that would facilitate private investment.

In past years, the lack of effective coordination in Italy between entities on the redevelopment side and those with remediation responsibilities has been a critical factor limiting cleanup of contamination and redevelopment of previously used land. If a site was legally contaminated, cleanup was required, but absent accompanying pressure for coordinated redevelopment, it sometimes took years for slow-moving administrative procedures to result in actual remediation; that is, because redevelopment typically could not even be considered until remediation was complete, it seldom provided sufficient pressure for remediation. Moreover, voluntary cleanups to speed the process were not facilitated. In addition, the prioritization of remediation by regional authorities occurred without the benefit of knowing whether redevelopment benefits at particular sites could offset remediation expenditures at those sites and extend the reach of remediation budgets further down the prioritized list.

In the last 10 years, brownfield governance through planning tools and the legal framework for environmental controls has moved toward more integration both within the separate development and remediation domains and, for developers who lack cleanup liability, jointly in remediation and redevelopment. On the development side in the 1990s, the Ministry of Public Works, with the Secretary General of the Housing Committee, instituted the so-called Complex Programs through a series of laws and decrees. These have effectively expanded the planning role in brownfields beyond zoning. The novelty of the approach is the integration between actors and financial resources, furthering collaboration, negotiation, and financial discussions among landowners, developers, and cities. Before plan approval, the major sources of finance must be acquired. In addition, local governments have started to play a larger role in sites not of national interest by cleaning up and redeveloping lesser contaminated properties under their own auspices through local land planning programs.

---

9   The Ministry of Public Works was the principal national-level regulator until 2001. The Ministry of Infrastructure and Transports and Ministry of Environment subsequently incorporated these ministerial functions.

On the contamination side, Law n. 13/2009 has provided a tool to accelerate remediation of sites on the National Priority List by allowing the Ministry of Environment to recover remediation charges incurred at such sites.[10] More strikingly, Legislative Decree n.4/2008 integrates remediation with redevelopment. The decree establishes that historical sites (those discovered prior to May 2006) in which the public has an interest—so-called Public Interest Sites[11]—be identified, and that programs be developed and actions taken to not only remediate them, but also to bring productive economic activities to them. These sites are selected jointly by the Ministries of Economic Development and Environment, and in agreement with the permanent conference for the relationship between state and regions and autonomous Italian provinces (Vanetti 2008). Because of its newness, the decree's legal interpretation and implementation are not settled, yet its existence implies a shift toward a development managerialist perspective. This may evince both a desire to hasten development for its own sake, as well as a hope of breaking the inertia of an unresponsive administrative apparatus whose inaction allows the effects of contamination to accumulate.

Table 3.2 summarizes the broad contours of the policy and governance regimes for the three countries, while Table 3.3 presents an appraisal of their relative strengths and obstacles to the redevelopment process.

---

10   The application of this rule for recovering environmental damages is contentious. Out of court settlements to recover the costs of remediation efforts that fall short of complete restoration extinguish claims for environmental damages associated with site contamination (Zaccheo 2009).

11   This includes sites not listed in the Remediation National Program established by Ministerial Decree n. 468/2001.

Table 3.2   Regulatory power, planning system and responsibility in the brownfield remediation and redevelopment processes

		United States	United Kingdom	Italy
	SET OF BROWNFIELDS (DEFINITIONAL BOUNDS)	USEPA and state definitions characterize brownfields in both remediation and redevelopment domains	Brownfields include both contaminated and non-contaminated sites in both remediation and redeveloped domains	Non-contaminated brownfields defined in redevelopment process and contaminated sites defined by national law; historical contaminated sites of public interest identified in both remediation and redevelopment domains
REDEVELOPMENT PROCESS	**REGULATORY POWER** (main authority)	Mostly local, with some state engagement in places	Nation-states	Nation through Ministry control, some sub-national engagement-
	PLANNING SYSTEM; RESPONSIBILITY	Decentralized; local governments are principal public player	Local authorities	Region, province and municipal levels all participate but play different roles
REMEDIATION PROCESS	**REGULATORY POWER** (main authority)	State agencies principal regulator w/ EPA in background; many sites in voluntary cleanup programs	Nation-states; some voluntary cleanup by liable parties to avoid Part 2A process	Nation, principally Ministry of the Environment; voluntary cleanups are rare
	PLANNING SYSTEM; RESPONSIBILITY	State agencies (often intra-state regional satellite office); local regulatory authorities for some aspects	Local authorities with centrally guided regime for managing contaminated sites	Contaminated sites of "regional interest" decentralized to region; province, municipality and regional environmental agency; for contaminated sites of National Priority List, Ministry control

Table 3.3   Obstacles and strengths in the brownfield remediation and redevelopment processes

	United States	United Kingdom	Italy
OBSTACLES			
INSTITUTION	Only properties w/ perception/presence of contamination are eligible for brownfield funding; voluntary cleanups common; local government reliance on local tax revenues; site-by-site approach	"Contaminated land" defined in very narrow terms; not considered contamination if pathway is not operative; limited public resources available for remediation; Part IIA system cumbersome and time consuming	Remediation and redevelopment processes typically separated in time, no national inventory of brownfield sites that are not of national interest; regional inventory of contaminated brownfield sites exists, but inventory doesn't contain information for redevelopment
PLANNING APPROACH	Site-by-site approach; frequently limited by lack of market demand; limited land use control powers; often subservient to economic development interests	Emphasis on site specific remediation at sites w/ significant or potentially significant threat to public health or the environment; planning system reactive	Separation in planning function; site-by-site approach for contaminated sites of national interest and for sites not contaminated or already remedied
STRENGTHS			
INSTITUTION	Financial resources available at federal and state level for brownfields; market rewards risk takers; risk-based approaches for setting remediation goals; statutory base for brownfields program	Strong national guidance and standards require surveys by local authorities; allows strategic priorities at two levels; controls costs by emphasizing "suitability for use" standards	System does not prohibit consideration of economic and social criteria in prioritizing contaminated sites of regional interest (areawide approach for contaminated sites in practice lacks integration of data, however)
PLANNING APPROACH	Responsive to local government revenue needs; redevelopment decisions often subject to public comment	Achieves privately funded treatment of contamination through public regulation (conditions on planning permissions)	Planning has potential larger role to play in targeting public resources more efficiently, reducing investor uncertainty, and incorporating community input

Broad-scale Sustainability

Sustainable development concerns have penetrated the brownfields policy regime and practice across the three countries, although in different degrees and forms. This variety in part reflects ambiguities in the term "sustainability" itself, but it also exists because the governance structures in the countries operationalize the concept in different directions.

The concept of sustainable development was popularized by the 1987 Brundtland Report (formally known as the World Commission on the Environment and Development) and endorsed by political leaders from across the globe at the Rio Earth Summit in 1992. Over the 1990s, the concept of sustainable development became embedded in the language of policy makers and academics to the point where it has been described as a new meta-narrative (Meadowcroft 2000: 370) or a "neo-renaissance idea" (O'Riordan and Voisey 1997: 4). It is an attempt to resolve the traditional tension in environmental politics between striving for economic growth and protecting the environment (see Meadows et al. 1972).

The discourse of sustainable development suggests that governments and their citizens can seek to promote economic growth, but that they must take greater responsibility for protecting the (global) environment from further damage. Furthermore, and particularly relevant for the purposes of brownfield regeneration, it suggests a greater regard for how future generations can enjoy similar resources and opportunities to the ones we presently enjoy. Hence, one of the key challenges for sustainable development is effectively to integrate plans that enable stable growth to complement social inclusion and the protection of the environment and its natural resources (Department of the Environment 2000). It is through this triad—economic, social and environmental—that various activities are commonly judged to be "sustainable."

Meadowcroft (2000: 374) argues that much of what governments have done in the name of sustainable development constitutes a repackaging of traditional environmental policies (for example, pollution control). However, this does not imply that we should dismiss sustainable development as an empty concept that has no core to it at all. Rather, governments have interpreted sustainable development in the light of current activities and have created new initiatives to accompany these efforts:

> There has been reform to structures and procedures, designed to integrate environmental problem-solving into the workings of the main branches of public administration. It has been accepted—at least in principle—that environmental policy cannot be operated as a post hoc corrective to normal (that is non-environmental) decision processes; rather the environmental dimension should be factored-in from the outset (Meadowcroft 2000: 344–5).

The principles of sustainable development applied to brownfields must, therefore, become a more integral factor in the development of a broader spectrum of public policies on brownfields and revitalization than has previously been the case, and not shunted onto the sidelines as just an optional consideration. With this understanding, economic policy on brownfields cannot be seen as a policy isolated from environmental or social considerations. Rather, these realms should be seen as inextricably linked. Whether they actually are taken as such depends on national context and the permeation of the concept into the style of national policy making (see Lafferty and Meadowcroft 2000).

In our view, two important gaps relevant to brownfields in the sustainability literature relate to the long-term prevention of brownfields and financing of their remediation when they are found, and to the scale of sustainability for economic development and the built environment (Pearce and Vanegas 2002). With respect to the former, in all three countries, the state can seek compensation from polluters for environmental damages. This can both provide revenues for compensation and additional remediation, and deter future damaging behavior by forcing cost internalization of damages in producer decision making (Cianflone and Wernstedt 2009). Such deterrence comprises a critical element of a regulatory apparatus, given the impossibility of devising a system that can universally detect damage after it has occurred. In all three countries, environmental damage also can provide an important lever to induce cooperation from a recalcitrant polluter. The threat of being forced in a court proceeding to pay both full remediation costs and full environmental damages may provide a compelling argument to negotiate an agreement with the state that remediation costs alone might not compel.

In terms of scale, brownfield redevelopment takes place in local, regional, and national contexts, yet decisions may not reflect concerns much beyond a site's border. From a sustainable land use perspective, however, it defeats the intent to increase viability at one site if it decreases viability in another. Some element of gains from any individual project may merely represent transfers of economic activity from other areas (Wernstedt 2004; Howland 2007). The scale of sustainability accounting should be consistent with the scale at which the net impacts from revitalization are considered important by society, implying that community- or even regional-level effects warrant consideration. In addition, the value of a brownfield redevelopment intervention depends on the degree to which labor, capital, land, and other resources are underemployed in an area, since the economic benefits of the redevelopment only comprise the marginal value of employing a resource in a more productive activity.

Sustainability in US Brownfields

The dominant model of brownfields development in the US rests on a site-by-site calculus, where investment decisions are made on individual properties to maximize gains from that property while often ignoring the cumulative effect on local or regional economies. This "tyranny of small decisions" (Kahn 1966) never explicitly addresses systems level problems—whether they be the ecological integrity of an area (Odum 1982), its economic components, or its social organization—nor does it typically take into account offsite benefits and costs in decision making (Meyer 1998). In addition, less desirable properties—ones that are too small or misshapen to be attractive for commercial enterprises or that lack transportation and other infrastructure (Howland 2004)—receive little market attention because as individual sites they generate insufficient rewards, even though they collectively may constitute the key to community-wide revitalization.

This does not mean that larger scale effects never receive consideration in brownfields decision making in the US The State of Wisconsin, for example, mandates that by 2010, local comprehensive plans required by and developed under Wisconsin's Smart Growth law need to include an Economic Development element that must "evaluate and promote the use of environmentally contaminated sites for commercial or industrial uses" (Wisconsin Department of Natural Resources 2002: 3). Development at a wider scale and area-wide brownfield planning, assessment, or cleanup efforts also have appeared in other states such as Alaska, Florida, Indiana, Minnesota, Missouri, New Jersey, New York, and Washington. In New Jersey, for example, the Brownfield Development Area Initiative (BDA) concentrates resources in a number of communities to help implement remediation and revitalization plans for clustered brownfield properties (van Hook, Shaw and Kloo 2004). New York State offers a similar concept in its Brownfields Opportunity Areas program. Tax increment financing (TIF) also has been used as a *de facto* area-wide strategy in some communities, and at least one state has formalized an environmental TIF for contaminated areas.[12]

However, the reliance of municipal governments on revenues from local property, sales, and income taxes, as well as user fees—these furnish 60% or more of public sector finances at the local level in the US (Tannenwald 2004)—may complicate broader-scale regeneration efforts.[13] Not only do greenfield areas compete with urban brownfields for development to secure local revenue streams, but those urban areas with brownfields often face short-term incentives to promote intense economic reuses, particularly ones with quicker revenue streams. From the private side, real estate capital in the US has been dominated by Wall Street debt and equity financing since the 1990s, and this typically focuses on a 5–7 year investment horizon, even though real estate nominally is a long-term asset (it has a 39 year depreciation schedule under federal tax rules). Brownfield redevelopment competes in this investment climate and therefore the development of longer-term, broader-scale sustainable land management practices with delayed returns

---

12   In addition, the new Leadership in Energy and Environmental Design for Neighborhood Development (LEED-ND) rating system could promote area-wide brownfields regeneration indirectly. This nationwide voluntary system aims to promote smart growth and more sustainable development at a wider neighborhood scale through a certification process to reward projects that meet a wide range of smart growth criteria. It includes credits for projects developed on brownfield sites as one of the elements of the Smart Location and Linkages section of the rating system.

13   This contrasts sharply with the UK situation, where British local authorities lack both the range of powers and rich local revenue veins of their American counterparts, raising and controlling only some 25% of their public sector expenditures (Simmie et al. 2006).

often face a competitive disadvantage, particularly in communities where a strong development managerialist discourse dominates and where the private sector has led brownfield regeneration efforts.

Finally, the ubiquity of voluntary cleanups in the US complicates efforts at broad-scale sustainability. EU member states also have voluntary efforts, but unlike Europe, such cleanups are pervasive in the US, the legacy of an adversarial enforcement culture that has influenced American brownfields practice. Nearly every state has developed a voluntary cleanup program to provide incentives for site principals to voluntarily address site contamination in exchange for less onerous requirements and certainty that state authorities will not indefinitely hold them liable for additional cleanup. Recent evidence from a survey of state voluntary cleanup program officials indicates that three-quarters of contaminated sites enter a remediation process through a voluntary rather than enforcement-led program (Wernstedt et al. 2009), and tens of thousands of sites in the US already have enrolled in such programs (U.S. Environmental Protection Agency 2008). Yet, although individual sites in these programs may have sustainable features (e.g., green roofs, low impact stormwater control, and energy efficiency), private and public parties seldom enroll such sites as part of a comprehensive effort to promote a sustainable system of revitalization at a local or regional scale. To the contrary, voluntary cleanup programs by their nature may promote a more opportunistic, less coordinated approach.

Sustainability in UK Brownfields

As noted above, when the UK government was developing the policy regime for contaminated land it was particularly eager to ensure that the resultant system would contain costs. This was made clear in Circular 02/2000 Contaminated Land (Department of the Environment 2000), which was couched in the language of sustainable development. The basic objectives were:

- to identify and remove unacceptable risks to human health and the environment;
- to seek to bring damaged land back into beneficial use; and
- to seek to ensure that the cost burdens faced by individuals, companies and society as a whole are proportionate, manageable and economically sustainable.

The policy regime that has actually emerged in the UK clearly reflects some balance of these three elements. However, Catney and co-authors (2006) argue that the pursuit of "development managerialism" has resulted in the weight given to these objectives following the reverse of their lexical ordering.

The "development managerialist" discourse was enhanced considerably by the move towards a plan-led system in the 1990s (Cullingworth and Nadin 2002; Rydin 2003) and the emergence of sustainable development as one of its key objectives. In combination, these features of planning have supported a markedly more restrictive stance towards the release of greenfield land for development. The national target for regional/local planning authorities to locate 60% of all development on brownfield land by 2008—a target reached by 2001, although aggregate housing completions across England had declined to historically low levels—clearly has increased pressure for development of contaminated sites (English Partnerships 2003: 6).

The consequence of the restriction on the release of greenfield land has been the marked increase in land values to a level sufficient to meet the costs of treating contaminated brownfield land without the cost of public subsidy. As a result, the reliance placed on the development process to remediate contaminated land has been made more secure. This is not to say that the latest evolution of the regime has not presented developers with problems of adjustment (see, for example, Adams 2004), or that no concerns have been expressed by other (non-planning) elements of the state, as evidenced by the debate surrounding the Barker Review (2004).

Sustainability in Italy Brownfields

The prioritization of health and environmental problems of brownfields in Italy has attenuated the development managerialist perspective to some degree, as noted earlier.[14] However, recent changes to the country's legal framework for environmental controls over contamination not only have started to accelerate remediation of contaminated sites on the National Priority List, but also have begun to integrate redevelopment and remediation at other sites. That is, developers who do not have cleanup liability at Public Interest Sites can now connect the remediation phase to the redevelopment phase of site revitalization. This change arguably has boosted the effectiveness of brownfield development in Italy and provided firmer grounding for sustainable brownfield practices consistent with a more overt development managerialism discourse that melds all three pillars of sustainability.

In addition, local governments have started to play larger roles in sites that are not of national interest by cleaning up and redeveloping lesser contaminated properties under local land planning programs.[15] The need for new urban spaces

---

14   The prioritization of environmental problems as policy tools dates back at least to 1986 (art. 18 of Law. n. 349) when, as the first Country in Europe, Italy introduced a fault and negligence legal system devoted to the reparation of "pure" environmental damages. These damages are defined strictly as an impairment of natural resources, regardless of the economic and moral damages suffered by private or public subjects.

15   Because they comprise different markets, remediation and redevelopment differ between bigger and smaller underutilized/inactive industrial sites in Italy. According to some observers, redevelopment has been dynamic at small sites (less than 10,000 square meters) that can host artisanal or small manufacturing activities. This results from supply

and infrastructure rehabilitation, as well as growing attention to the daily quality of life in urban areas, undergird much of this trend, leading toward a locally-driven integration of the two spheres of brownfields revitalization and collaboration among public and private actors that historically has been lacking.

Thus, on both the planning and remediation sides, the trend toward integration arguably has begun to promote more brownfield redevelopment in Italy. This has occurred at the local level principally in two dimensions of sustainability—the economic and social—while leaving the environmental dimension to existing regional authorities that have primary responsibility to compel remediation.

Looking to the future, although not yet fully implemented, the obligation for Italian regional authorities to prioritize contaminated sites of regional interest in their remediation plans could furnish a viable planning framework for brownfield interventions that would incorporate all three dimensions of sustainability. In particular, integrating these plans with economic and social information so as to prioritize interventions on contaminated sites not only by health and environmental criteria but also by socio-economic ones could result in more efficient targeting of public resources, a reduction in investor uncertainty, and the inclusion of community input in sustainable brownfields practice.[16]

Conclusions

In this chapter we have sought to analyze the extent to which the policy and governance regimes for recovering contaminated land for re-use have been guided by the discourse of "development managerialism" in Italy, the UK, and the US Development managerialism frames the issue of land contamination primarily in economic terms—as an obstacle to economic progress and urban (re)development—and structures policy responses primarily through the existing administrative apparatus of planning and economic development processes. The emphasis within the discourse is on minimizing urban blight, protecting economic interests and harnessing market-led development processes to bring contaminated

---

side factors, such as low acquisition costs (usually 50% lower than new industrial real estate with similar area); relatively simple sale and acquisition procedures; and the unavailability of alternative land to redevelop in strategic areas. In contrast, the regeneration of large sites is still limited because of high costs and the complex organizational efforts necessary to broker agreements between public and private stakeholders (Liberatore 2004).

16   Ranieri and Di Marco (2004) suggest that Regional Remediation Plans need to include more information, such as socio-economic, demographics, employment and economic activity characteristics. In addition, they argue that these plans should prioritize contaminated sites not only on environmental but also on socio-economic grounds to provide incentives to private investors. Cianflone and Di Marco (2007) express the same view and propose the development and implementation of a common Brownfield Information System in European countries to define and monitor brownfield redevelopment policies and interventions.

land back into productive use. Pragmatism and cost effectiveness have been recurrent themes of the discourse.

In both the UK and US, the approach adapted to contaminated land has been pragmatic and ad hoc. Cost-effectiveness has been an important driver in the regulatory system particularly in the UK, where contaminated land has not been treated primarily as an environmental or health issue requiring comprehensive action. The history of brownfields redevelopment in the US and its placement in programs run by the EPA or state environmental agencies has attenuated this emphasis, although formal regulation and treatment of contaminated land clearly have evolved in the country to facilitate economic gains, and cleanups often are made conditional and relative to those needs. Both the UK and US have sought to harness the strength of the development industry to recycle contaminated land with as little recourse to the public exchequer as possible. In contrast, Italy has maintained a stronger emphasis on environmental planning, public health protection and state involvement in remediating contaminated sites. As we discussed in this chapter, Italy's traditional separation of remediation from redevelopment has started to weaken as a greater focus on redevelopment as a means of recycling brownfield sites has begun to elevate the importance of economic objectives. We suggest that although it is weaker in Italy than in the US and UK, development managerialism operates here as well as it strives to increase the effectiveness of brownfield policy through market instruments. In short, the policy regime in Italy is trying to find more cost-effective means of dealing with its legacy of contamination, a key principle of development managerialism.

The rise of sustainable development as a discourse has done little to check the growing dominance of development managerialism in debates in all three countries. Sustainability discourses have focused on the need to develop more compact urban forms by containing urban sprawl and pushing development away from greenfields. In the US, this is promoted through the use of an array of publicly funded incentive schemes, while the UK principally utilizes a restrictive land-use planning system (with more modest incentives than the US). In Italy, more complex planning and direct state interventions (though this is now in flux) are the norm. Apart from pushing a more compact urban form, however, sustainability in the context of contaminated land policy is currently interpreted in all three countries through an economic and cost-effective prism, making public health and environmental protection more secondary concerns.

References

Adams, D. 2004. "The Changing Regulatory Environment for Speculative Housebuilding and the Construction of Core Competencies for Brownfield Development." *Environment and Planning A* 36(4): 601–24.

Barker, K. 2004. *Review of Housing Supply—Delivering Stability: Securing our Future Housing Needs, Final Report & Recommendations.* Wetherby, UK: ODPM Literature.

Catney, P., J. Henneberry, J. Meadowcroft and J.R. Eiser. 2006. "Regulating Contaminated Land through 'Development Managerialism'." *Journal of Environmental Policy and Planning* 8(4): 331–56.

Catney, P., J. Henneberry, N. Lawson and S. Shaw. 2008. "Deliberating Environmental Risk on Contaminated Land: The Importance of Local Context." *Land Contamination and Reclamation* 16(2): 113–24.

Cianflone, T. and G. Di Marco. 2007. *The Integration of the Brownfield Remediation and Redevelopment Processes: The Relevance of the Information.* Paper read at Proceedings of the 2nd International Conference on Managing Urban Land at Stuttgart & Bonn.

Cianflone, T. and K. Wernstedt. 2009. "Reducing Uncertainty in the Monetary Assessment of Environmental Liabilities from Waste Landfilling." In *Waste and Environmental Policy*, edited by M. Mazzanti and A. Montini. Milton Park, UK: Routledge.

Cullingworth, B. and V. Nadin. 2002. *Town and Country Planning in the UK.* 13th edn. London: Routledge.

Department of the Environment, Transport and the Regions. 2000. Environmental Protection Act 1990: Part IIA Contaminated Land. DETR.

Dixon, T. 2007. "Heroes or Villains? The Role of the UK Property Development Industry in Sustainable Urban Brownfield Regeneration." In *Sustainable Brownfield Regeneration: Liveable Places from Problem Spaces*, edited by T. Dixon, M. Raco, P. Catney and D.N. Lerner. Oxford: Blackwell.

Dryzek, J. 2005. *The Politics of the Earth: Environmental Discourses.* 2nd ed. Oxford: Oxford University Press.

English Partnerships. 2003. *Towards a National Brownfield Strategy, Research Findings for the Deputy Prime Minister.* London: English Partnerships.

Ferber, U., D. Grimski, K. Miller and P. Nathanail. 2006. *Sustainable Brownfield Regeneration: CABERNET Network Report.* Nottingham, UK: University of Nottingham.

Fischer, F. 2003. *Reframing Public Policy: Discursive Politics and Deliberative Practices.* Oxford Oxford University Press.

Fortney, M.D. 2006. "Devolving Control over Mildly Contaminated Property: The Local Cleanup Program." *Northwestern University Law Review* 100(4): 1863–906.

Heberle, L. and K. Wernstedt. 2006. "Understanding Brownfields Regeneration in the U.S." *Local Environment* 11(5): 479–97.

Henneberry, J., W. Wehrmeyer, J. Meadowcroft, P. Catney and K. Pediaditi. 2005. "Interlocking Processes? Monitoring and Policy-making for Urban Brownfield Regeneration." In *1st SUBR:IM Conference*. London.

Howland, M. 2004. "The Role of Contamination in Central City Industrial Decline." *Economic Development Quarterly* 18(3): 207–19.

———. 2007. "Employment Effects of Brownfield Redevelopment: What Do We Know from the Literature?" *Journal of Planning Literature* 22(2): 91–107.

ISPRA. 2009. *Annuario dei Dati Ambientali 2008*. Rome: APAT.

Kahn, A.E. 1966. "The Tyranny of Small Decisions: Market Failures, Imperfections, and the Limits of Economics." *Kyklos* 19(1): 23–47.

Lafferty, W.M. and J. Meadowcroft (eds). 2000. *Implementing Sustainable Development: Strategies and Initiatives in High Consumption Societies*. Oxford: Oxford University Press.

Liberatore, P. 2004. "Diffusione e Tipologia dei Siti Inquinati." In *Metodologie, Tecniche e Procedure per il Supporto degli Interventi di Valorizzazione dei Siti Inquinati*, edited by A. Ranieri and G.D. Marco. Rome: APAT.

Luo, Q., P. Catney and D. Lerner. 2009. "Risk-Based Management of Contaminated Land in the UK: Lessons for China?" *Journal of Environmental Management* 90(2): 1123–34.

Meadowcroft, J. 2000. "Sustainable Development: A New(ish) Idea for a New Century?" *Political Studies* 48(2): 370–87.

Meadows, D.H., D.L. Meadows, J. Randers and W.W. Behrens, III. 1972. *The Limits to Growth*. London: Earth Island.

Meyer, P.B. 1998. "Accounting for Differential Neighborhood Economic Development Impacts in Site-Specific or Area-Based Approaches to Urban Brownfield Regeneration." In *Center for Environmental Management, University of Louisville Working Paper*. Louisville, KY.

Meyer, P.B., R.H. Williams and K.R. Yount. 1995. *Contaminated Land: Reclamation, Redevelopment and Reuse in the United States and the European Union*. Cheltenham, England: Edward Elgar.

O'Riordan, T. and H. Voisey. 1997. "The Political Economy of Sustainable Development." *Environmental Politics* 6(1): 1–23.

Odum, W.E. 1982. "Environmental Degradation and the Tyranny of Small Decisions." *BioScience* 32(9): 728–9.

Pearce, A.R. and J. Vanegas. 2002. "Defining Sustainability for Built Environment Systems: An Operational Framework." *International Journal of Environmental Technology and Management* 2(1–3): 94–113.

Ranieri, A. and G. Di Marco (eds). 2004. *Metodologie, Tecniche e Procedure per il Supporto degli Interventi di Valorizzazione dei Siti Inquinati*. Rome: APAT.

Rydin, Y. 2003. *Conflict, Consensus, and Rationality in Environmental Planning: An Institutional Discourse Approach*. Oxford: Oxford University Press.

Simmie, J., J. Carpenter, A. Chadwick, R. Martin and P. Wood. 2006. *State of the English Cities: The Competitive Economic Performance of English Cities*. London: Department for Communities and Local Government.

Tannenwald, R. 2004. *Are State and Local Revenue Systems Becoming Obsolete?* Washington, DC: National League of Cities.

U.S. Environmental Protection Agency. 2008. *State Brownfields and Voluntary Response Programs: An Update from the States.* Washington, DC: U.S. EPA.

van Hook, D.E., J. Auer Shaw and K.J. Kloo. 2004. "The Challenge of Brownfield Clusters: Implementing a Multi-site Approach for Brownfield Remediation and Reuse." *New York University Environmental Law Journal* 12(1): 111–52.

Vanetti, F. 2008. *TU Ambientale: Il Possibile Sviluppo dei Siti Contaminati Dismessi in Italia Alla Luce delle Novità Introdotte dal D.Lgs. n. 4/2008*, in giuristi ambientali.

Walker, S. 2002. *The Politics of Contaminated Land: A Political History of UK Contaminated Land Policy 1975–2002.* University of Newcastle, Newcastle.

Wernstedt, K. 2001. "Devolving Superfund to Main Street: Avenues for Local Community Involvement." *Journal of the American Planning Association* 67(3): 293–313.

——. 2004. *Overview of Existing Studies on Community Impacts of Land Reuse.* Washington, DC: U.S. EPA, National Center for Environmental Economics.

Wernstedt, K., A. Blackman, T.P. Lyon and K. Novak. 2009. *Volunteering for State-Level Cleanup Programs: Perspectives from the States and Volunteers.* Blacksburg, VA: School of Public and International Affairs, Virginia Tech.

Wisconsin Department of Natural Resources. 2002. *Brownfields and Comprehensive Planning.* Madison: Wisconsin Department of Natural Resources.

Yount, K.R. 2003. "What are Brownfields? Finding a Conceptual Definition." *Environmental Practice* 5(1): 25–33.

Zaccheo, V. 2009. "Le Transazioni del Danno Ambientale, in Giustizia Amministrativa." *Rivista di Diritto Pubblico* 4.

Chapter 4

Redevelopment Strategy of Brownfield Sites in the Czech Republic

Petra Rydvalova and Miroslav Zizka

Introduction

The term "brownfields" signifies devastated and unused or very little-used localities and properties of an industrial, agricultural, or public character, and non-industrial sites that are abandoned, having lost their original function. This research analyzes causes of brownfields in the Czech Republic, showing some differences from countries that did not go through a period of a centrally planned economy. Another goal of this chapter is to propose a systemic approach to accelerating the redevelopment of brownfields. It is obvious from the experience of developed economies that the problem of brownfields would have occurred even without economic reforms. In transition countries, the changes in economic structures came into effect in a very short period of time, intensifying the problem of brownfields. A brief summary of the causes of brownfields within the context of the historical development of the economy in the territory of the present Czech Republic is provided. For foreign readers the discussion includes characteristics of Czech Republic economic reforms at the beginning of the 1990s. The second part of the contribution quantifies the extent of brownfields in the Czech Republic and presents the strategy, organization, and financial arrangements of redevelopment processes. The last part of the chapter suggests implications of effective brownfield redevelopment in the Czech Republic.

Brownfields in the Context of Historical Development

We can look at brownfields from various points of view. The one that is obvious to all residents, investors, developers, or municipalities is from the point of view of environmental change (in landscape and surroundings). In Europe, the environment suffers from a lack of sufficient open space. This situation can be extreme. In the Czech Republic the population has remained about the same since 1925 (see Figure 4.1) but a lack of open space considerably exceeds the situation of 100 years ago. Thus the redevelopment of brownfields is a means for coping with a lack of space, at the same time decreasing negative externalities brought about by unused, devastated, and depressing zones. As Kraft (2005a: 40) asserts, the reason for the existence of externalities is market failure. This assertion is important since non-market intervention, which means regulated elimination of brownfields by the state, can be justified only if there is no other effective market solution to the problem. Non-market solutions should be used only in those regions where a potential demand for brownfields is so low that it does not eliminate the present or future supply of brownfields. Market and non-market solutions to the problem of brownfields are discussed in detail in the article by Kraft (2005b: 28–33).

The attack on the environment and the expansion of human activities outside original settlements in central Europe started during the industrial revolution. As we can see in Figure 4.2, a considerable increase of population in the area of the Czech Republic corresponds with this period; a dramatic increase can be observed from 1800 to 1900. The number of inhabitants increased with a gradual improvement of living conditions and health care and with the development of new technologies. Larger populations logically brought greater demands for space. Herein is the genesis of today's brownfields. According to the process of industrialization, brownfields are divided into the following three categories:

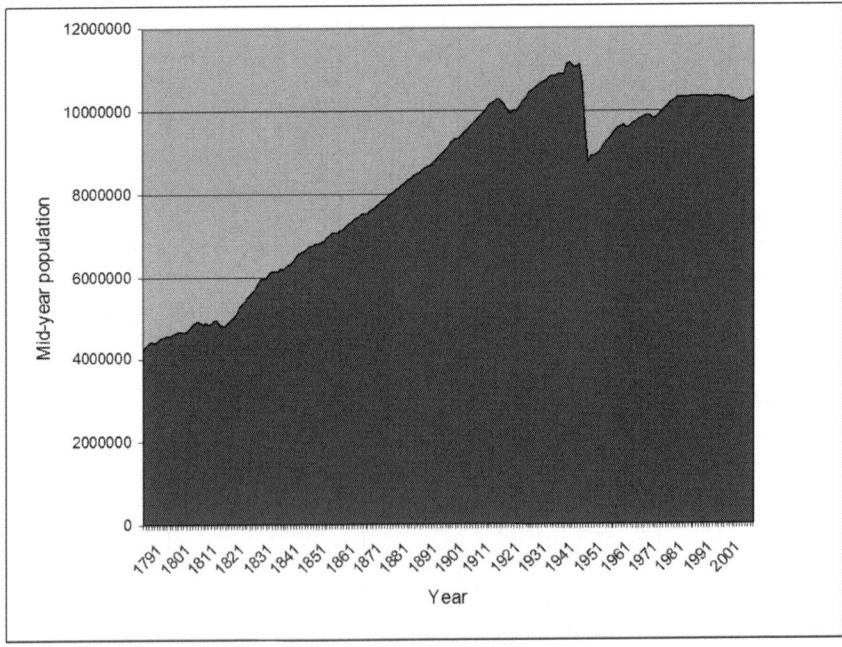

Figure 4.1   Development of population in the Czech Republic
Source: Based on the data of the Czech Statistical Office.

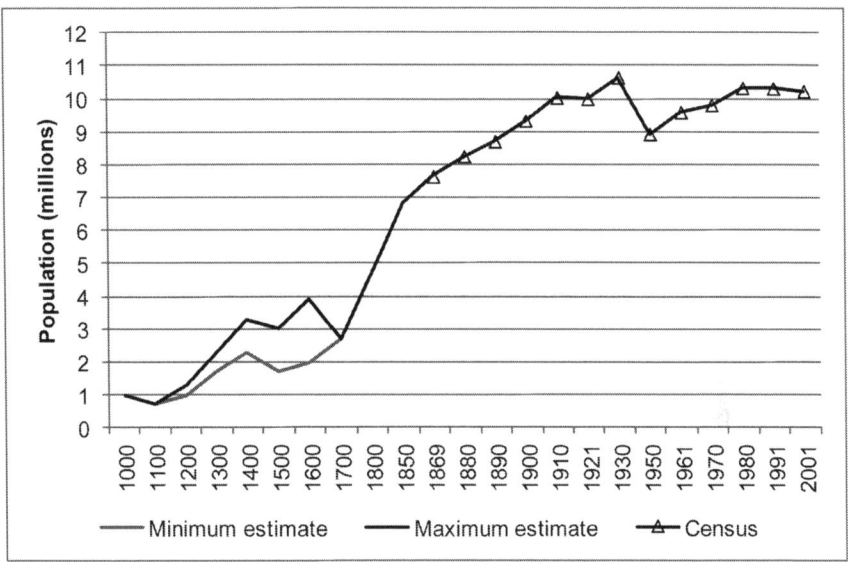

Figure 4.2   Population on the territory of the Czech Republic
Source: Based on the data of the Czech Statistical Office.

The first category: The remains of the first industrial revolution, which began with the invention of the steam engine (approx. the 18th century);

The second category: The remains of the second scientific and technical revolution, which began with the invention of electric current (the turn of the 19th and the 20th centuries); and,

The third category: The remains of the third scientific and technical revolution, which began with the development of information technologies and the Internet (the turn of the 20th and the 21st centuries).

Manual work was typical of the period up to the 18th century; after *the first wave of the industrial revolution* this work was automated. In the 18th and 19th centuries roads were improved and later the railroad was built. As a consequence of the steam engine, the productivity of work increased with the effects visible mainly in textile production. In the first half of the 19th century technological and economic progress sharply increased and we can speak of the transition into *the second wave of the industrial revolution*. This progress was enhanced by innovations in the area of energy resources (electric energy, gas) and with the invention of a combustion engine. The innovations were transferred across nations in several ways—migration of employees, stays abroad (typical of the textile industry in Bohemia, where Liebig, the textile industrialist, gained experience and bought machines in England and then had his technologies patented), or indigenous research and development (typical of the glass industry in Bohemia, e.g. the Riedl family). From the first wave of industrialization in Bohemia, brownfields were mainly left along water courses (in the Liberec region after the textile industry) or in forests (the glass industry). By the turn of the 19th and 20th centuries, living and enterprise were no longer dependent on rivers, forests, and agriculture. The industrial environment then spread into other fields of human activity. Yet in the 19th and at the turn of the 20th century the capitalist lived for his factory, building his house and a public park near it, often investing in the development of the town in the form of construction of railroads, roads, bridges, post offices, and also galleries, museums, baths, and viewing towers. These improvements were the responsibility of an entrepreneur, but they often now constitute abandoned infrastructure left at the mercy of nature. If redevelopment was unsuccessful, a plaque with the information about a given significant building or factory might be erected and the site registered on a cartographic map as a technical monument. We can find such examples in the Czech Republic, for example the Kladno steelworks.

The 1930s brought the development of suburban industry. The original face of a landscape often changed into disordered land uses. The world wars, a period of nationalization, and a forced relocation of population in Central and Eastern Europe brought a gradual alienation of people from their work, life, traditions, towns, and surroundings, and destroyed unity with the environment. Disrespect for property rights caused a disconnect between the development of a municipality and the life of its inhabitants and a lack of interest in social values. There were attempts to obliterate differences between the country and the city, such as building centers for health care, supermarkets, and cultural venues in even the smallest villages. In such places the demand for these services was not sufficient for various reasons (demographic development or local tastes and values). This contributed to the development of today's brownfields. Village life was continuously devastated, although the collectivization of agriculture had the worst impact. The reason for living in a village and for its existence (farming and small local crafts) ceased. Another cause of village depopulation was migration to cities and changing life styles. So-called "cottage-dwelling," when people from cities escaped to nature on weekends, also changed the style of village life. Technological progress had a significant influence too, through nationalization and separation of property rights from entrepreneurial values. In places where it was possible to harmonize technical development with the landscape and thus support municipal patriotism, unique values remain up to the present.

Another breaking point in the development of Central and Eastern Europe occurred during the 1990s, with economic reforms that took place after the centrally planned economies collapsed. At the turn of the 20th and 21st centuries the third wave of industrial revolution arrived. It was characterized by the growing influence of information technologies, the Internet, and the creation of a virtual environment. Today's style of business reflects globalization and changes in values. An owner of a multinational company need not be familiar with branch plant locales. Certain types of industrial sectors, such as automotive, favor rapid construction of factories in remote industrial zones, with little connection to a particular municipality. Investors primarily want to avoid any problems with landowners, and to have an inexpensive, comfortable infrastructure preferably prepared in advance as a part of investment incentives. The focus is only on the time of project duration, without broader social considerations. Another phenomenon is big shopping centers offering "all in one services," ranging from food and shoes to "culture" such as multicinemas. Localization of such centers requires construction of huge car parks that are still not sufficient, construction and expansion of motorways, petrol stations, and large reloading areas. Will this leave future generations with the potential for technical monuments, or only devastated land? Under certain conditions some zones might be used for redevelopment as a foundation for construction of solar panels or car parks, but current development poses the threat of brownfields for future generations. The landscape can be divided into valuable and valueless, but the measures are often misleading. The landscape must be valued according to its diversity and ecological stability. It is necessary to restrict new exploitation of greenfields and use legislative and economic tools to motivate investors to reconstruct developed and abandoned localities. One example of such a tool is the method used in Germany, based on the evaluation of a potential developed area from the point of view of usefulness for investors, the municipality, and from the broader society (see Rydvalova and Zizka 2006).

Description of the Transformation Process in the Czech Republic

Between 1948 and 1989, the Czechoslovak economy was centrally planned. Another characteristic feature was a significant degree of monopolization in the economy caused by efforts to simplify management of individual enterprises. For this reason production was concentrated into a small number of giant enterprises. In effect, there was no private sector; in 1989 only 1.2% of the work force was employed in the private sector (Zidek 2006: 8). Foreign trade was focused on the countries of the Council for Mutual Economic Assistance (CMEA) whose participation in import and export was as much as 60%. The Czechoslovak economy was largely industrialized with a prevalence of industry over services. Holub (1998: 791) states that in 1989, the industry share of GDP was 59%, compared to services with 31%, and agriculture with 11%. The Czechoslovak economy was thus classified as an ISA type (I—Industry, S—Services, A—Agriculture), while developed countries at that time were in the SIA phase, i.e. with a dominance of services in GDP and employment.

After the "velvet revolution" in November 1989, a market economy was gradually introduced. The main reforms took place between 1991 and 1993; a preparation phase was implemented in 1990. The main attributes of the economic reform were the end of a centrally planned economy, price deregulation, foreign trade liberalization, privatization, private enterprise renewal, tax reform, macro-economic stabilization, and social transformation. It is necessary to note that development in Czechoslovakia was not separate from other countries in Central and Eastern Europe, where similar changes occurred. In June 1991, the Council for Mutual Economic Assistance was cancelled, a fatal blow to the majority of companies, which lost their guaranteed sales. Trade in CMEA countries was not based on the individual contracts of companies but on intergovernmental bilateral agreements. The loss of Eastern markets, over-equipped production capacity, high fixed costs, obsolete and low quality production, and inability to adapt to new more demanding markets led to the failure of many companies. In association with events at the beginning of the 1990s, it is necessary to recall that the federative republic was divided into two independent countries—the Czech and Slovak Republics—on January 1, 1993. This division obviously brought significant economic costs (costs of currency separation, government division, and a drop in mutual trade) and had a negative impact on foreign capital inflow.

Great economic and social changes obviously had an influence on the development of a central macro-economic indicator—gross domestic product. In the initial phase of transformation a significant drop in GDP occurred (see Figure 4.3). This drop cannot be evaluated in a negative way only. In many cases it consisted of a termination of production that consumed scarce resources without being useful. The market revealed this sort of production and eliminated it. As Dyba (1998: 280) states, GDP before 1990 was a rate of consumed resources, not a rate of useful production. Therefore the Czech economy in the period between 1990 and 1992 is described as a transformation recession (Vostrovska 2005: 7). As Table 4.1 illustrates, other countries of the former CMEA went through a similar phase of economic development.

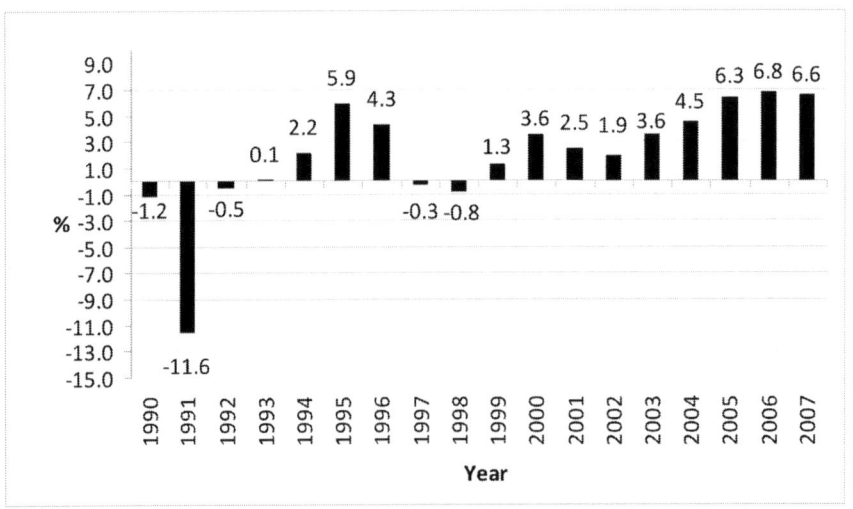

Figure 4.3   Real changes in GDP
Source: Based on the data of the Czech Statistical Office.

Table 4.1   Transformation recession in the former CMEA countries (GDP changes in %)

Country	1991	1992	1993
Bulgaria	-11.8	-7.4	-1.5
Czech Republic	-11.6	-0.5	0.1
Slovakia	-14.6	-6.5	-3.7
Hungary	-11.9	-3.1	-0.6
Poland	-7.0	2.6	3.8
Romania	-12.9	-8.8	1.5
Slovenia	-8.9	-5.5	2.8
Russia	N/A	-19.0	-12.0

Source: Mikula (2008: 4).

To understand the reason for brownfield creation, it is necessary to analyze the changes that occurred in branches of the national economy. If the year 1989 is considered as a point of reference, the share of the economy in industry decreased from nearly 59% to 42% by 1993; that is, by 17 percentage points in four years. Likewise, the decrease in agricultural production from 11% to 7% for the same period can be regarded as unprecedented. These changes are more visible in the services sector whose share increased from 31% to more than 51%, i.e. by more than 20 percentage points. Holub (1998: 791) notes that this phenomenon is quite unique in the country's economic history. Similar structural changes of this type occur only in other transforming economies. A strategy of transformation was necessarily connected to an absolute decrease in GDP, caused by an absolute decrease in industrial and agricultural production and a corresponding absolute increase in the service sector. In the subsequent period, mutual sectoral proportions were changing very slowly, as Figure 4.4 shows. Further, a relative decrease in the significance of the agricultural sector continued (to 2% in 2007), the share of industry on GDP creation stabilized at 38% and the services sector at 59%. It was notable that there was more rapid growth in industrial production than in services in the period up to 2004. This was caused by foreign investments, mainly in the automotive industry, and government support of industrial zones. In the last quarter of 2008 there was a decrease in industrial production, mainly in the automotive industry (by more than 20%)—according to CZSO (Czech Statistical Office 2009). The question remains whether the economic policy of the state, oriented mainly toward the support of one industrial branch, was correct. Clearly there is a potential danger for the creation of new brownfields in industrial zones that were only recently developed on greenfields.

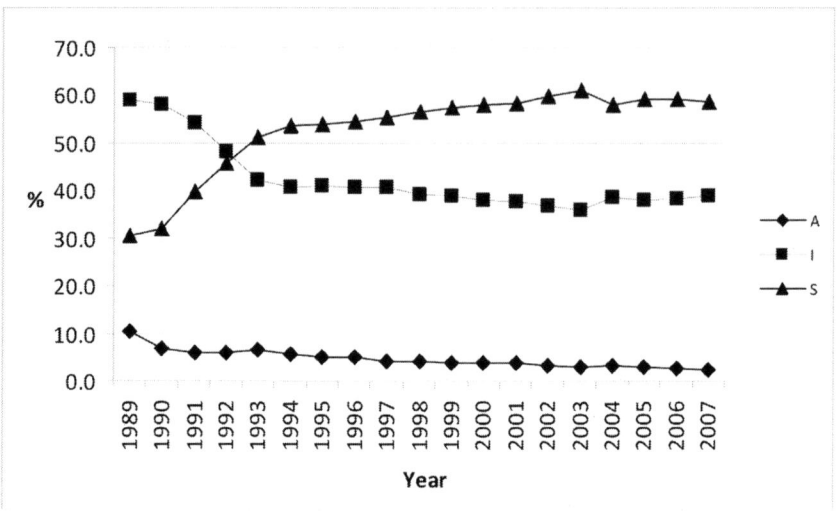

Figure 4.4   Industry share on GDP creation
Source: Based on the data of the Czech Statistical Office.

Transformation of the economy and the restructuring of companies are the main cause of brownfield creation in the Czech Republic, following on the abandonment of historical industrial sites and surplus infrastructure in the periphery of villages previously described. In the times of a centrally planned economy, brownfields were not a crucial problem for two reasons. First, facilities were artificially kept in operation, even those that did not make a profit. Second, responsibility for the environment was not a priority of central offices at that time. The claim that brownfield creation is closely connected with the process of transformation was proved correct by research carried out during 2004 and 2005 in one of the regions in the Czech Republic (Rydvalova and Zizka 2005: 467–76). Two hundred and forty-seven brownfield sites were identified, out of which nearly 40% had not been used for more than 10 years, while only 11% were not utilized for more than 15 years. Only 27% of sites were not used for up to five years (see Figure 4.5). The research further showed that the majority of sites were previously used in agriculture (32%) and industry (30%). These results correspond with a later survey of the CzechInvest agency made between 2005 and 2007 to locate brownfields in the Czech Republic. A comparison of the two studies is presented in Table 4.2. The differences in some categories result from the fact that the research of Rydvalova and Zizka was limited to the Liberec Region only. Both studies show that the dominant cause of brownfields is the remains of economic activity in primary and secondary sectors of the economy.

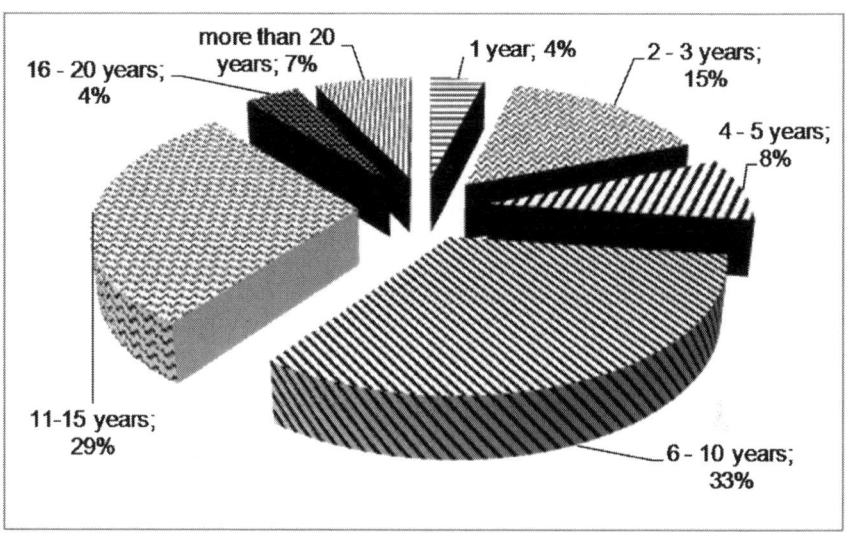

Figure 4.5   Non-utilization of brownfields

Table 4.2   The structure of brownfields according to their original use (in %)

Former use of a locality	Rydvalova, Zizka (2005)	CzechInvest (2008b)
Agriculture	31.6	34.9
Industry	29.6	33.3
Civic amenities	9.7	12.9
Army, military areas	0.8	6.4
Housing	8.5	4.0
Tourism, health resorts, hotels	18.6	0.9
Others	1.8	7.5
Total	**247**	**2,355**

Source: Rydvalova and Zizka (2005: 469), CzechInvest (2008b: 3).

Method of Brownfields Survey in the Czech Republic

Despite the fact that the majority of brownfields emerged at the beginning of the 1990s, the state began to deal systematically with the problem much later. The chronological succession showing the policy cycle for brownfields in the Czech Republic is apparent from Figure 4.6. The first survey dealing with brownfields was carried out in 2004 as a part of the project called "National Strategy of Brownfields" conducted from October 2003 to October 2004 by the state organization CzechInvest, established by the Ministry of Industry and Trade of the Czech Republic. As a result, a pilot database was created covering only two regions of the Czech Republic (approximately 14% of the total area). In the two regions 2,000 sites were identified with the total area of 11.5 thousand ha (see Table 4.3).

Table 4.3   Number of brownfields in CZ (estimate in 2004)

Region	No. of localities	Area in ha
North Bohemia	1,000	3,500
North Moravia	1,000	8,000
Other regions	4,000	14,000
CZ total	**6,000**	**25,500**

Source: Chlebna (2004: 2).

Based on this research, the number of brownfields in the Czech Republic was estimated at 6,000 sites with a total area of 26,000 ha. This estimate did not include brown coal surface mines (Chlebna 2004: 2). At the same time, state administrative authorities participated in brownfields regeneration in different ways. The Ministry of Industry and Trade was responsible for the creation of the brownfields regeneration program, and the CzechInvest Agency provided technical help and organized program implementation. The Ministry of the Environment addressed the problem of ecological burdens endangering the environment. The Ministry of Regional Development was dealing with the problem of regional planning and the use of brownfields for housing purposes. Finally, the regions had the task of developing regional strategies of redevelopment and integrating them into their larger development strategies and plans. Under the terms of the national brownfields strategy, financial sources that could be used for redevelopment were delimited: these were structural funds of the EU; the Fund of National Property intended for ecological-damage redevelopment drawn from the privatization of state property; special programs directed by CzechInvest; funds of the Ministry of the Environment; regional and municipal budgets; and private sources.

It was determined that it would be useful to divide the brownfield issue into two parts: the redevelopment of large brownfields (with an area of more than two hectares) with a development potential for processing industries and connected services, and the regeneration of smaller non-industrial sites suitable for small and medium enterprises or for the cooperation of a private and public sector enterprise. The Ministry of Industry and Trade was engaged in the category of large brownfields through CzechInvest.

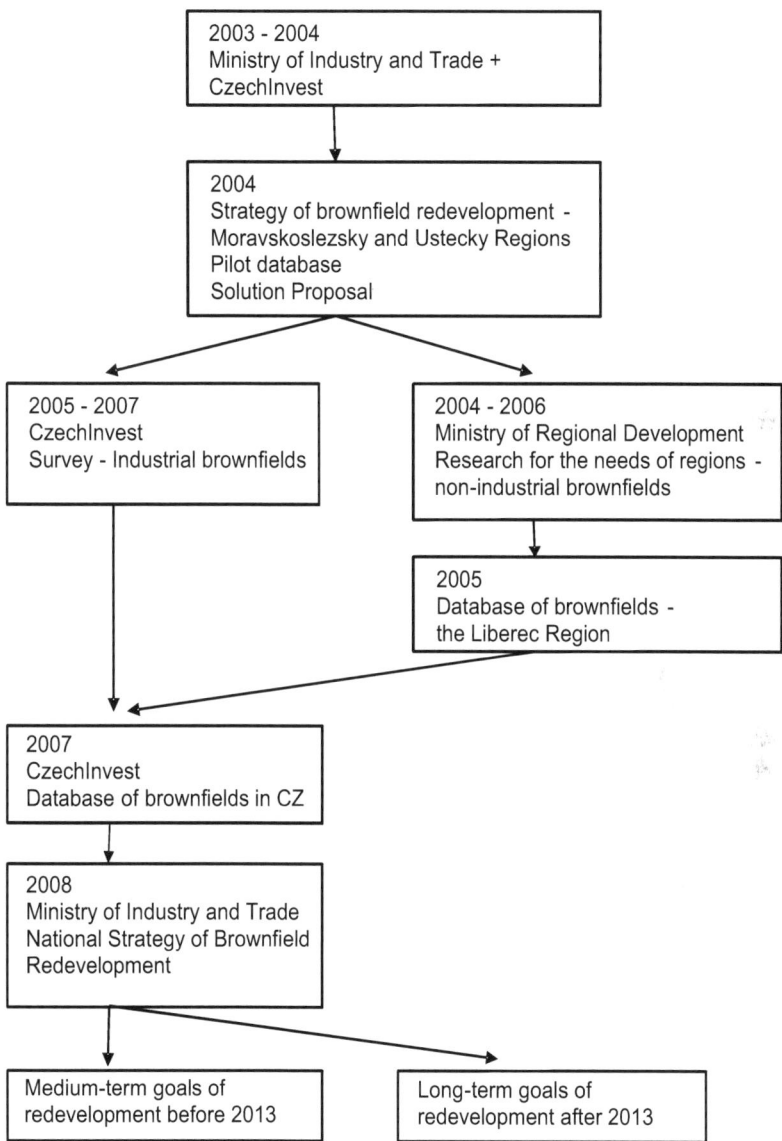

Figure 4.6   Solution of brownfield redevelopment

The National Strategy of Brownfields Redevelopment was carried out by the Ministry of Industry and Trade on the basis of the Government Resolution of August 31, 2005. Its goal was to make investment preparations for the areas in question for the location of strategic industrial zones. Within this framework, between 2005 and 2007, CzechInvest (in cooperation with all regions) prepared a survey of brownfields in the Czech Republic. It is necessary to point out the interconnection of this study with the research activities carried out by the Ministry of Regional Development that focused on non-industrial brownfields. Both research programs logically overlapped from the point of view of contents and time since, during the mapping phase, it was necessary to find all brownfields in the Czech Republic regardless of character and size.

A national strategy of brownfields redevelopment was recognized by the government on July 9, 2008. There were 2,355 brownfields with a total area of 10,326 ha. This incorporated all regions except Prague (see Table 4.4) and included brownfields that were one hectare and more, except the so called "mining brownfields." The rough estimate of the redevelopment costs of these sites is CZK 200 billion (approximately US $9 billion). The total number of brownfields is estimated to be 11.7 thousand sites with a total area of 38 thousand ha (MPO 2008: 3), however. Work on the database will therefore continue in the upcoming years.

It is obvious from Table 4.4 that there are large regional differences in the original use of brownfields, particularly in ecological issues. Even if there is a majority of brownfields in agriculture and industry, in some regions (for example in the Hradec Kralove and the Zlin Regions) there are also large former military areas that are much bigger than other brownfields. As Kocourkova (2008) states, an average military brownfield has an area of about 16 ha, an abandoned factory 5.5 ha and a decayed farm 2.2 ha. In the Vysocina and the Karlovy Vary Regions there is a higher number of brownfields originally used as civic amenities. The most problematic sites, with existing or assumed ecological burdens, can be found in the Usti, the Hradec Kralove, and the Liberec Regions. There areas were contaminated by the army or by chemical or textile production. On the other hand, the Olomouc and the Zlin Regions are the least ecologically overburdened. Among brownfields, private lands and objects prevail over public ownership. Brownfields in public ownership can be found most frequently in the Vysocina Region. These are mainly abandoned cultural venues and former country stores. It is apparent from the list of the original use of sites under investigation that it was not feasible to strictly divide monitoring of industrial and non-industrial brownfields as was initially planned in the survey study.

Table 4.4   Number of brownfields in the regions of the Czech Republic

Region	Estimated no. of localities	Prevailing original use	Percentage of localities with ecological burden	Type of ownership	
				Public	*Private*
Pardubice	600	agriculture 44%	60%	17%	77%
Hradec Kralove	750	agriculture 33% industry 32% army 19%	66%	20%	71%
Liberec	650	industry 40% agriculture 33%	65%	19%	79%
Zlin	880	agriculture 49% industry 33% army 14%	6%	14%	86%
Olomouc	750	agriculture 50%	5%	19%	65%
Vysocina	630	agriculture 43% civic amen. 22%	34%	38%	62%
South Bohemia	750	agriculture 32% industry 29%	57%	16%	73%
Plzen	750	agriculture 45% industry 29%	38%	22%	78%
Karlovy Vary	420	industry 33% civic amen. 25% agriculture 17%	26%	17%	72%
Usti and Labem	760	agriculture 39% industry 34%	67%	11%	71%
Central Bohemia	1,500	industry 38% agriculture 37%	16%	23%	74%
South Moravia	1,300	industry 40% agriculture 35%	41%	28%	69%
Moravia-Silesia	1,350	industry 50% agriculture 22%	47%	27%	73%
CZ in total	**11,100***	**agriculture 35%** **industry 33%**	**46%**	**20%****	**73%****

Source: CzechInvest (2008b).

Notes: *The total number of brownfields does not include the area of Prague, the capital of the Czech Republic. **The total in the "type of ownership" does not equal 100% as in some cases the ownership is mixed or it is owned by the church.

The second group of non-industrial brownfields came under the jurisdiction of the Ministry of Regional Development within four programs (see Damborsky 2005: 29):

- Collective regional operation program: regeneration and redevelopment of selected towns—focused on the redevelopment of urbanized town areas and historical centers of towns;
- Support of the development of areas significantly affected by the demobilization or reduction of military facets: focused on support of regional planning, housing and related infrastructure;
- Unified document Prague: redevelopment of damaged and poorly utilized areas, landfills and ecological burdens in the area of the capital of Prague; and,
- Research on the needs of regions with a priority for non-industrial depressed zones: focused on processing of instructions for the integration of non-industrial brownfields into social and economic plans used by municipal authorities.[1]

The Technical University of Liberec team implemented the project "Redevelopment of Non-industrial Brownfields as a Part of Regional Development Strategy" (project reg. no. WB-13-04) within the "Research on the Needs of Regions" research program mentioned above. The project included monitoring of brownfields in the Liberec region, compiling a brownfields database, evaluating individual sites from the point of view of attractiveness for potential investors, and including selected sites in the catalog of investment opportunities (see Rydvalova and Rydvalova 2005).

The brownfields database was created in Microsoft Access and consists of three mutually interconnected subsystems:

- monitoring,
- microenvironment,
- macroenvironment.

The municipality where a problematic locality could be found was selected as a basic classificatory feature of a brownfield.

Monitoring: the first subsystem contains basic information about individual brownfield sites: name; cadastral territory number; the area of the brownfield divided into building area and adjacent areas; type of ownership; building limitations; availability of technical infrastructure and transportation approach; brownfield specification; regulations according to planning documentation; an original, the latest and an intended usage (if known); period of non-utilization;

---

1   The above programs were implemented between the years 2004–2006.

location in relation to a municipality; spatial surroundings delimitation; photo documentation; owner contact information; and value of investor's criterion.

Microenvironment: this subsystem includes information about the municipality where the brownfield is located. Potential investors can gain information about the number of inhabitants, potential for tourism, public transport, public revenue collection (divided into individual kinds of taxes such as subsidies, exclusive taxes, shared taxes, motivation taxes, and untaxed income, for the last three years), and typical local industry (defined by the coefficient of localization according to the number of employees and added value).

Macroenvironment: the third subsystem characterizes an authorized superordinate municipality in the area from the point of view of unemployment, structure of job candidates, legal forms of entrepreneurial subjects, and a number of economic variables according to classification of economic activities. All information was provided for the last three years (Rydvalova and Zizka 2006: 643). The database was submitted to the representatives of the Ministry of Regional Development, the Liberec Region, emergency services, and other parties, and is a model for creation of other brownfield databases at the national level and for individual regions.

Strategic Measures for the Support of Brownfield Redevelopment

Based on the forgoing research, the *National Strategy of Brownfields Redevelopment* has been implemented. Its goal is to create a suitable motivating environment for quick and effective implementation of redevelopment projects and the prevention of new brownfields. The goals are divided into medium- and long-term. Medium-term goals are set in such a way that their achievement is possible by 2013. By that time the 2007–2013 programming period of the European Union will have ended, along with the possibility of drawing financial resources from structural funds. It is expected that after this period most regions in the Czech Republic will reach 75% of the European Union GDP average, and the ability to finance redevelopment of brownfields from structural funds will be limited.

Medium-term goals include (see MPO 2008: 5):

- Maximum involvement of available European sources for the redevelopment of brownfields in the programming period 2007–2013.
- Possibility of brownfield redevelopment for other than industrial uses (e.g. mixed city functions, civic amenities, agriculture, and housing).
- Development of the educational system in the field of brownfield redevelopment and professionalization of a state administration for this issue.

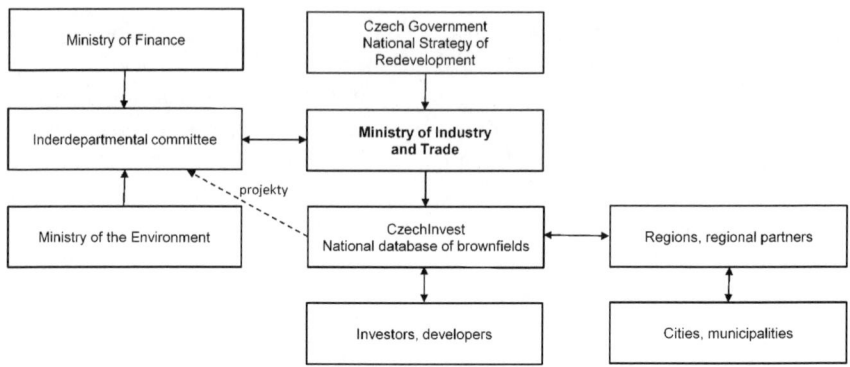

Figure 4.7   Scheme of cooperation

Long-term goals, after 2013, emphasize preventive measures in order to use the areas of existing brownfields for redevelopment and to limit the creation of new brownfields.

As far as the organizational structure is concerned, the Ministry of Industry and Trade of the Czech Republic is fully responsible for solving the problem of the redevelopment of brownfields (see Figure 4.7). In June 2008, the *National Strategy of Brownfields Redevelopment* was introduced to and further recognized by the government of the Czech Republic. Since there is no longer a division of power and sources between the Ministry of Industry and Trade and the Ministry of Regional Development, the agenda was delegated to the CzechInvest agency established by the Ministry of Industry and Trade in 2004.

The CzechInvest agency is charged with the following tasks:

- To identify and map suitable sites for implementation of entrepreneurial projects in individual regions of the Czech Republic; these are then registered in the National Database of Brownfields,
- To engage in active participation in the National Strategy of Brownfield Redevelopment in the Czech Republic,
- To provide information about programs to support brownfield redevelopment financed from the state budget and from structural funds of the EU (Real Estate program, see further),
- To prepare and organize the so-called "road shows" (tours of available real estate),
- To prepare seminars and conferences on the redevelopment of brownfields,
- To submit projects for evaluation to an interdepartmental committee,
- To look for potential investors, and
- To administer subsidy programs supporting brownfield redevelopment (CzechInvest 2008a; MPO 2008: 7).

The interdepartmental committee consists of representatives of the Ministry of Industry and Trade, the Ministry of Finance, and the Ministry of Environment. Its task is to evaluate projects for public support. The cooperation of regions and municipalities is also important. That is why regional coordination groups will be introduced by CzechInvest in 2009 (Adamkova 2008). There are CzechInvest offices in all regional cities. The regions have already cooperated with CzechInvest on the survey study; it is now essential to assure that the support of brownfield redevelopment is included as one of the priorities of regional development programs. Large brownfields and the enormous financial demands of redevelopment require the use of multi-source financing. Despite this, it is estimated that by 2013 a maximum of 25% of the total of brownfield sites will have been redeveloped: 750 sites. At present it is projected that the majority of remaining sites will have been redeveloped by approximately 2025 (Lorenzova 2006). For this reason, priorities have been set for brownfield redevelopment. Preference will be given to sites with ecological burdens and historical buildings; that is where immediate economic gain can be expected. As far as the area is concerned, the preferred sites will range from middle-sized to larger, formerly used in industry, agriculture, housing, civic amenities, and as former military areas (see Table 4.5).

Table 4.5   Priorities in brownfields redevelopment realization

Localization of area	1 Priority ++	2 Priority +	3 Priority -
A—Urban area (Intravilan) of cities and villages compact, urban areas of cities and villages	*Big areas:* - industry - housing - civic amenities	*Middle areas:* - industry - housing - civic amenities	*Small areas:* - industry - housing - civic amenities *Big and middle areas:* - agriculture
B—Extravilan of cities and villages urbanized but a suburb part of settlements, scarcely built-up	*Big areas:* - industry - housing - army	*Middle areas:* - industry - housing - army	*Small areas:* - industry - housing - civic amenities *Big and middle areas:* - agriculture - civic amenities
C—Interurban areas except settlements in the main region, free or loosened areas	*Big and middle areas:* - airport - agriculture	*Small areas:* - industry, - housing - agriculture	*Big and middle areas:* - army - former mines

Source: Alexova (2007: 7), adapted.

The brownfield redevelopment strategy supposes three basic ways of remediation and redevelopment:

- *Reconstruction* means changing an unused and neglected building to a new use provided the technical conditions of the building allow it;
- *Recultivation* means cleaning the area of contamination, remediating the site, and thus approximating greenfield conditions;
- *Revitalization* is a follow-up to remediation and represents a new use of the site for entrepreneurial and other purposes.

In practice, the above-mentioned possibilities can be combined; for example, part of a neglected site can be reconstructed and another part can be used after remediation and revitalization.

Financial Resources for Brownfield Redevelopment

The costs of brownfield redevelopment in the Czech Republic are enormous and exceed the possibilities of the state budget. The costs of brownfield redevelopment in mapped localities are estimated at CZK 200 billion; the average costs of one hectare of redevelopment are around CZK 20 million. If we take into account the fact that the total area of brownfields in the Czech Republic is 38,000 ha, then we arrive at a cost of CZK 760 billion, which is 21% of GDP of the Czech Republic. It is obvious, then, that the redevelopment of brownfield sites must be spread over a longer timeframe, marking out priorities and involving other financial resources. In the previous programming period, 2004–2006, resources mainly from structural funds via the REALITY program (CZK 2.1 billion) and from the national "Program for the support of industrial zones development" (CZK 1.3 billion) were used for financing brownfield redevelopment. For the years 2007–2013 several main resources are projected (MPO 2008: 7):

- European funds: mainly the Operational Program for Entrepreneurship and Innovations (as a part of the Real Estate Program), Operational Program for the Environment, Program for the Country Development, Operational Program Prague Competitiveness, and Regional operational programs;
- State budget grants: mainly programs of the Ministry of Industry and Trade, the Ministry of Regional Development, and the Ministry of the Environment;
- Private resources: developer investments; partnership of public and private sectors;
- Debt financing: international and Czech financial institutions.

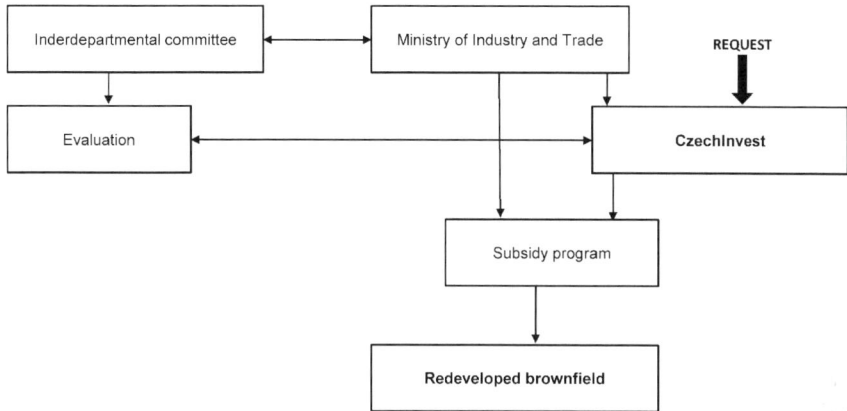

Figure 4.8   Scheme of brownfield redevelopment

A specification of individual public resources is presented in Table 4.6. CzechInvest is responsible for helping potential investors and developers select suitable subsidy sources (See Figure 4.8).

The biggest funding source is the Real Estate program that is a part of the *Operational Programme for Entrepreneurship and Innovations*. This program is not focused on brownfield redevelopment but encourages preparation of and investment in new business facilities or construction of new facilities for rent. As far as brownfields are concerned, the program also supports their transformation into a business facility or reconstruction of a facility for business purposes. Self-governing territorial units, towns, regions, and entrepreneurs can apply for the subsidy; the maximum amount of subsidy is variable (see Table 4.6). Apart from the size of a company, the amount of a subsidy for enterprises also depends on the region in which the project will be implemented. The rule is that in less developed regions the support can be higher than in economically more developed regions. Some industrial branches are excluded from the subsidy including: food, steel, and coal, production of artificial fibres, ship building, agriculture, and fishing (these branches are so-called "sensitive" branches and are under a special mode in the EU). Another condition is that the business facility must have a minimum of 500 m^2 and will be used for activities in the fields of manufacturing, technology, and strategic services.

Operational Program—Environment, managed by the Ministry of the Environment, addresses the problem of brownfields mainly in the Priority Axis Subsidies for Waste Management and Eliminating Old Environmental Burdens. Financing from this program can be used for remediation of old dumps, removal of illegal dumps in protected areas, inventory of contaminated and potentially contaminated sites, classification of priorities for selecting the most seriously contaminated localities for redevelopment, research projects, risk analysis, and redevelopment of seriously contaminated localities. Further, the Priority Axis

Protection and Regeneration of Urbanized Landscape concerns the problem of brownfields and is focused on small (up to 10 ha), economically utilizable areas, and former military training areas. Recultivation of such localities includes planting natural vegetation.

Program of the Country Development of the Czech Republic for the period 2007–2013, conducted by the Ministry of Agriculture, offers broad possibilities for utilization. Brownfields belong to the first and the third axis. Axis I deals with modernization of agricultural companies and supports reconstruction of agricultural buildings (especially brownfields). This axis is mainly aimed at entrepreneurs in agriculture. Axis III addresses the issue of renovation and improvement of villages, facilities, services, and protection and development of country cultural heritage. Within the third axis are projects for reconstruction of buildings (such as modernization) and areas with public facilities in the field of education, health care, social services, child care, culture, shopping infrastructure, sports activities, and projects for reconstruction of historical buildings. The projects can be submitted by villages, associations of villages, non-profit organizations, churches, and associations of legal entities. There is a condition that the project must be implemented in a village with up to 500 inhabitants. The amount of subsidy varies according to the economic level of the given region.

Operational Program "Prague Competitiveness" is only valid for the region of the capital of Prague, which also holds the office of a managing body. The reason for the independent Operational Program for the capital is that Prague is a developed region and could not draw financial resources appointed for other economically weak regions. In Prague there are many abandoned and devalued areas representing ecological danger. They emerged mainly after the conversion of production and also include vast areas of railroad. There is high demand for land in the city center and at the same time extensive building "on the greenfield" (OPPK 2008: 17). The second axis of this operational program—Environment, Revitalization and Area Protection—also includes brownfields. This axis is focused on the regeneration of unattended, damaged, and inappropriately used areas with the goals of increasing their attractiveness for investors, citizens, and visitors, and strengthening their "city-forming" function and economic potential. One of the priorities is a renewal of unused, historically valuable buildings and sites that are a part of a revitalized area and that have been neglected so far (e.g. periphery areas). Another priority is the redevelopment of areas that represent a current environmental danger regardless of the future use of such a site. Projects can be submitted by the capital, its districts, organizations set up by the capital or its quarters, and non-governmental non-profit organizations.

Another *Operational Program* is the *Transport Operational Program*. Its sub-program called "Redevelopment of Rail Sidings in the Czech Republic" can be used for the regeneration of brownfields. A project can be submitted by both legal entities and physical persons that have property rights to land located by rail sidings. The aim of the project is to renew rail sidings that have been decommissioned or are out of order, improve the accessibility of railroad networks in industrial zones by construction of new rail sidings, and increase the use of environmentally-friendly haulage. According to program documentation (MD 2008: 10), in 2006 there were a total of 495 abandoned rail sidings (about 27% of their total number) in the Czech Republic. The amount of financial support depends on the economic development of a region where the project will be implemented.

Other significant resources can be found in *Regional Operational Funds* (ROP) that are created at cohesion region levels (regions on the level NUTS II according to Nomenclature of Territorial Unit for Statistics used by Eurostat). There are seven such programs, which cover all areas of the Czech Republic except the capital of Prague. Every ROP is an independent document directed by a regional council. The structure of these documents is similar and all ROPs have one common priority; the development of urban and rural regions, thus addressing the issue of reconstruction and redevelopment of brownfields. ROP NUTS II South-East, including the Liberec, Hradec Kralove, and Pardubice regions, is a good example. Here the brownfield issue is addressed within two priority axes— Country Development and Enterprise Development. The first axis supports the redevelopment of sites owned by public authorities. The second axis encourages redevelopment and revitalization of brownfields intended to be used to develop entrepreneurial activities. The areas used for manufacturing industry, tourism, and abandoned agriculture production are excluded from this support. The reason is that ROPs mainly concentrate on sites not eligible for other existing support programs (for example the Real Estate Program or the Country Development Program).

Besides the Real Estate Program, which is co-financed from structural EU funds, there is also the national *Program for the support of entrepreneurial real estates and infrastructure* financed from the state budget of the Czech Republic. The Ministry of Industry and Trade administers this program as well. Unlike the previous program, this grant is intended for strategic investors in the field of manufacturing and for projects that cannot be supported from the Real Estate Program. The aim of this support is to contribute to the process of restructuring industry in the Czech Republic, replacing old production plants with modern facilities with low resource and energy dependence, high added value, and significant export potential; to support restoration of the environment and removal of contamination; and to create new jobs (MPO 2007: 19). The program can finance the preparation and implementation of a project focused on transformation of brownfields into an entrepreneurial park or a building. When the area of the entrepreneurial park exceeds 100 ha it is a "strategic project," and the amount of the subsidy can be up to 100% of eligible costs. Smaller projects are financed partly from the "cost gap," which is the difference between the costs of project implementation and the yield from its sale or rental. The minimum area of the entrepreneurial zone must be 5 ha. The program was ratified for the years 2006–2010.

Table 4.6 Potential public resources for brownfields redevelopment for the period of 2007–2013

Source	Managing body	Applicants	The amount of subsidy	Available money*
Real Estate Program	Ministry of Industry and Trade	Self-governing territorial units, entrepreneurial entity	From the total of eligible costs: Municipalities: max. 40% Small entreprises: max 50%–60% Medium entreprises: max 40%–50% Large entreprises: max 30%–40% CZK 1–500 mil./project	CZK 16.1 billion
OP the Environment	Ministry of the Environment	Self-governing territorial units, entrepreneurial entity	max 85% of eligible costs min CZK 0.5 mil./project	CZK 7.2 billion
Country Development Program in the Czch Republic	Ministry of Agriculture	*Modernization of agriculture companies:* - entrepreneur in agriculture - entrepreneurial entity (basic industry, primary producer) - young farmer	From the total of eligible costs: Young farmer: max 50%–60% Other farmers: max 40%–50% min CZK 100 th./project max CZK 30 mil./project	CZK 16.2 billion
		Revitalization and development of villages, facilities and services: - village - village association - non-profit org. - church - association of legal entities	From the total of eligible costs: Small entreprises: max 50%–60% Medium entreprises: max 40%–50% Large entreprises: max 30%–40% min CZK 50 th./project	CZK 1.9 billion
		Protection and development of country cultural heritage: ditto	max 90% of eligible costs min CZK 50 th./project	
OP Prague Competitiveness	The capital of Prague	The capital of Prague, city districts, organizations set up by the capital or its districts, non-profit organizations	max 85% eligible costs min CZK 1 mil./project	CZK 1.7 billion CZK 1.6 billion

Table 4.6   Continued

OP Transport	Ministry of Transport	Legal and physical entities	max 30–40% eligible costs	CZK 0.4 billion
Regional operational programs NUTS II: - North East - South East - South West - Central Bohemia - North West - Moravia-Silesia - Central Moravia	Regional council of cohesion regions	Regions, municipalities, voluntary village associations set up by the region and municipalities, non-governmental non-profit organizations, chamber of trade, associations of entrepreneurial entities, entrepreneurs—mainly small and medium entreprises SME	Projects not applying for public support: max 85% up to 92.5% eligible costs In case of public support, from total of eligible costs: Small entreprises: max. 40%–60% Medium entreprises: 40%–50% Large entreprises: max. 40% min CZK 1–5 mil./project	CZK 28.7 billion
Program for support of entrepreneurial real estates and infrastructure**	Ministry of Industry and Trade	Self-governing territorial units, developing company, entrepreneurial entities	Strategic projects: up to 100% of eligible costs Municipalities: max 80% eligible costs Entrepreneurial entities: max 36%–40% cost gaps, max 1,500 CZK/m²	CZK 3.3 billion

Notes: * for exchange rate used EUR 1 = CZK 28. ** assumed program finishing in 2010.

Source: the authors' own.

Table 4.6 lists possible sources for financing of brownfield redevelopment from public budgets. In total there is CZK 77 billion available. If we take into account the fact that the costs of redevelopment of currently mapped brownfields are estimated at CZK 200 billion, then public resources cover about 38% of total costs. It is necessary to stress, however, that grant titles in Table 4.6 are not exclusively intended for brownfield redevelopment and can be used for the support of building on greenfields. The volume of financial means for brownfield remediation is thus much lower. Based on statistical data from the Real Estate Program, 73% of the total amount of subsidy was used for brownfield redevelopment. Disposable financial resources (a presupposed volume of financial resources from public budgets that will be used for the regeneration of brownfields) can then be estimated to be approximately CZK 56 billion until 2013. From these data it is obvious that private capital must be involved in the process of brownfield redevelopment in the Czech Republic in order to improve the current situation.

Implications for the Support of Brownfield Redevelopment

In the Czech Republic there is an annual demand for 240 ha of land for development (Kovarik 2005: 17). Theoretically, it would be possible to meet the whole demand with redeveloped brownfields; the current map of brownfields would satisfy the interests of entrepreneurs for many years. The reality is quite different, however. Investors are mainly interested in sites in industrial zones created from overzoned greenfields. For rental buildings, only 8% of investors look for brownfields (Kocourkova 2009; see Figure 4.9). In the last two years, some change in the structure of demand for entrepreneurial real estate has occurred. According to Kocourkova (2009), there is less interest in large spaces in industrial zones (inter-annual decrease by 9%), however, there is an increasing interest in rental spaces (inter-annual increase by 6%) and also slight interest increases in brownfield sites (increased by 2%). The explanation for such a change can be found in statistical data about new investments in the Czech Republic provided by the CzechInvest agency. Instead of large production projects, the Czech Republic is more attractive for smaller technology and plants. According to the general manager of the CzechInvest agency (Rudysarova 2007), research and development or services formed only 32% of investment projects; in 2008 it was 63%. The increasing interest in rental spaces shows how the Czech economy approximates old European member states where a majority of new investments focus on rental spaces (Kocourkova 2009). Despite this positive development regarding investors' interest in brownfields, it is still insufficient to use up the stock. There are several reasons including:

- complicated property rights,
- devastated buildings,
- contamination, and
- depopulated surroundings.

For these reasons investment in brownfields appears to be much more time-consuming, with higher costs and many risks. The interest in brownfields can be seen mainly in attractive locations with a lack of available land, for example in urban areas.

Figure 4.10 shows the stimulation process for a higher use of brownfields. A first step is a thorough mapping of all brownfield sites and description of the surrounding areas. Types of information gathered during the monitoring process were mentioned previously. Information about the locality itself, the macro- and micro-environment, is important for potential investors; therefore it is appropriate to make it available to the public. Information from the database can be displayed on a website and in catalogues of investment opportunities. At the same time the National Database of Brownfields is available on websites of the CzechInvest agency (available on brownfieldy.cz) and other databases run by regional offices.

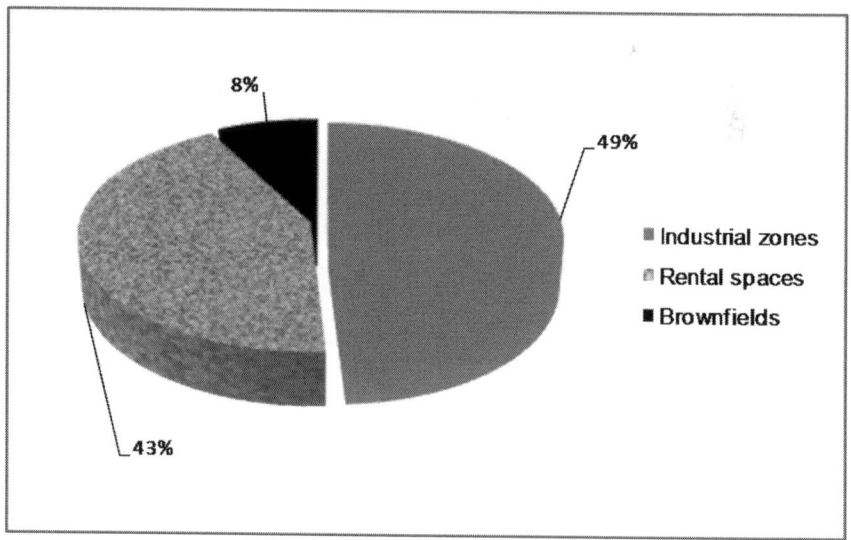

Figure 4.9   Structure of demand

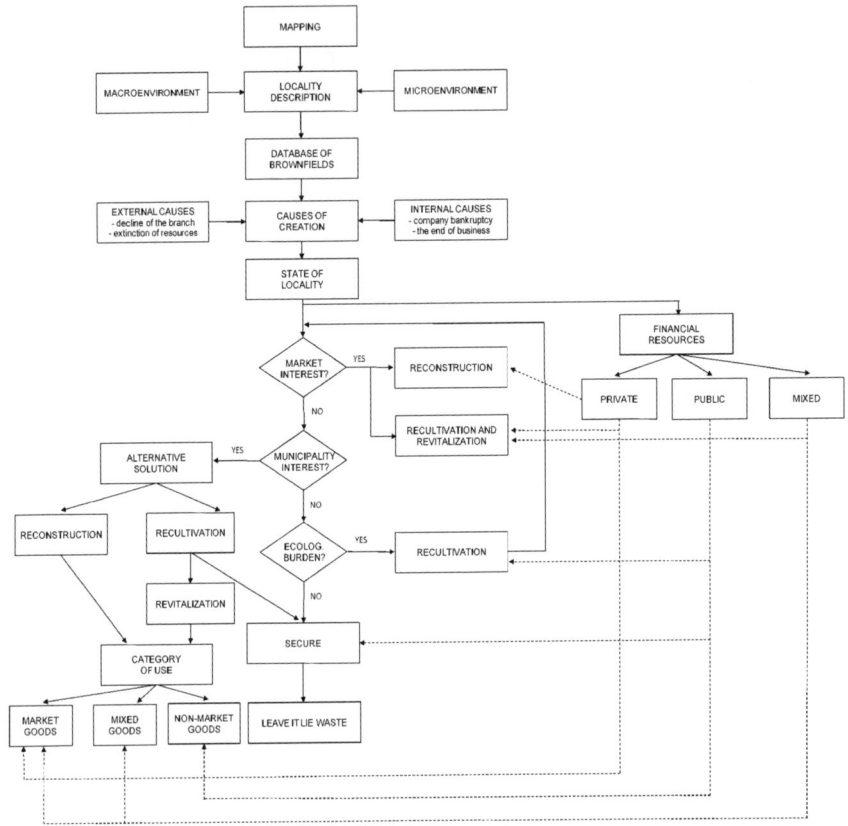

Figure 4.10   Process of brownfield redevelopment

It is important to determine the reasons why existing brownfields were created in order to seek the best possibilities for future use. Brownfields result for a variety of reasons, for example as a consequence of a bankruptcy or voluntary termination of business. Such situations lead to the creation of smaller sites that can be redeveloped more easily. If a brownfield was created by external influences, e.g. the decline of whole industries or extinction of resources, the situation is more complicated. In the Czech Republic such brownfields can be found in former textile factories, glassworks, or mines. Since the decline of the industries occurred in a relatively short time period and concerned thousands of businesses, the number of such sites is very high. The buildings of textile factories (spinning factories and weaving mills) can serve as an example of brownfields unlikely to be usable for anything other than textile purposes. They have many bearing columns that occupy quite a lot of potentially usable space. Apart from that, textile production in such factories damaged the subsoil, which is contaminated by chemicals. Former mining activities are also large in land area and are associated with contamination.

They often are found in scantly populated locations, which is the consequence of towns and villages becoming extinct due to the mining activity.

The result of previous research is a detailed description of site conditions that can serve as a guideline for further decision-making. Attractive sites, which form only a small part of the total number of brownfields, may succeed on the market. Therefore it is a prerequisite to strengthen the interest of investors in such sites, for example by co-financing redevelopment and revitalization from public funds. An example of such redevelopment can be seen in Prague-Liben, where a multipurpose entertainment and ice-hockey hall was built on the site of the former CKD, a large engineering company. The complete remediation of ecological damage in the amount of CZK 680 million was financed from public resources (Tylova 2004); the construction of the building itself was financed from private sources. The total costs of the construction reached CZK 8 billion between 2002 and 2004 (Burian 2005).

If investors are not interested in a neglected site for a long time, it is possible for a given municipality to undertake redevelopment on its own with the support of national and European grants. Reconstructed and renewed buildings can be offered to both private investors and non-commercial entities with the cooperation of the public and private sector. The resulting business activity is for-profit and is usually financed from private capital. In certain cases, entrepreneurs can gain the support of public resources, for example where the building is on the site of a former brownfield. The amount of public support depends on a grant title usually ranging between 40% and 60% of total costs. The rest of the costs must be covered by an entrepreneur.

Non-commercial activities are not conducted in order to make a profit. Former brownfields are used for education, social care, information centers, and so forth. In these cases redevelopment is usually financed from public resources (state budget, municipalities, and European funds). The cooperation of the public and private sectors (Public Private Partnerships; PPP) emerged from the idea that the involvement of the private sector will ensure more efficient use of public resources. This cooperation also brings diversification of risk among a larger number of subjects and utilization of private companies' experience. A project beneficiary is always the public sector; the private sector is a supplier or services provider. The alliance of the PPP is used quite often abroad, for example when building and operating traffic infrastructure, accommodations, sports centers, or in the area of prisons. In the Czech Republic there were only a few projects of the PPP type. An example in brownfield redevelopment is the current construction of a sports center, including follow-up housing development, administration, and shopping spaces, in Brno-Ponava, to be created by the reconstruction of an old and unused football and ice-hockey stadium. The project beneficiary is the statutory city of Brno, the project duration is estimated to 25 years, and the total costs at CZK 3 billion. Other examples are the construction of a building in former barracks in Jihlava, the Vysocina Region, and housing development in Prague-Bechovice on the site of a former brownfield (Kubistova 2008).

In practice, however, there are many cases when either private investors or a municipality are not interested in a brownfield site. Then it is necessary to consider the efficiency of redevelopment from public resources. The key parameter is the presence of contamination which can endanger the environment or health of people. In any case, it is necessary to remediate identified ecological burdens. Apart from the removal of a potential threat, the chances for renewal by private investors or a municipality are increasing. In a case where the site is in a remote location, without contamination, but without any long-term interest for private investors or a municipality, there is the possibility of "reconciliation" with its present state; that is, to ensure no unauthorized entries, leaving it to lie waste for future generations. At the first blush, such an approach may seem irresponsible. It must be stressed however, that from the point of history, economic activities followed people, not the other way around. There is no point in supporting economic activities in areas that do not possess even basic prerequisites. Abandoned mines and quarries are one example of such sites. A study by Cilek (2005: 90–93) presents inspiration for how to address them. The author deals with the question of how to prevent the creation of brownfields on sites of abandoned quarries. The suggested method can be generalized for other types of brownfields.

Quarries can be "beautiful," if they adapt to local conditions before their opening, not the contrary. The process of redevelopment must be elaborated before building or operations commence. This approach is required by departments of the environment all over Europe. It cannot be said in advance what will occur during the quarry functioning, therefore regulations for quarry abandonment should be accepted upfront. In the first place, all buildings and technical equipment should be removed. An essential part of a quarry should be left as a vertical wall because of vegetation, birds, or as a demonstration of a geological profile.

Redevelopment should be carried out continuously, even during quarry functioning. The average lifespan of a quarry is about 30 to 50 years. Approximately one-half of a quarry from the edge to the center and further could be naturally revitalized. When deciding which part of a quarry should be left as a stand for birds (close to rock steppes) it is advisable to cooperate with natural scientists.

An aesthetic integration of a quarry into the surrounding landscape is also a vital rule. This is the work of a landscape architect (in the Czech Republic there are only garden architects, but their services can also be used). The beauty of a quarry is in its diversity, especially when it overgrows with vegetation.

Nature itself will come. Native plantings under the control of a plant scientist can help spontaneous reproduction.

Beauty is in its simplicity. The quarry must be cleaned and abandoned. This is better for the environment than planting it artificially and then building a road, train, or cable car for tourists. Especially with smaller quarries it is also necessary to consider whether it is worth hiring a landscape architect. Here, nature will help itself; it is necessary, however, to prohibit dumping. Tourists may be allowed to pass through.

Quarries with buildings do not assimilate well with the landscape. In places where garages and storehouses were left after abandonment, there is little to be done. Therefore when approving the quarry, it is necessary to require that all buildings and machinery be liquidated and taken away after its termination. Before a complete abandonment of a quarry, it is also necessary to secure the area. A fence is not a solution, since it can be destroyed. On the contrary, it is suitable to prepare one safe access, such as a footpath. Other access roads can be blocked by piles of soil or stone. Then, it is up to adventurers to decide to climb the quarry walls, risking injury, or not.

When approving a new quarry it is necessary to require documentation of the redevelopment track record of mining companies. For example, requiring photo documentation showing what they did with other old quarries is useful.

Conclusion

The intention of this contribution was to analyze a range of problems connected with brownfields in the Czech Republic, uncover the cause of their creation in the historical context of economic and social development, and make a proposal for successful redevelopment. It must be stressed that even if existing brownfields represent a serious problem from the point of view of economic, ecological, and public interests, the prevention and identification of potential brownfields is much more important. In this respect, a key role is attributed to uncovering historical causes of brownfield creation. Present brownfields often have roots in history and past events or in the technological lock-in of certain industries. The identification of the dependence of industry on its historical development (so called "path dependency") might help with brownfield forecasting in the future. The situation in the Czech glass industry, which has begun to decline in recent years, is an example of such dependence. The main causes of the decline are historical—the majority of glass products in the Czech Republic are hand-made, and glassworks are proud of this tradition. A final product is of premium quality, but the costs of hand-made production are very high compared to industrial production. Czech glass can hardly compete with that of Asian producers, whose automated production is much cheaper and eventually will achieve the same quality. This technological "lock-in" can be overcome, but it requires a lot of time and high costs, factors that many glass companies cannot afford.

The threat of new brownfields can be foreseen in the present economic policy of the government (not only the central level, but also regional and municipal levels) which may lead to lower demand for brownfields. Brownfield redevelopment is generally supported, but mainly on a theoretical as opposed to practical level. As was stated previously, financial resources intended for the development of entrepreneurial property can be used either for brownfield redevelopment or for building on greenfields. The government or the regions, then, invest in building infrastructure in a greenfield in order to create new job possibilities. As a result, the supply of new sites for doing business is increasing and logically the interest in investing in brownfields is decreasing. It can be said that the demand for brownfields is higher in built-up city areas where there is a low supply of vacant sites. In these cases, investors are willing to accept the risk and pay the costs related to brownfield redevelopment. On the other hand, there is only slight interest in doing business in remote brownfields. More and more vacant landscape is being built up and nature cannot fulfil its functions. Further, land prices are generally lower outside city centers (the question of land price and scarcity). As Cilek et al. (2004: 25) state, land in the Czech Republic is the cheapest in the last two hundred years. The reason lies in the fact that there is an abundance of agricultural production in all European countries caused by the common EU agriculture policy. Low land prices logically lead to waste. The problem can be solved by better utilization of financing from the public sector and the use of disposable financial resources mainly for brownfield redevelopment. The focus of individual support programs must be considered. They often favor investments in manufacturing or services closely connected with this sector. That means a gradual relative decrease of the secondary sector for the benefit of tertiary and quaternary sectors. Statistical data confirm this tendency.

This text was written in relation with the results of the project WB-13-04 with the support of the project WD-30-07-1 of the Ministry of Regional Development.

References

Adamkova, A. 2008. *Kdy se ceske brownfields zazelenaji?* Available at asb-portal. cz (accessed: January 6, 2010).

Alexova, M. 2007. "Regenerace brownfields." *Planeta* 15(3): 2–20.

Burian, A. 2005. *Bude se u Sazky Areny stavet?* Available at estav.cz (accessed: January 6, 2010).

Chlebna, H. 2004. *Brownfields—dosavadni vysledky projektu "Strategie regenerace Brownfields".* Prague: University of Economics and IREAS, CR.

Cilek, V. et al. 2004. *Vstoupit do krajiny. O prirode a pameti strednich Cech.* Prague: Dokoran, CR.

Cilek, V. 2005. *Krajiny vnitrni a vnejsi.* Prague: Dokoran, CR.

CSU. 2009. *Treti mesic mezirocniho poklesu prumyslove vyroby.* Available at www.czso.cz (accessed: January 6, 2010).

CzechInvest. 2008a. *Brownfieldy.* Available at czechinvest.org/brownfieldy (accessed: January 6, 2010).

——. 2008b. *Zakladni statisticke vysledky Vyhledavaci studie brownfieldu.* Prague: Ministry of Industry and Trade of the Czech Republic.

Damborsky, M. 2005. "Verejna politika k brownfields." In: Zizka, Miroslav (ed.) Sborník prispevku ze 4. mezinarodniho sympozia Ceske podnikatelstvi v evropskem prostoru 2005. Liberec: Technical University of Liberec, 26–31.

Dyba, K. 1998. "Makroekonomicka politika a tempo hospodarskeho rustu v obdobi transformace ceske ekonomiky." *Politicka ekonomie* 46(2): 279–83.

Eurostat. 2008. *National Accounts Aggregates and Employment by Branch (NACE).* Available at epp.eurostat.ec.europa.eu (accessed: January 6, 2010).

Holub, A. 1998. "Strukturalni odvetvove zmeny v procesu transformace ceske ekonomiky (s ohledem na mezinarodni aspekty)." *Politicka ekonomie* 46(6): 788–804.

Kocourkova, Lucie. 2008. *Na 750 opustenych arealu v Kralovehradeckem kraji nutne potrebuje investora.* Prague: CzechInvest, CR. Available at CzechInvest. org (accessed: January 6, 2010).

———. 2009. *Statistika: novi investori chteji v CR podnikat v najmu.* Prague: CzechInvest, CR. Available at CzechInvest.org (accessed: January 6, 2010).

Kovarik, P. 2005. *Brownfield versus Greenfield.* Prague: DTZ Central and Eastern Europe, CR.

Kraft, J. 2005a. "Brownfields—negativni externality, ztracena prilezitost a pozitivni externality." In: Zizka, M. (ed.) Sborník příspěvků ze 4. mezinarodniho sympozia Ceske podnikatelstvi v evropskem prostoru 2005. Liberec: Technical University of Liberec, 39–48.

———. 2005b. "Uloha trhu pri systematicke revitalizaci brownfields." *E&M Ekonomie a Management* 8(4): 28–33.

Kubistova, M. 2008. *Vyuziti PPP pri regeneraci brownfieldu.* Prague: PPP Centrum Czech Republic. Available at arr.cz (accessed: January 6, 2010).

Lorenzova, K. 2006. *Od poskozeneho uzemi k atraktivni nemovitosti: Projekt "Brownfields 3000".* Prague: CzechInvest, CR. Available on CzechInvest.org.

MD. 2008. *Dokumentace programu "Podpora revitalizace zeleznicnich vlecek".* Prague: Ministry of Transport, CR.

Mikula, S. 2008. *Komparace vyvoje HDP v CR a zemich EU.* Brno: Masarykova univerzita, CR.

MPO. 2007. *Program na podporu podnikatelskych nemovitosti a infrastruktury.* Prague: Ministry of Industry and Trade of the Czech Republic.

———. 2008. *Narodni strategie regenerace brownfieldu.* Prague: Ministry of Industry and Trade of the Czech Republic.

MZE. 2008. *Program rozvoje venkova Ceske republiky na obdobi 2007–2013.* Prague: Ministry of Agriculture of the Czech Republic.

MZP. 2008. *Prirucka pro zadatele o dotace z Operacniho programu Zivotni prostredi.* Prague: Ministry of the Environment of the Czech Republic.

OPPK. 2008. *Implementacni dokument Operacniho programu Praha— Konkurenceschopnost.* Prague: Magistrat hlavniho mesta Prahy, CR.

Rydvalova, P. and Rydvalova, R. 2005. *Lokality typu brownfields.* Liberec: Agentura regionalniho rozvoje and Technicka univerzita v Liberci, CR.

Rydvalova, P. and Zizka, M. 2005. "Analyza neprumyslovych deprimujicich zon ve vazbe na hospodarsky slabe oblasti Libereckeho kraje." In: Marsikova, K. (ed.) Sborník prispevku z VII. mezinárodní konference Liberecké ekonomicke forum 2005. Liberec: Technicka univerzita v Liberci, 467–76.

——. 2006. "Ekonomicke souvislosti revitalizace brownfields." *Politicka ekonomie* 54(5): 632–45.

Tylova, E. 2004. Sazka Arena—vyhra, nebo prohra? *Hospodarske noviny*, 2004-04-02. Available at hn.ihned.cz (accessed: January 6, 2010).

Vostrovska, Z. 2005. *Transformace ceske ekonomiky. Teoreticka vychodiska.* Prague: University of Economics, CR.

Zidek, L. 2006. *Transformace ceske ekonomiky 1989–2004.* Prague: C.H. Beck, CR.

PART II
Implementation and Evaluation

Policy is made as it is being administered and administered as it is being made.
(Anderson 1975, quoted in Hill and Hupe 2002: 1)

There has been ongoing debate in the policy literature on several points related to implementation. At root, much has focused on whether, indeed, implementation can be understood as a distinct part of the policy making process, differentiated from policy creation in particular, but also from evaluation, feeding as it does back into the policy-making cycle (Wildavsky 1979). Over time, scholars of implementation have presented the implementation endeavor as top-down or bottom-up, or some synthesis of these two general approaches (see Hill and Hupe 2002 for an excellent summary of this literature). In short, top-down models present implementation as being one step in a rational decision-making process that begins with goals set by the political process. Then a variety of agencies, influenced by each other and the broader political, economic, and social environment, interact to implement policies to achieve these goals (Van Meter and Van Horn 1975). Significant variation between initial goals and policy and what is implemented, or between what is achieved and what is expected—viewed alternately as the implementation gap (Dunshire 1978), implementation failure (Morgan 1993), or even the implementation fiasco (Bovens and 't Hart 1996)—is undesirable and should be controlled. This world view has resulted in a great deal of additional research supporting recommendations for controlling bureaucrats and thus lessening the implementation gap (Hill and Hupe 2002).

Other scholars, however, see policy implementation as a bottom-up endeavor where bureaucrats (or more specifically street-level bureaucrats) create coping mechanisms or standard operating procedures, and policy emanates from those actions (Lipsky 1971). In short, implementation brings together a number of different actors from which policy is "recombined" (Hjern and Hull 1982). Variations between goals/policies and what is implemented are thus to be expected as a natural part of the policy-making process. Indeed, some have argued that such bottom-up policy processes not only conform to reality but are better in a normative sense, in that street level bureaucrats will be more in tune with environmental realities regarding what is possible and with the desires and needs of their clients (Lipsky 1971; Maynard-Moody and Musheno 2003).

Synthetic approaches combine the elements of top-down and bottom-up models, not only emphasizing the desirability of achieving goals but also acknowledging the realities when discretion and the "real world" intervene leading to an implementation process that is variable, political, and networked, and yet still represents a viable and distinct part of the overall policy process which can be studied, evaluated, and potentially improved (Hill and Hupe 2002). The chapters in this section show many elements of these debates. The cases vary in the unity of brownfield policy and the goals contained therein; they are similar in that they show significant discretion left to bureaucrats and the larger environment, so that initial policies and goals are indeed changed. Implementing brownfield policy clearly involves a network of public and private organizations, citizens, and levels of government, which ensures that implementation is a dynamic and complex act.

Several authors have suggested that implementation "inevitably takes different shapes and forms in different cultures and institutional settings" (Hill and Hupe 2002: 1) and thus, there are different national policy styles. Clearly, the US, UK, Chinese, and German cases presented here vary significantly in all of these aspects. Yet, as should become clear, there are many more similarities among these implementation stories than there are differences. The facilitators and barriers to the successful implementation of brownfield policies are surprisingly uniform even across very different national settings. Why might this be the case? Is has been suggested that cross-national policies tend to be more similar if the policy area in question is more technical, requiring specialized knowledge and training to create and implement (Wolman and Page 2002). Thus, policy transmission between countries is more common (Reese 2006). This would be true of brownfield policy where there are relatively uniform scientific methods for measuring contamination, remediation standards, and clean-up processes. And despite efforts at and the desirability of involving citizens and neighborhoods in the planning process, there are certain aspects of brownfield policy that, almost of necessity, are the purview of scientific specialists. These factors may explain why there appears to be greater similarity in implementation experiences in the case of brownfield policy than there may be in other policy areas.

In short, while the legal, policy and incentive contexts vary by country, and in the cases of federalist systems, by state, there are significant commonalities when it comes to implementing and evaluating brownfield remediation and revitalization efforts. The chapters in this section point to barriers and facilitators of effective policy implementation across national contexts: the US, England, Germany, and China.

Chapter Summaries

The chapter by Knapp and Hollander provides central recommendations for the implementation of brownfield redevelopment policy. Specifically, they recommend the application of "community benefits agreements" typically used in planning and economic development policy for brownfield remediation efforts. A community benefits agreement, or CBA, is defined as a:

private contract between a developer and a community coalition that sets forth the benefits that the community will receive from the development. Common benefits include living wages, local hiring and training programs, affordable housing, environmental remediation and funds for community programs (http://communitybenefits.blogspot.com/).

Such agreements are particularly important given the market-driven nature of most state voluntary cleanup and redevelopment processes. As the inherently punitive federal policies have given way to incentive-based or voluntary state policies, market forces tend to determine development sites to a much greater extent than health or social equity needs. As shown in the proceeding policy section, this may well have increased the number and extent of brownfield remediations, but the market approach has tended to leave urban sites in poor areas off the table. Public sector incentives along with CBAs can potentially serve to mediate this balance.

The application of CBAs to brownfield redevelopment projects has a number of things to recommend it. First, because many brownfields are located in the most financially stressed areas of central cities, CBAs engage residents most in need of environmental justice, political representation, and economic development. Second, they not only ensure community participation in the planning of redevelopment but also foster community participation in and acceptance of redevelopment outcomes. Third, by focusing on social equity, community organizing, economic development, land use planning, and environmental health, CBAs link often disparate goals and processes. The community benefits delineated in the development agreements can result in much-needed investment in poor inner-city neighborhoods including: neighborhood development funds, employment, affordable housing, amenities such as parks, infrastructure investment, and so on. More specifically, the benefits of CBAs for brownfield redevelopment include:

- Physical community building and the elimination of development gaps created by sitting brownfields
- Social capital building through the planning and development process
- Improved planning through the inclusion of community stakeholders.
- Increased community revitalization due to benefits delivered from developers.
- Better balance to the gentrification effects of many urban redevelopment projects.
- Encouragement of public/private partnerships as a means of implementing policy.

While the benefits of CBAs are clear, there are a number of implementation challenges that must be addressed for successful use, and the jury is still out on 1) whether they are actually a new implementation methodology (development agreements and linkage programs have been around for a long time) and more importantly, 2) whether they are effective. Implementation barriers discussed in the chapter can be summarized as follows:

- There are often multiple and conflicting goals for brownfield remediation and redevelopment, particularly in the diverse neighborhoods where urban sites are often located. While health protection and economic development probably represent the primary potential goal conflict, residents and developers may have very different goals (community benefit versus profit), and residents themselves are unlikely to weigh all potential benefits—jobs, infrastructure, amenities, services—equally.
- CBAs require that historic distrust between developers, city officials, and citizens be overcome, since effective planning and community participation require a measure of mutual trust. Development and funding processes require that these parties be able to work with city officials.
- CBAs require a set of skills within the community or neighborhood that are often in short supply: planning, coalition building, and negotiating, for example. Significant resource constraints within most city governments limit their ability to supplement or develop neighborhood skills.
- Successful brownfield redevelopment and CBAs in particular require sustained, long term commitment from all parties to the agreement.
- Because urban brownfields are often located in areas where the market will not clear and redevelop land, development will typically require local, state, and/or federal subsidies. Identifying and leveraging public resources for inner-city brownfield redevelopment are, again, often beyond the skills sets of many community groups.
- Cities are often fearful of CBAs, thinking that they will deter developers from investing in brownfield sites, and thus they do not push for such agreements. This barrier is more perceptual than real, however, based on assessments of linkage programs in economic development that indicate that they are not anathema to developers who are more interested in certainty and predictability in the redevelopment process than in getting the cheapest site (see Molina 1998; and Reese, 1998 for a discussion of this literature).

These barriers are clearly challenging, and evaluations of brownfields developed with CBAs need to be expanded to provide public officials with a more complete sense of the costs and benefits of this approach.

The chapter by Williams focusing on brownfield remediation in England provides an excellent quantitative assessment of the effectiveness of programs in that country as well as a set of barriers that still need to be addressed. The goal of British brownfield policy (which encompasses contaminated and previously development sites) is both to reduce the number of brownfields and also to promote sustainable development through reduced use of greenfields, densification of housing, and green/sustainable development principles. Implementation of policy related to previously developed land (PDL) has been facilitated in the British context by several factors noted below. The common thread in these facilitators is strong national commitment to the redevelopment of PDL, while clearly absent are the unavoidable challenges of fragmentation and multiple authorities present in federalist systems:

- Lack of land for housing requiring increased density and re-use of inner-city sites and more positive market forces for inner-city housing sites;
- Specific national targets for brownfield remediation;
- National push to preserve green space;
- Comprehensive national land use database with records on all PDL sites.

As a result of these facilitators, the evaluation of PDL in Britain indicates that brownfield land has been reduced significantly, exceeding national targets; derelict land and buildings have been reduced; and the targets for new housing on PDL have been met. Although data are not as clear, it appears that the amount of greenspace in the country overall has increased. Housing has become denser, greenfield land preserved, and urban areas regenerated.

On the other hand, there are still some challenges to policy implementation. There is an abundant (and apparently growing) amount of land that is designated for redevelopment through the PDL program still being used for a former purpose. This is typically the case because the land is making a profit under its original, and generally less than optimal, purpose. Further, there is some unevenness in the amount of redeveloped land, with some areas of the country—including London—experiencing increases. The qualitative sustainability of new developments is more difficult to measure, and national databases have not addressed these measures. While overall green land cover has increased as noted, housing has not always included all new urbanism or green features, and transportation connectivity and energy efficiency remain problematic in some cases. As housing has become denser (and smaller), traffic congestion and gentrification have occurred in desirable areas, while land in other, less desirable locations remains undeveloped. Even with optimal national goals and a real need for housing, markets do not operate to clear all brownfield land available, particularly in poor areas, leading to the sorts of conditions that make CBAs desirable in the US.

The stimulus for brownfield redevelopment policy in Germany, as explained by Grimski, Dosch, and Klapperich, is much the same as in Great Britain: the presence of derelict industrial facilities, a need to stop development on greenfields, and the desire for housing densification, all driven by specific national targets. The primary challenges that make the German case different are the lack of a centralized brownfield database and the federalist system: two issues of similarity between Germany and the US. "Hardcore" brownfield sites are particularly challenging as they are located in areas with declining populations, making them undesirable from a market perspective, and there is insufficient public money to provide incentives and support redevelopment in the case of such market failures.

For such sites, interim uses are more feasible than full-scale redevelopment, and Germany provides a useful case for US urban policy-makers in this regard. Potential interim uses include: green cover, farms or urban agriculture, temporary event spaces, container nurseries, drive-in movies, sports facilities, and art spaces (Urban Collaborators, http://spdc.msu.edu/LinkClick.aspx?fileticket=eql7%2bGs %2fU2I%3d&tabid=365).

The challenges posed by federal systems are probably the most evident in the German case. While federal brownfield policy in the US was initially too directive and punitive, with too little discretion left to the states, the opposite tends to be true in Germany. Federal policy is in separate pieces, across several federal acts and agencies, and implementation is left to regions or local governments. Too little federal coordination limits the effectiveness of cleanup and redevelopment processes. This fragmentation and lack of communication and coordination among stakeholders are probably the most signficiant barriers to successful policy implementation in Germany. In addition, other common challenges include:

- The general reliance on the market to stimulate interest and funding for brownfield remediation and reinvestment. C sites or reserve lands that are not profitable for the market to redevelop and which thus require governmental action are particularly challenging.
- A general lack of cleanup financing either from the public sector or from banks.
- Lack of coordination between the local planning process and land development and the need for more flexible local regulations for remediation and site use.
- Insufficient technology and information about brownfields generally, as well as implementation processes.

After being somewhat slow to take up the issue of brownfield remediation and redevelopment, according to Hongyun and Liange, China has many environmental policies and regulations on the books. Implementation of these rules and regulations, however, is extremely problematic. Using a game-theoretic framework, these authors clearly show how information asymmetry and the lack of a general brownfield reporting system are the primary barriers to successful brownfield policy implementation. In a fragmented and decentralized system, the main challenge is to devise a system where regulators effectively regulate polluters within a state system not designed for monitoring. Although often viewed from the outside as a highly centralized system, at least in the area of environmental policy, China exhibits many of the barriers found in federal systems such as the German case just described. The federal Ministry of Environmental Protection (MEP) must rely on supervisors at local or regional monitoring stations for information about levels of pollution as well as actions to remediate and control that pollution. Firms have an incentive to pay off these supervisors, so that reporting to higher levels of government is positive. Due to lack of information and other governance challenges, it becomes very easy for local-level regulators to be bribed. The challenge then becomes how to reduce the proclivity for bribes while working through the existing governance system. Vertical fragmentation and high levels of discretion for monitoring supervisors lead to less than optimal implementation of environmental regulations.

The authors show the decision calculus for both abuse of authority and collusion, with the latter being more easily deterred. Abuse of authority occurs when monitoring supervisors tell firms that they will not report positive information about remediation efforts unless a tribute is received, while collusion is where the polluting firm offers a bribe to the supervisor to avoid reporting negative information. Abuse of authority is more costly to deter, in part because the polluting firms reap the greatest benefits, yet it must be controlled for environmental legislation to be implemented. In short, both collusion and abuse of authority are less likely to occur if there is more effective and independent environmental monitoring at higher levels of government. A corollary option would be to increase the pay of monitoring supervisors so as to reduce the incentives for taking bribes and tributes.

Other barriers to the implementation of environment regulation in China include:

- The involvement of multiple levels of government;
- Insufficient monitoring causing information uncertainty at higher levels;
- Discretion left to supervisors allowing for abuse of authority;
- Conflicts of interest inherent in relying on local government to monitor the environment because it often sponsors or has other interests in the polluting firms.

The final chapter in the implementation section, by Gillem and Schreifer, highlights generally successful cases of housing and mixed-use development on former brownfields. The cases show how environmental, economic development, and social capital benefits can accrue from focusing on housing on former brownfield sites. Again, because brownfields are prevalent in inner cities, their conversion to housing can improve health, increase walking, foster development in inner cities, and use existing urban infrastructure more effectively. As with the British case, this assessment emphasizes the quality of development in terms of smart growth or sustainability, and not merely that the land is developed at all. The four cases discussed present generally positive examples of housing redevelopment, although two more effectively achieve sustainable principles.

While the other implementation chapters emphasize barriers, this one has much to say about factors that lead to success. In looking across the four cases, what distinguishes the two most successful projects, Glenwood Park and Hercules, includes: housing and mixed-use development, strong public-private partnerships, serious attention to new urbanist principals, a focus on live-work spaces, access to public transportation, and strong involvement of citizens in planning at all stages of the development. The access to transportation and the extent of integration between the larger city and the new development are the critical components that differentiate between a sustainable project and a more problematic one. In the cases of Atlantic Station and Bay Street, while many of the positive features are present, the presence of big box or generic retailers, poor transportation connectivity, a continuing emphasis on car-related transit, and lack of integration with surrounding communities (leading to a sense of "fakeness") limit the long-term sustainability of the projects. In all cases, however, there appear to have been positive outcomes in terms of increased use of urban infrastructure already in place, reduction in external car trips, improved health, and a potential reduction in the use of alternative greenfield sites.

Again, some common barriers were present, including:

- Gentrification and increased housing costs, particularly in the less successful cases that included lifestyle malls;
- Difficulty in connecting the new development to the rest of the urban area;
- Challenges in obtaining necessary zoning changes and variances for new urbanist development within central cities (typically related to lot size, setbacks, and mixed-use zoning);
- The difficulty in creating wholesale a sense of "community."

Common Challenges

The cases in this section point to several common brownfield policy implementation challenges across countries. And, while the cases vary in the level of success in both implementation and outcomes, even the British assessment provided by Williams and the profile of successful housing and mixed redevelopment projects described by Gillem and Schreifer, show where pitfalls lie in the effort to achieve maximal success of outcomes. These common barriers are as follows:

Lack of comprehensive datasets: Clearly the lack of comprehensive data on the existence and nature of brownfields and efforts at remediation are a central problem in implementing brownfield policy in China and Germany. Even in the British case, however, the collection of data on greenfields and on sustainable development criteria would make implementing policy and assessing its results more effective.

Federalism: The unitary system in Great Britain shows the advantage of a centrally created and administered brownfield policy. Yet, the mere existence of policy developed at the national level does not ensure successful implementation. Indeed, many aspects of brownfield policy in the US were problematic until significant discretion and delegation occurred for the states. This is a delicate balance, however. In Germany and China, national policies implemented at the local level lead to lack of coordination and unhealthy discretion left to local bureaucrats and governments with little interest in enforcing remediation regulations. This latter issue raises another common barrier: lack of official commitment to achieving brownfield remediation and redevelopment goals. A large part of the success rate in the British case results from strong national commitment to the goals of brownfield redevelopment and housing densification. Without such commitment among officials at all levels, implementing brownfield or any policy becomes much more difficult. While the China case may be extreme in this regard, implementation success varies across state and localities in the US based on differing local commitment to policy goals.

Market failure: In all the cases brownfield policy implementation is inhibited or sub-optimal when the market is left to identify projects and provide funding. Many of the most serious brownfields are located in areas most in need of redevelopment. These often inner-city locations tend to be the least desirable for market interests, however. Thus environmental and economic justice goals are unlikely to be achieved when the market is left to drive the redevelopment process. While community benefits agreements are one way to address this, it is clear that some brownfields will require public sector stimulus and funding before redevelopment occurs. In cases where the public and private sectors have come together, along with citizen groups, projects have achieved greater levels of success in both their economic and distributive aspects, as clearly shown in the US housing cases.

Lack of clean-up financing or incentives: A corollary of the market failure barrier noted above is the general paucity of public funding of brownfield remediation and redevelopment. Where the market will not support private projects, public funding is required. Many US states have come up with funding or inventive mechanisms that address this barrier. In China and Germany, however, lack of public funding creates almost insurmountable barriers to brownfield redevelopment in cases deemed undesirable or unprofitable by the market.

References

Bovens, M.A.P. and P. t'Hart. 1996. *Understanding Policy Fiascoes*. New Brunswick, NJ: Transaction Publishers.

Dunshire, A. 1978. *The Execution Process, Volume 2: Control in a Bureaucracy*. Oxford: Martin Robertson.

Hill, M. and P. Hupe. 2002. *Implementing Public Policy*. Thousand Oaks, CA: Sage Publications.

Hjern, B. and C. Hull. 1982. "Implementation Research as Empirical Constitutionalism," in B. Hjern and C. Hull (eds), *Implementation beyond Hierarchy*. Amsterdam: Elsevier.

Lipsky, M. 1971. "Street-level Bureaucracy and the Analysis of Urban Reform." *Urban Affairs Quarterly* 6: 391–409.

Maynard-Moody, S. and M. Musheno. 2003. *Cops, Teachers, Counselors: Stories from the Front Lines of Public Service*. Ann Arbor: University of Michigan Press.

Molina, F. 1998. *Making Connections: A Study of Employment Linkage Programs*. Washington, DC: HUD.

Morgan, G. 1993. *Imaginization: The Art of Creative Management*. Newbury Park, CA: Sage.

Reese, L.A. 2006. "The Planning-Policy Connection in US and Canadian Economic Development." *Government and Policy* 24: 553–73.

——. 1998. "Sharing the Benefits of Economic Development: What Cities Utilize Type II Policies?" *Urban Affairs Review* 33: 686–711.

Van Meter, D. and C.E. Van Horn. 1975. "The Policy Implementation Process: A Conceptual Framework." *Administration and Society* 6(4): 445–88.

Urban Collaborators. 15 July 2010. Available at http://spdc.msu.edu/LinkClick.as px?fileticket=eql7%2bGs%2fU2I%3d&tabid=365 (accessed: May 10, 2010).

Wildavsky, A.B. 1979. *Speaking Truth to Power: The Art and Craft of Policy Analysis*. Boston: Little, Brown, and Company.

Wolman, H. and E. Page. 2002. "Policy Transfer among Local Governments: An Information Theory Approach." *Governance: An International Journal of Policy and Administration* 15: 477–501.

Chapter 5

Exploring the Potential for Integrating Community Benefits Agreements into Brownfield Redevelopment Projects

Courtney E. Knapp and Justin B. Hollander

Brownfield redevelopment is increasingly becoming a strategy of urban development, as planners and policymakers who are concerned with the revitalization of older, deindustrialized communities turn to their local vacant and underutilized land as an opportunity to reverse blight, attract investment capital and boost their local economy. For more than a decade, federal and state programs have responded to the growing concerns about brownfields by providing incentives—including liability protection, grant and loan opportunities, tax credits, and other technical resources—to developers who are willing to invest their dollars in the revitalization of these potentially contaminated, underutilized, and risky sites (USEPA 2007b). This investment has produced significant concrete benefits, but not without criticism. Environmental justice critics argue that brownfield redevelopers should be held more accountable to the residents living in the communities where they operate, both in terms of the participatory process and in deciding on the end-uses of sites (Greenberg and Lewis 2000; Brachman 2004; DePass 2006; Gute and Taylor 2006; Cummings 2007; Eisen 2007). Despite public participation being a central goal of brownfields programs, scholars, practitioners, and government officials repeatedly articulate the need for developing new strategies to promote a more meaningful participation process, one wherein "community groups can be involved in real decision making, not just feedback" (EPA 1999: 18).

The US Conference of Mayors annual report *Recycling America's Land,* a survey focused on brownfield redevelopment in cities around the country, has consistently revealed that a lack of cleanup funds is the number one factor inhibiting site revitalization (US Conference of Mayors 1998–2007). Thus, in addition to the accountability issue, there is a political-economic critique against brownfield programs in the way they tend to prioritize sites for redevelopment funds according to their anticipated profitability. Programs tend to privilege private developers and target scarce funds primarily into those communities and projects thought to have the most potential from a private market perspective. This reinforces a system whereby communities that need financial support the most are overlooked and de-prioritized by both private and public actors (Brachman 2004; Evans 2004; Howland 2004; Eisen 2007).

Given the scarcity of public funds and the primarily market-driven criteria for awarding public resources, it becomes clear how those communities suffering most from deindustrialization, both economically and environmentally, might fall through the cracks of federal and state programs. It is for this reason some critics argue for a new, equity-driven approach to programming and funding brownfield redevelopment projects (Gute 2006; Gute and Taylor 2006; DePass 2007; Eisen 2007). In this chapter, we explore one such avenue for an equity-driven approach to brownfields: Community Benefits Agreements. We begin describing Community Benefit Agreements, and present an analysis of brownfield redevelopment policy in the United States over the past 30 years and the ways in which policies have privileged traditional economic benefits over the more social local benefits that CBAs seek to secure. From there, a discussion about the limitations of integrating the two as well as the political, social, and economic conditions needed for successful integration is offered. Ultimately, this discussion is theoretical in nature; it is the beginning of a conversation that will ideally continue among community development stakeholders on the local level.

Toward Accountability and Equity: Community Benefits Agreements

Community Benefit Agreements (CBA) are one value-recapture tool that might hold the potential for transforming the brownfield redevelopment movement. CBAs are a community development and organizing mechanism used to promote equitable and participatory community development. A CBA is a legally-binding contract, negotiated between developers and community coalitions, wherein developers commit to providing a discrete set of public goods—determined by the community—in exchange for the public support of the development project (Gross, Leroy and Janis-Aparicio 2005; DePass 2007; Lavine 2008). "Support" in this sense refers not only to residents' approval of development plans, but in many cases also takes the form of regulatory approval and public subsidies, as well.

Given that the brownfield redevelopment arena relies heavily on public subsidies, another reason for integrating CBAs into projects becomes clear. Cities often compete with one another over development projects because of the increased tax bases that result from a vibrant local economy; large-scale projects in particular hold the potential to provide many long term economic benefits. Municipalities try to attract these developments into their communities by offering competitive development subsidies, of which brownfield funds are one type.

The subsidization of development projects may bring jobs and revenue to revitalizing communities. However, there is no guarantee that the end products of development will benefit existing residents over time. According to the authors of *Community Benefits Agreements: Making Development Projects Accountable*, "While economic development projects are often heavily subsidized by taxpayer dollars, they produce decidedly mixed results for city dwellers" (Gross, Leroy and Janis-Aparicio 2005: 4).

Displacement caused by gentrification, for example, is a very real consequence of some urban revitalization efforts; others create large numbers of "low wage retail and service sector jobs, leaving low income families ... mired in an endless cycle of poverty" (4). In this sense, without taking certain precautions to ensure that existing residents benefit from the development projects that enter their communities in meaningful and sustainable ways, there is a serious risk that taxpayers will inadvertently fund projects that will either displace them from the community or reinforce their unstable socioeconomic positions. Community benefits agreements, insofar as they produce procedural and concrete advantages for existing residents, seem as though they would offer protection from some of the negative effects that revitalization efforts can have on poor communities.

Benefits achieved through CBAs take a variety of forms, and the specific amount and types of benefits negotiated depend largely on the scale of the development project in question as well as the local economic climate of the community where the project is being proposed. Popular benefits include neighborhood funds, first source hiring programs, provision of living wage jobs, affordable housing development, public facilities and amenities, transportation infrastructure improvements, and environmental abatement measures.

In addition to measurable benefits, CBAs promote equitable community development vis-à-vis the negotiation/participation process. No longer are residents forced to sit back and merely react to contracts articulated between cities and developers. Instead, they play a proactive role in setting the parameters of the development agreement and shaping the futures of their neighborhoods (Gross et al. 2005; DePass 2007; Lavine 2008).

Thus the justification for integrating CBAs into brownfields projects becomes even clearer. As previously mentioned, public participation has increasingly become a goal of federal and state brownfield redevelopment programs, as more and more public agencies espouse the rhetoric of participatory planning and Smart Growth. Developers are also realizing the importance of bringing the community into planning discussions; one survey of 158 brownfield developers, for example, discovered that "community support" was the most frequently cited indicator of a project's success (Lange and McNeil 2004: 103). And members of the environmental justice and equity fields have long argued that a more inclusive, participatory process is critical to achieving sustainability in brownfield redevelopment (U.S. EPA 1999; Greenberg, Lowrie, Solitare and Duncan 2000; Solitare 2001; Faber, Jennings and Loh 2004; Brachman 2004; Gute 2006). Yet, as many critics have pointed out, achieving meaningful citizen participation is easier said than done. The need for developing special tools to ensure that public participation is meaningful leads back to community benefits agreements as a strategy for promoting equitable brownfield redevelopment.

Based on this discussion, community benefits agreements and brownfield redevelopment programs appear to have similar goals: civic engagement, political empowerment, local economic development, and smart growth/environmental sustainability. Given the compatibility of interests, it seems logical that CBAs could be employed as tools to help brownfield programs achieve social and political empowerment goals. To that end, research assessing the potential to integrate the two models is guided by three primary questions.

- Are Community Benefits Agreements a useful tool for opening the lines of communication between residents and developers and for promoting meaningful community participation with brownfield projects?
- Are Community Benefits Agreements a viable strategy for ensuring that brownfield redevelopment programs and projects advance equitable community development?
- What are the political, economic, and social conditions needed for community benefit-driven brownfield redevelopment to succeed?

These research questions are important for a number of reasons. First, as previously mentioned, brownfield redevelopment initiatives are increasingly being integrated into local economic and community development plans. Simultaneously, state and federal programs continue to provide incentives to developers willing to take on risky sites. Given the quantity of brownfields in the United States and the push by both policy makers and developers to treat these sites as unique real estate opportunities, communities increasingly need tools to ensure their voices are heard in the re-envisioning and redevelopment of their neighborhoods. Because they depend on communication and collaboration between residents and developers to produce concrete public benefits, CBAs seem to be an especially promising tool for advancing social equity.

The efficacy of CBAs, however, is highly speculative—primarily because of the lack of evaluative literature concerning community benefits agreements. Although many progressive development advocates speak about the promises of CBAs, evaluations of the model are only now beginning to emerge. The little evaluation that does exist confirms mixed results (Gross et al. 2005; Pratt Center 2005; DePass 2007; Lavine 2008). This has led to speculation about the effectiveness of CBAs as tools from prominent voices within the sustainability and community development fields.[1]

According to critiques, there is insufficient evidence to claim that development projects possessing a CBA advance equity or benefit communities significantly more than projects without agreements. Furthermore, cooperation (or development) agreements have long been drafted between cities and developers, wherein promises are made to provide certain public goods in exchange for development rights on a site (Levine 1989; Gross et al. 2005; Shragge 2005). Thus, community

---

1   This statement is based on informal discussions held with faculty members from Tufts University's Department of Urban and Environmental Policy and Planning and Harvard University's Graduate School of Design.

benefits are merely a new name for an old concept—a concept that, historically, has been criticized because municipal officials often do not (or cannot)—hold developers accountable for their promises over time.

To this end, the second reason research focused on these topics is valuable is to address a significant hole in the equitable community development literature: namely, the effectiveness of the Community Benefits Agreement model. In addition to thinking about CBAs directly in relation to brownfields, it is important to understand the promises and challenges of employing CBAs as a development and empowerment tool in general.

The (D)Evolution of US Brownfield Redevelopment Policy and its Effect on the Equitable Revitalization of Communities

The brownfield redevelopment arena has evolved around three prominent discourses: administrative rationalism, economic rationalism, and environmental justice (Kirkwood 2007). While the administrative rationalist discourse is primarily concerned with policy and regulatory aspects of redevelopment, economic rationalism frames the issue in terms of economic revitalization and real estate development. The third discourse—environmental justice—is less concerned with economic growth and more interested in the legacy of land contamination and deindustrialization and its disproportionate environmental, social, and economic impacts upon low income, politically disempowered communities.

Over the past decade, brownfields literature has increasingly pointed to a growing interconnection between administrative and economic discourses, as policies are implemented to facilitate the profitable redevelopment of contaminated and underutilized land (US Department of Housing and Urban Development 1999; American Institute of Architects 2000; Bowman and Pagano 2000; Gattuso 2000; Meyer and Lyons 2000; Brachman 2004; Evans 2004; Gardner 2004; Howland 2004; Eisen 2007; Hazardous Waste Consultant 2007). Reactions to this shift vary according to an authors' position along the political-economic spectrum. Although virtually everyone agrees that current policy conditions are having a positive impact on urban redevelopment in terms of the numbers of sites being bought and sold, there remain a number of critiques against the ways these policies privilege the private sector and do not necessarily benefit existing communities.

Not surprisingly, the business and economic development-oriented literature tends to commend the surge in public-private partnerships, much of it focusing on ways that federal and state programs mitigate some of the legal and financial risk associated with brownfield investment. Page and Rabinowitz (1996) argue that assessment and remediation costs are a great disincentive for private developers, who expect promised returns on their investments. Using four development projects as case studies, the authors develop a risk threshold model and assert that the potential for a site's redevelopment corresponds inversely to expected return on investment. By subsidizing cleanup and assessment costs, public subsidies can level the playing field between the cost of developing on brownfields versus greenfields, effectively returning the level of risk to zero in the eyes of private investors.

The authors note that brownfields as a category is not monolithic: some sites will be marketable enough on their own not to need public funds to encourage redevelopment, while others will likely remain underutilized and vacant irrespective of their contamination status, because they are located in declining or stagnant local markets. They argue that the most efficient use of public subsidies would be to target sites "within or just below the redevelopment threshold," while continuing to develop public policies to ensure returns on private investment.

In 2000, the Competitive Enterprise Institute, a conservative think tank dedicated to "the principles of free enterprise and limited government" (Competitive Enterprise Institute 2007), issued a report titled *Revitalizing Urban America: Cleaning Up the Brownfields*. The report's author, Dana Joel Gattuso (2000) criticized the Environmental Protection Agency for trying to exercise oversight in matters that are essentially local and private in nature. Instead, she argues that the federal government should "get out of the business of brownfields altogether" (15), leaving redevelopment programming to the discretion of states, municipalities, and private developers. In her concluding remarks, Gattuso writes,

> In many ways, the states have succeeded where the federal government has failed ... Despite the failings of CERCLA, its regulations still govern state and local brownfield programs, enforcing rigid cleanup standards and threatening lengthy, costly legal challenges (15–16).

Many other studies confirm that brownfield redevelopment is best dealt with as a real estate issue, and approached through the development of public-private partnerships. The American Institute of Architects (AIA), for example, released a report in 2000 discussing the potential that brownfield redevelopment has to boost local economies while promoting livable communities and Smart Growth: "Brownfields are the new market frontier, bursting with community capitalism" (AIA 2000: 3). Instead of articulating the issue of vacant, (potentially) contaminated land as a great risk, AIA encourages policy-makers to think about redevelopment as a "value statement" (1) and an opportunity to make sustainable economic, environmental, and social change.

In more recent years, the issue of risk has persisted in the discourse, although there is increasingly a debate about the extent to which costs and liability issues are the primary variable inhibiting redevelopment. Wernstedt, Meyer and Alberini (2006) surveyed over 300 developers about the factors they consider to be the greatest incentives to redeveloping brownfields. Liability relief was a top answer; however, the researchers discovered significant heterogeneity among respondents in terms of the values they placed on public monetary incentives. A study of an industrial area in Baltimore (Howland 2004) confirmed that many other factors— including outdated infrastructure and zoning restrictions—have an impact on the redevelopment of contaminated sites. Howland maintains that by focusing almost exclusively on liability and cost concerns, state and local officials miss opportunities to tackle other, more manageable redevelopment hurdles.

But while most agree that the public-private approach is the most promising model, some remain critical of the way these partnerships privilege certain types of projects. Green and Coffin (2005) examine how real and perceived contamination impacted property values in Cleveland and Atlanta, concluding that brownfields do have a negative impact on neighborhood property values. Moreover, depressed property values due to land contamination is usually only one issue among many in "persistently poor neighborhoods" (Green and Coffin 2005: 277). Others have criticized a focus on high-market value properties by bringing attention to the worst-of-the-worst brownfields and the unique challenges they bring to their communities (Greenberg et al. 2000; Hollander 2009).

Thus from an environmental justice perspective, communities suffering the hardest should be prioritized for the scarce public funding that exists, and public policy should be reoriented to encourage private investment not only in marketable communities, but in and for the benefit of poor neighborhoods as well.

Other scholars have made similar arguments about the need to refocus policy on overall community development instead of individual developers. Joel Eisen (2007) deconstructs what he terms the "successful brownfields story" by examining the development and implementation of state programs in New Jersey. He argues that a "second generation" of state policies are needed, ones that develop new measures to promote citizen involvement, equity and public health, not merely a profitable return on investment. Such a policy transformation would move beyond the "parcel by parcel [approach to] development, and bring brownfields revitalization closer to the idea of "smart growth"" (756).

The Origins of Brownfield Policy in the United States

Throughout the 1970's and 1980's, Congress passed a number of laws intended to hold polluting industries accountable for the waste they produced and contamination they caused to the air, water, and land (U.S. EPA 2007a). The most significant law to effect the redevelopment of brownfields was the Comprehensive Environmental Response, Compensation, and Liability Act (CERCLA), or Superfund, enacted by the EPA in 1980.

Superfund stipulations affected brownfield redevelopment in three important ways. First, it established a pollution tax against dirty industries, whereby revenue was generated to be used to fund cleanup of the most heavily contaminated and dangerous sites. The EPA reports that in the first five years following the implementation of Superfund, $1.6 billion dollars were collected from the industrial sector to fund assessment and cleanup. Related to this point, Superfund established the National Contingency Plan—a set of federal guidelines for dealing with redevelopment of hazardous sites—and created the National Priorities List (NPL), a database of the "national priorities among the known releases or threatened releases of hazardous substances, pollutants, or contaminants throughout the United States and its territories" (EPA 2007a). The NPL was significant not only because it allowed policymakers to prioritize redevelopment funds, but also because it created a level of public transparency around environmental hazards in communities.

The third and most significant element of CERCLA was the liability statute it established. Section 107A of the law asserts that under CERCLA, any "potentially responsible party"—including owners, operators, generators, and transporters—can be held liable for the costs of cleaning up and restoring a polluted site in an action brought by EPA, the state, or a private party (U.S. EPA 1980). In other words, an owner of contaminated land can be held legally and financially responsible for any and all cleanup costs, regardless of whether and the extent to which the contamination was of his or her own doing. For sites with a legacy of industrial uses, the specter of unforeseen and astronomically high environmental costs created a culture of deterrence within the real estate development field. Rather than risk discovering contamination, developers opted to abandon the urban core and build on greenfields, a practice that requires less up front capital and is much less risky from a liability perspective.

The result of this exodus was an epidemic disinvestment in urban industrial land. Banks and other financial lending institutions compounded the problem by practicing "brownlining" in communities and neighborhoods thought to be too environmentally risky (Leigh 2004: 123). In the absence of private capital, even those developers who *wanted* to help often could not. Ironically, Superfund began as an effort to direct resources into the revitalization of environmentally burdened sites, yet created such a legal disincentive that developers actually avoided redeveloping them more than they had prior to the implementation of CERCLA (Howland 2004; Li and Sattler 2007).

Federal policymakers soon realized that Superfund was not achieving its goals, and in fact, was having a negative impact on urban revitalization as a whole. Six years after the implementation of Superfund, the EPA amended the law with the Superfund Amendments and Reauthorization Act (SARA) (U.S. EPA 1986). Essentially, SARA upheld the liability clause articulated in the 1980 version but attempted to reverse some of the disincentives felt by the private sector by transferring power into the hands of individual states and increasing the amount in the federal cleanup/redevelopment trust to $8.5 billion. Additionally, on some level SARA was an attempt to reorient planners and policymakers toward the original environmental justice brownfield discourse vis-à-vis its emphasis on public health and encouraging greater public participation in the redevelopment process (U.S. EPA 2007). SARA is important in so far that it explicitly reemphasized the role that individual states should play in the redevelopment process, effectively marking the beginning of a federal devolution process that now characterizes US brownfield redevelopment policy as a whole.

Today, although the EPA continues to play a critical role in the process, its responsibility has shifted mainly to that of funds distributor. The truly difficult tasks—planning, program development, implementation, and enforcement—have been largely passed on to states, and subsequently to municipalities. Like many other areas of community development policy marked by devolution in the late 1970s and 1980s, a major challenge to brownfield redevelopment has been local (in)capacity for handling complex redevelopment initiatives, as well as inter-municipal competition for scarce public funds (Gardner 2004; Eisen 2007).

From National to State Contingency Plans

In 1996, the EPA formally recognized the need for a program to target those brownfield sites not hazardous enough to be included on the National Priorities List but still perceived as risky and therefore in need of additional financial and environmental support. They launched the Federal Brownfields Program, an initiative with a mission "to empower states, communities, and other stakeholders in economic redevelopment to work together in a timely manner to prevent, assess, safely clean up, and sustainably reuse brownfields" (EPA Brownfields Program 2007). Primarily, this mission is achieved through the allocation of funding: the program offers assessment, cleanup, job training, and revolving loan fund grants on an annual basis through a competitive application process (EPA 2007).

The first few years of the initiative were marked by successful pilot programs. Simultaneously, this era witnessed an emergence of state brownfield programs (US Conference of Mayors 1998–2002; Department of Housing and Urban Development 1999; Leigh 2004). In 1999, the Department of Housing and Urban Development (HUD) conducted an analysis of state-initiated programs—a movement it characterized as a "widespread effort on the part of state legislatures to respond to local redevelopment barriers posed by past pollution in relation to CERCLA requirements" (HUD 1999: 1)—and discovered that 90% of states had developed some form of Voluntary Cleanup Program (VCP) as a way to encourage private investment in the absence of public subsidy and faced with Superfund liability.

By 2002, every state but North and South Dakota had a program in place (Leigh 2004). Although individual state programs vary in terms of *where* they allocate funds, most are similar in *how* they design their plans. Voluntary Cleanup Programs are one of the most popular strategies; others include property tools such as Tax Lien Foreclosures and Eminent Domain; redevelopment programs, including Enterprise, Empowerment, and Brownfield Redevelopment Zones; and finance tools such as Development Authority, Tax Increment Financing, and Business Improvement Districts (Leigh 2004: 114–15).

The impact that these state programs have had on brownfield redevelopment is significant. With the introduction of federal grants and state financing incentives, the environmental risks and economic costs associated with brownfield redevelopment suddenly become less daunting. Although the literature previously discussed reveals that liability issues persist as one of the main deterrents to redevelopment (at least among private developers), brownfield redevelopment has become an industry—transformed into a game of creative financing (Brachman 2004; Penn 2007).

Combined with the passage of the Small Business Liability Relief and Brownfields Revitalization Act of 2002—a federal law intended to further reduce risk for individual investors by loosening some of the liability restrictions articulated under CERCLA—the brownfield redevelopment "industry" has exploded. In 2006, for example, the EPA's annual Brownfields Conference drew more than two thousand participants from throughout the real estate, public policy, community development, engineering, and waste management fields (Brownfields National Conference 2006).

Although it is clear that many people are benefiting in the brownfield redevelopment arena, the issues previously discussed raise serious questions about equity, most significantly as it relates to local capacity and municipalities' ability to leverage state and federal funding opportunities. In her article "Environmental Devolution and Local Capacity" Sarah Gardner (2004) illustrates how local issues such as social and political capital, local real estate markets, and civic capacity greatly determine the success of a city's brownfield redevelopment program.

Gardner compares four cities in New Jersey—Newark, Camden, Patterson and Trenton—to make the point that a municipality need not be strong in all three of these areas to be capable of implementing an effective local redevelopment program. At least some form of local capacity is necessary to be able to get off the ground and access state and federal resources, however. Gardner's analysis of Paterson reveals that in the absence of any local driving force, little to nothing gets done. Of the 167 confirmed contaminated sites within the city, only four have been fully redeveloped (152).

From an equity perspective, the case of Paterson reveals the central problem that arises when public policy is devolved to the local level while scarce public funding is allocated according to a competitive process. Cities that possess the resources required to launch an effective grant writing and financing campaign are awarded money, while those already stretched too thin are unable (or unwilling) to compete in the process. Moreover, because multiple rounds of funding are often awarded to communities with a proven track record, the inequitable distribution of funds is exacerbated over time.

Considering the Future of Brownfield Redevelopment in the US

Essentially, the theoretical and policy framework proposed here is concerned with integrating social equity and community organizing agendas into community economic development and land use planning, to ensure that revitalization efforts respect and benefit existing residents. The underlying debate—both in relation to brownfield redevelopment specifically and community revitalization efforts in general—is whether social equity can be achieved when programs and projects are almost exclusively oriented toward economic growth and private development.

Brownfield policy and discourse has shifted away from thinking about idle, contaminated land as an environmental and social justice problem and toward thinking about it as a private market opportunity.

In particular, since the implementation of the EPA's Brownfields Program, equity concerns have been de-emphasized within the dominant administrative and economic (i.e. policy) discourses. The Agency states, "Brownfields revitalization presents an opportunity for environmental justice to be achieved through community involvement in cleanup and reuse decisions and activities and through the leveraging of new investment and jobs in distressed communities" (U.S. EPA 2007b). While the Minority Working Training Grants touted as the federal program's great environmental justice success are certainly a start, the fact remains that many of the most vulnerable communities never receive a share of federal funding. Thus the EPA gets away with not having to tackle environmental justice issues in any meaningful way. There are a few exceptions to this rule, however. Federal agencies have sponsored research focused on brownfields' impact on neighborhood property values, local tax bases, and the extent to which potential Title VI (Environmental Justice) complaints dissuade private investors from redeveloping contaminated sites (HUD and EPA 1997; EPA 1998; HUD 1999).

Similarly, community economic development and urban revitalization discourse has increasingly been co-opted by neo-liberal discourse. While this is not to say that progressive advocates have been entirely silenced, laws and regulations increasingly frame redevelopment issues in terms of the free market and private investment, effectively pushing plans that deal with comprehensive community development and equity planning to the political and economic margins (Shragge 1997; Goetz 2003; Fainstein 2007; Markusen 2007).

Re-emphasizing the Third Discourse:
Brownfields and Environmental Injustice

Ironically, it was environmental justice and public health activists who first brought the issue of idle, contaminated urban land to the attention of policymakers, demanding that a strategy be developed to revitalize the industrial and commercial blight in urban neighborhoods. Today, although environmental justice concerns are understood and addressed by federal and some state policymakers, the resources allocated specifically to equity-driven projects pale in comparison to market-driven ones.

Ellerbusch (2006) calls attention to the many different ways that risk can be conceptualized by stakeholders within the brownfields redevelopment arena. While developers often think about the risk of losing a return on investment, property owners risk being held liable for environmental contamination. Residents in neighborhoods impacted by brownfields may worry about the risk to their properties and health if land is left vacant and unremediated for long periods of time. Ellerbusch maintains that a brownfield program truly committed to equity and community development must be willing to take all of these perspectives into consideration when forming strategies and plans.

Articles by Gute (2006) and Gute and Taylor (2006) argue that the key to sustainable revitalization of cities suffering the impacts of brownfields is greater public participation and collaborative environmental decision-making. An analysis of redevelopment projects in Bridgeport, CT, revealed that a resident-driven planning and risk assessment process has been central to the municipal program's success. Many other environmental justice activists and planners have emphasized this idea as it relates to brownfield redevelopment and community development in particular (Greenberg et al. 2000; Greenberg and Lewis 2000; Wernstedt et al. 2006).

Eric Shragge (1997) offers a political economic critique of neo-liberal approaches to community economic development, arguing that urban poverty is in large part the result of poor people not controlling resources in their communities. Shragge argues that local economic revitalization is a component of social empowerment, but cannot be expected to solve problems of urban poverty alone. Instead, market efforts should be focused on improving social conditions, not facilitating individual profit.

By and large, progressive critics of brownfield redevelopment and community economic development programs are not against private development; nor do they wholly discourage public-private partnerships (Levine 1989; Shragge 1997; Grogan and Proscio 2000; Bennett and Giloth 2007). Rather, they see these issues in relation to overall community empowerment. From this perspective, private investment is encouraged as long as investors are willing to work with and on behalf of the communities where they operate. This rethinking of local economic development strategies requires,

> The establishment of new social partnerships with the shared understanding that a neighborhood is more than a market,that the local economy must be based on a solid social foundation reflecting the life experience and needs of its residents (Mendel and Evoy in Shragge: 115).

Insofar as brownfield redevelopment is an increasingly important facet of community economic development, this same principle holds true.

Community Benefits Agreements hold—at least in theory—the potential to revolutionize the brownfield redevelopment arena. CBAs are powerful for two reasons: first, because they demand a more inclusive, collaborative and meaningful public participation process, and secondly, they target much needed investment to areas of community development that might normally be considered outside the scope of individual development projects.

In fact, at least one other researcher has pointed directly to the natural link between brownfield redevelopment and CBAs. Michelle DePass (2007), a Program Officer with the Ford Foundation, recently published an article calling for a convergence of the two movements. She writes,

> In a practical sense, the success of brownfield redevelopment efforts has created a rare and robust opportunity for environmental and community economic development goals to converge ... Leveraging valuable redevelopment opportunities to generate a broader sweep of community benefits may be the next frontier for brownfields (602, 604).

Like DePass and other progressive critics, this chapter is most concerned with questions of social, economic, and environmental equity in so far that they relate to community development and brownfields. To that end, research should ask questions such as: How might the incorporation of CBAs into brownfield redevelopment projects advance the equitable and sustainable revitalization of communities? What opportunities and challenges exist for integrating the two together? And what might a collaborative movement look like?

Conclusion

This analysis has provided a foundational understanding of both the promises and problems inherent with integrating community benefits agreements into brownfield redevelopment projects. Older, deindustrialized cities, while possessing a number of demographic similarities, are ultimately very different places, with unique political, economic, and social environments. As such, the model needed for effectively linking CBAs to brownfield redevelopment projects in individual communities will vary according to the particular socio-political and environmental context. In short, there is no standard model for integrating the two.

Despite their unique contexts, however, a number of notable trends do emerge, suggesting that there are some common questions, challenges, and opportunities which should be further explored in the pursuit of developing a more general framework for incorporating CBAs into brownfield projects. This section explores these common themes and concludes with a set of recommendations about how both the community benefits agreements and community-based brownfield redevelopment movements might join forces and move forward together in the future.

Economic Development

Community development specialists understand the potential for local economic benefits to be achieved vis-à-vis the integration of brownfield redevelopment programs and community benefits agreements (Knapp 2008). Clearly, the redevelopment of idle, underutilized, and potentially contaminated urban land into productive uses makes sense for a number of economic and social reasons. Jobs are created, the local tax base is improved, perceptions about the quality of the community change (which can spur additional private investment and activate more widespread revitalization), and overall, the quality of life for people living in a neighborhood is improved.

In practice, however, questions about how to define "local economic benefits" emerge. Representatives from the community economic development field recognize that training, jobs, and affordable housing are critical for equitable urban revitalization (Bartalameo 2008; Damaisi 2008). Without taking certain measures to ensure that economic opportunities exist for poor and working class residents and affordable housing is built and preserved, gentrification on redeveloped sites might occur.

In some cities, municipal officials work with community groups, state and federal agencies, and private developers to link environmental remediation goals with economic revitalization, particularly in neighborhoods and among residents who need sustainable solutions the most. Their systems are not perfect- not all training opportunities end in a permanent placement, public resources are scarce, coalition building is challenging, and targeted sites often must wait to be redeveloped until economic conditions are favorable. Nevertheless, there are models that show how different types of stakeholders might work collaboratively to link community-based environmental, economic and social development projects together through brownfield redevelopment. And with progressive, committed nonprofits working in a city, the potential for engaging some of the most socioeconomically vulnerable residents in green collar jobs and sustainability movements seems to be a very promising way to link brownfield redevelopment to economic and social sustainability goals and ultimately succeed in moving beyond a "permanent underclass" in urban centers (Razzaq 2008).

In communities that have begun to feel gentrification pressures, but have not gotten to a point economically where market variables outweigh environmental ones in terms of influencing redevelopment, cities continue to suffer both economically and environmentally from postindustrialization. City officials often remain skeptical about making community benefits demands from developers, while community development activists remain skeptical of municipal leaders' commitment to promoting the public interest.

This cycle of distrust creates a political situation that could make integrating CBAs into brownfield redevelopment projects extremely challenging. Further, given the current economic recession, a lack of resources has stalled many community-based planning projects. Municipal officials increasingly feel they must cater to large-scale private projects as a way to boost the city's tax base and fill in resource gaps. These projects hold great community economic benefits, but these benefits will only be realized when community groups and local officials are willing to put real pressure on developers.

Building Community and Re-knitting Neighborhoods

Brownfield redevelopment has the potential to physically reknit communities that have suffered the damaging effects of deindustrialization by returning idle, contaminated, and often inaccessible land to active and productive uses. Brownfield sites create physical barriers in communities, severing neighborhoods from one another and deterring reinvestment.

By linking the community benefits agreement movement to brownfield redevelopment, residents and neighborhood stakeholders have the opportunity to join forces and have political influence over the end uses of sites. Arriving at the decision-making table with first-hand knowledge about the neighborhood, engaged stakeholders can make collaborative recommendations about how best to use sites to achieve community goals.

In addition to physical community building, CBA-driven brownfield redevelopment has the potential to foster social sustainability. The coalition building process required for the negotiation of a community benefits agreement provides an opportunity for people with a variety of stakes in the neighborhood to sit around a table, have a frank discussion about specific goals, and work together to devise a solution that addresses concerns comprehensively. This process has the potential to build community and social capital among individuals and groups who might not have otherwise joined forces, ultimately strengthening the community's political voice and making it possible to put real pressure on public officials and developers.

People get very excited about neighborhood revitalization, especially in communities that have suffered legacies of disinvestment. In cities already experiencing gentrification, however, there may be tension around what sorts of benefits are most needed. Poor and working class residents may hope for affordable housing, job training, and affordable child care, while the more affluent may want to emphasize benefits such as open space and community funds. In cities reaching out to developers in hopes of instigating revitalization, the challenges may not lay so much in agreeing on common goals for the community, but in having the human and economic capacity required to sustain coalition-building and community benefits negotiating processes over time. Although city staff may have the capacity to leverage state and federal funds for cleaning up sites, administrative capacity is only one piece of the puzzle with respect to a successful community-centered brownfields program. In communities where the majority of residents are strapped for time and money, community organizers must figure out ways to engage people with projects that may take years and millions of dollars to complete.

Furthermore, what is missing in some cities is not the ability of community groups to work together and collaboratively recommend public-oriented end uses for brownfield sites. Nor is the main challenge an issue of organizational sustainability. The primary issue is the city's unwillingness to put pressure on developers and demand that brownfield redevelopment projects be oriented toward equitable community revitalization. Without municipal support for the community benefits negotiation process, leveraging real and meaningful public benefits is next to impossible (at best).

Final Thoughts

A consideration of CBAs is important for understanding how decisions are made about the distribution of the benefits and burdens of environmental and economic revitalization. It is also important to activists and scholars who are interested in understanding strategies that work for engaging diverse stakeholders around redevelopment initiatives, and how unified visions for community development can transcend cultural and socioeconomic divides.

In sum, the theoretical framework offered here suggests that great potential exists for integrating community benefits agreements with brownfield projects to promote just, equitable community revitalization. Both the community benefits agreement and community brownfield redevelopment movements hold great promise, and both operate successfully in a number of cities across the United States. The two have similar goals: both aim to create opportunities for meaningful public participation among politically and economically marginalized citizens, ultimately bringing people to the table to make decisions about and share in the benefits of community and environmental revitalization efforts.

In promoting brownfield redevelopment, care must be taken to ensure that the social consequences of revitalization efforts are equitable and fair. As older, postindustrial cities are reborn environmentally and economically, special precautions will be necessary to ensure that there is a fair distribution of both the benefits and burdens of reinvestment; that environmental and economic development goals coincide with social equity goals, that include issues such as labor and employment, education, affordable housing, transportation improvements, public health, and community building.

It is here that the community benefits agreement movement holds the most potential with respect to brownfield redevelopment projects. If the same multidisciplinary, participatory approach that coalitions have applied to large scale development projects where CBAs have been negotiated are applied to brownfield projects, a truly community-based form of brownfield revitalization could be achieved.

The task of linking them together, however, is much easier to do in theory than in practice. Although community development specialists generally agree that an integration of the two models would help secure the equitable revitalization of communities, there is hardly an easy solution or blueprint for linking them together.

There are clear opportunities for integrating CBAs and brownfield projects as a way to create meaningful and sustainable environmental, economic, political, and social opportunities in neighborhoods recreating themselves in the wake of deindustrialization. And although their integration is not assured, in the sense that the model is subject to myriad social and political tensions, with the right planning and capacity building, it is a model that could advance the development of just and sustainable cities in the future.

References

American Institute of Architects. 2000. *The New Market Frontier: Unlocking Community Capitalism through Brownfield Redevelopment.* Available at http://www.usmayors.org/brownfields/brownfieldsreport0801.pdf (accessed: October 15, 2007).

Bartalameo, J. Telephone Interview. March 6, 2008.

Bennett, M. and R. Giloth. 2007. *Economic Development in American Cities: The Pursuit of an Equity Agenda.* Albany: SUNY Press.

Bowman, A. and M. Pagano. 2000. "Transforming America's Cities: Policies and Conditions of Vacant Land." *Urban Affairs Review* 35(4): 559–81.

Brachman, L. 2004. "Turning Brownfields into Community Assets: Barriers to Redevelopment." In Rosalind Greenstein and Yesim Sungu-Eryilmaz (eds), *Recycling the City: The Use and Reuse of Urban Land.* Lincoln Institute of Land Policy: Cambridge, MA.

Cummings, B. "Advocates Rally before Steel Point Hearing." *Connecticut Post Online* (Bridgeport, Connecticut), October 30, 2007.

Damaisi, B. Telephone Interview. March 14, 2008.

DePass, M. 2006. "Brownfields as a Tool for the Rejuvenation of Land and Community." *Local Environment* 11(5): 601–6.

Eisen, J. 2007. "Brownfields at 20: A Critical Reevaluation." *Fordham Urban Law Journal* 34: 721–56.

Ellerbusch, F. 2006. "Brownfields: Risk, Property, and Community Value." *Local Environment* 11(5): 559–75.

Evans, A. 2004. "The Economics of Vacant Land." In Rosalind Greenstein and Yesim Sungu-Eryilmaz (eds), *Recycling the City: The Use & Reuse of Urban Land.* Lincoln Institute of Land Policy: Cambridge, MA, 53–63.

Faber, D., J. Jennings and P. Loh. 2002. "Solving Environmental Injustices in Massachusetts: Forging Greater Community Participation in the Planning Process." *Projections: MIT Journal of Planning* 3: 109–32.

Fainstein, S. 2007. Class Lectures. *Urban Politics and Land Use Policy.* Harvard Graduate School of Design.

Gardner, S. 2004. "Environmental Devolution and Local Capacity: Brownfield Implementation in Four Distressed Cities in New Jersey." in Rosalind Greenstein and Yesim Sungu-Eryilmaz (eds), *Recycling the City: The Use & Reuse of Urban Land.* Lincoln Institute of Land Policy: Cambridge, MA, 53–63.

Gattuso, D.J. 2000. *Revitalizing Urban America: Cleaning up the Brownfields.* Competitive Enterprise Institute: Washington, DC.

Goetz, E. 2003. *Clearing the Way: Deconcentrating Poverty in Urban America.* Washington, DC: The Urban Institute Press.

Green, N. and S.L. Coffin. 2005. "Modeling the Relationship among Brownfields, Property Values, and Community Revitalization." *Housing Policy Debate* 15(2): 257–80.

Greenberg, M. and J. Lewis. 2000. "Brownfields Redevelopment, Preferences, and Public Involvement: A Case Study of an Ethnically Diverse Neighborhood." *Urban Studies* 37(13): 2501–14.

Greenberg, M., K. Lowrie, L. Solitare and L. Duncan. 2000. "Brownfields, TOADS, and the Struggle for Neighborhood Development." *Urban Affairs Review* 35(5): 717–33.

Greenberg, M., K. Lowrie, H. Mayer, K. Tyler Miller and L. Solitaire. 2001. "Brownfield Redevelopment as a Smart Growth Option in the United States." *The Environmentalist* 21: 129–43.

Greenstein, R. and Y. Sungu-Eryilmaz (eds). 2004. *Recycling the City: The Use and Reuse of Urban Land.* Lincoln Institute of Land Policy: Cambridge, MA.

Grogan, P. and T. Proscio. 2000. *Comeback Cities: A Blueprint for Urban Neighborhood Revival.* Boulder, CO: Westview Press.

Gross, J., G. LeRoy and M. Janis-Aparicio. 2005. *Community Benefit Agreements: Making Development Projects Accountable.* Published by Good Jobs First and California Partnership for Working Families.

Gute, D.M. 2006. "Sustainable Brownfields Redevelopment and Empowering Communities to Participate More Effectively in Environmental Decision-Making." *Local Environment* 11(5): 473–8.

Gute, D.M. and M. Taylor. 2006. "Revitalizing Neighborhoods through Sustainable Brownfields Redevelopment: Principles Put into Practice in Bridgeport, CT." *Local Environment* 11(5): 537–58.

Hazardous Waste Consultant. 2007. "Brownfields Tax Incentive Encourages Site Revitalization." *Hazardous Waste Consultant.*

Hollander, J. 2009. *Polluted and Dangerous: America's Worst Abandoned Properties and What Can be Done about Them.* Lebanon, NH: University Press of New England.

Howland, M. 2004. "Is Contamination the Barrier to Inner City Industrial Revitalization" in Rosalind Greenstein and Yesim Sungu-Eryilmaz (eds), *Recycling the City: The Use & Reuse of Urban Land.* Lincoln Institute of Land Policy: Cambridge, MA, 89–109.

Kirkwood, N. January 26, 2007. Class Lecture: *Brownfield Redevelopment Practicum.* Harvard University Graduate School of Design.

Knapp, C. 2008. "Assessing the Potential for Integrating Community Benefits Agreements into Brownfield Redevelopment Projects: Case Studies from Three Northeastern Cities." Master's thesis, Tufts University.

Lange, D. and S. McNeil. 2004. "Clean It and They Will Come? Defining Successful Brownfield Development." *Journal of Urban Planning and Development* (June): 101–8.

Lavine, A. Telephone Interview. March 28, 2008.

Leigh, N.G. 2004. "Survey of State-Level Politics to Address Urban Vacancy Land and Property Reuse" in Rosalind Greenstein and Yesim Sungu-Eryilmaz. *Journal of Urban Planning and Development* (June): 101–8.

Leigh, N.G. and S. Coffin. 2005. "Modeling the Relationship Among Brownfields, Property Values, and Community Revitalization." *Housing Policy Debate* 16(2): 257–80.

Levine, M. 1989. "The Politics of Partnership: Urban Redevelopment since 1945." In G.D. Squires (ed.), *Unequal Partnerships*. New Brunswick, NJ: Rutgers University Press.

Li, D. and R. Sattler. February 2, 2007. Class Lecture: *Brownfield Redevelopment Practicum*. Harvard Graduate School of Design.

Markusen, A. 2007. *Reigning in the Competition for Capital*. Kalamazoo: W.E. Upjohn Institute for Employment Research.

Meyer, P. and T.S. Lyons. 2000. "Lessons from Private Sector Brownfield Redevelopers: Planning Public Support for Urban Regeneration." *Journal of the American Planning Association* 66(1): 44–55.

Page, G.W. and H. Rabinowitz. 1994. "Potential for Redevelopment on Contaminated Brownfield Sites." *Economic Development Quarterly* 8(4): 353–63.

Penn, B. April 15, 2007. Class Lecture: *Brownfield Redevelopment Practicum*. Harvard Graduate School of Design.

Pratt Institute Center for Community and Environmental Development. 2005. *Slam Dunk or Airball? A Preliminary Planning Analysis of the Brooklyn Atlantic Yards Project*. Brooklyn, NY.

Razzaq, T. Telephone Interview. February 18, 2008.

——. Personal Interview. March 7, 2008.

Shragge, E. 1997. *Community Economic Development: In Search of Empowerment*. New York: Black Rose Books.

Solitare, L. 2003. "Participation in Brownfields Redevelopment." Available at http://www.bos.frb.org/commdev/c&b/2003/winter/Brownfields.pdf (accessed: October 15, 2007).

United States Conference of Mayors. 2006. *Recycling America's Land: A National Report on Brownfields Redevelopment*.

——. 2008. *Recycling America's Land: A National Report on Brownfields Redevelopment*.

United States Department of Housing and Urban Development. 1999. *Assessment of State Initiatives to Promote Redevelopment of Brownfields*.

United States Environmental Protection Agency (U.S. EPA) 1980. *Comprehensive Environmental Response, Liability, and Compensation Act*, P.L. 42 U.S.C. 9601, 1980 (amended 1986). Available at www.epa.gov/superfund/policy/cercla.htm (accessed: September 10, 2007).

——. 1986. SARA Overview. Available at www.epa.gov/superfund/policy/sara.htm (accessed: March 29, 2010).

——. 1999. *Brownfields Title VI Case Studies: Summary Report.*

——. 2005. *Brownfields Fact Sheet: ISLES, Incorporated.* Available at http://www.epa.gov/brownfields/05jtgrants/jt_isles.htm (accessed: March 29, 2010).

——. 2006. *2006 U.S. EPA Brownfields Conference.* Available at http://www.brownfieldsconference.org (accessed: March 29, 2010).

——. 2007b. *Brownfields and Land Revitalization.* Available at http://www.epa.gov/brownfields/ (accessed: December 5, 2007).

——. 2008. *Brownfields Fact Sheet: Bridgeport, CT.* Available at http://www.epa.gov/brownfields/cities/bridgeport.htm (accessed: March 29, 2010).

Wernstedt, K., P. Meyers and A. Alberini. 2006. "Attracting Private Investment to Contaminated Properties: The Value of Public Interventions." *Journal of Policy Analysis and Management* 25(2): 347–69.

Chapter 6

The Quantitative and Qualitative Impacts of Brownfield Policies in England

Katie Williams

Introduction

The redevelopment of brownfield sites (or Previously Developed Land, PDL) has been a concerted planning objective in England for over a decade now. Brownfield land is defined in England as "land that has been previously subject to physical development (other than agriculture) and where its reuse may be complicated by one or more factors, which may include contamination" (English Partnerships 2006: 1). Therefore, in contrast to many other countries in Europe, and common usage in the USA, English brownfield sites are not necessarily contaminated, but have been developed (CABERNET 2004). A range of policy initiatives, which are described below, have cumulatively attempted to steer development on to PDL, mainly in urban areas. The policies have, broadly, been devised to meet the twin objectives of: 1) Reducing the total amount of brownfield land in England; and 2) Contributing to sustainable development and urban regeneration. These two policy aims can be seen as "quantitative," i.e. related to the amount of land redeveloped, and "qualitative," focusing on the types of places and wider social, economic and environmental benefits that can be achieved via the development of brownfield land (Ganser and Williams 2005).

This chapter explores the extent to which English policies have succeeded, in both these "quantitative" and "qualitative" ambitions. First, it sets out the context of brownfield development in England, discussing why the issue rose to policy prominence from the late 1990s to the present day. Second, it sets out the government's response, tracking the development of a series of initiatives, policy statements, and targets addressing PDL, housing, regeneration, and sustainable urban development. Third, the extent of PDL redevelopment is analyzed, with the aim of assessing the success or otherwise of the policies outlined at improving land recycling rates. Fourth, a qualitative analysis of the types of development that have been delivered in this policy context is offered. Specifically, issues such as the sustainability of the developments and their contribution to urban regeneration are evaluated. The chapter finishes by drawing some conclusions about England's experiences in attempting to tackle the "brownfield problem."

The Origins of the Contemporary Policy Focus on
Brownfield Redevelopment in England

A specific set of circumstances in England from the mid-1990s to the present day have combined to bring about a new focus on brownfield development by government and a number of key agencies working in planning, regeneration and countryside protection. These circumstances do not have a neat chronology, but have coincided to produce a new and targeted policy response to PDL. The policy activity grew out of the emergence of a number of meta-narratives that influenced political discourse, including "sustainable development" and "holistic regeneration" (Raco and Henderson 2006).

The origins of the renewed emphasis on brownfield redevelopment, by the Labour Government from 1997 onwards, is set firmly within the context of growing interest in sustainable development. The concept became a strong objective in UK, and European policy throughout the 1990s and 2000s, and is intrinsically linked to brownfield policies (see Dixon 2008, for a full account of the development of the concept of "sustainable brownfield regeneration"). In 1994, the UK Government became the first to develop a national strategy on sustainable development, and subsequent strategies and frameworks have developed the concept more deeply in policy ever since (ibid.: 241). The planning system was seen as one of the key mechanisms for delivering a more sustainable future, and the planning profession welcomed the new challenge and purpose that the sustainability agenda offered. From the mid-1990s onwards, planning policies have been revised continually to embed and operationalize sustainability into all land use planning decisions, from large-scale strategic developments to minor changes.

The sustainability agenda is important to the brownfield development story because it informed and accelerated national and international debates about desirable spatial development patterns. Brownfield sites came to form part of the debate about the capacity and opportunity to achieve sustainable urban forms through urban compaction (see for example Breheny and Rookwood 1993; Jenks et al. 1996; Power and Mumford 1999; Rudlin and Falk 1999; Williams et al. 2000; Adams and Watkins 2002). Redeveloping urban brownfield sites was seen as an opportunity to raise densities. Proponents of more compact cities argued that high-density, mixed-use living enhances sustainability because it reduces car use and pollution, leads to urban vitality, encourages social interaction, provides support for the local economy and facilities, and diverts development from greenfield sites (CEC 1990; Adams and Watkins 2002).

Although there was far from consensus on the sustainability of the compact city model, with some research evidence suggesting it was neither necessarily sustainable nor feasible, the government still embraced the concept (see, for example, Breheny 1992; Jenks et al. 1996; Williams 1999; Adams and Watkins 2002). It stated frequently the dual benefits of contributing to regeneration and staving off development in the countryside (DETR 1998a, 1998b, 2000a, 2000b; DTLR and CABE).

In addition to the transition to sustainability, in the mid-1990s the English government was also facing political pressure in response to the development patterns of the 1980s and early 1990s. This period had seen a weakly regulated development industry build extensively on greenfield sites, particularly in the South East of the country. This had led to dispersed settlement patterns and fast peripheral growth of urban areas. In due course, adverse environmental and social effects became obvious (Breheny 1996; Stretton 1996; Williams 1999; Williams et al. 2000), and consequently, the development of greenfield sites became a central political issue. National and local pressure groups were effective in their campaigns to raise awareness of the impacts of greenfield development: the loss of countryside, increased car use, and social exclusion were all highlighted (Breheny 1999; CPRE 2001). This fuelled arguments that a more integrated, and some argued, regulated, approach to land development was required.

Yet in the late 1990s, massive housing growth was still required to meet demand. This was driven partly by natural population increases (i.e. more births than deaths) and net in-migration, but also by changes in household sizes. On average, English households are becoming smaller due to social and demographic trends such as aging, higher divorce rates, and a growing number of single households. This situation left the Government with the challenge of finding new land for housing in an already densely populated country (Williams 2009).

At the same time, PDL was also seen as a genuine catalyst and mechanism for much needed "holistic regeneration." Like much of Western Europe in the post-war period, England had experienced decline in industries such as coal, steel, and textiles, and had witnessed the closure of large-scale infrastructure facilities (Grimski and Ferber 2001). Many cities, particularly in the North of England, were in economic decline, suffering from the effects of recessions in the 1970s and 1980s. Many were experiencing counter-urbanization, with populations declining dramatically in the largest English cities. The worsening social conditions, combined with dereliction of sites and their negative impacts on neighboring properties and wider urban areas, contributed to the policy focus on brownfield regeneration.

Against this backdrop of decline in many cities, continual pressure for new housing and the growing "sustainability" and "regeneration" agendas that favored clustering development in existing built-up areas, the issue of urban brownfield development came to the fore. In 2001 it was estimated that there were around 66,000 hectares of PDL in England, and a focused policy response was seen as necessary to bring this land back into beneficial use.

An Overview of Brownfield Policies and Targets in England

> The redevelopment of land is not an end in itself: rather it should maximize
> benefits for new and existing communities through the provision of housing,
> retail, employment, and amenity space, as well as sustaining the living
> environment (English Partnerships 2007: 4).

A general policy of urban containment has existed in England throughout the post-
war period. A distinctive element of the English planning system is the Green
Belts that surround most major cities and restrict development in the countryside.
However, as set out above, from the mid-1990s onwards, the impetus for compact
city policies including development on urban PDL gained in strength.

As far back as 1996, the government made the bold move of setting an
"aspirational" target for 60% of new housing in England to be built on PDL.
This target was stated in the Green Paper, Household Growth: where shall we
live? (DoE 1996). The Labour Government, which came to power in 1997, then
incorporated the target into national planning guidance, Planning Policy Guidance
Note 3: Housing (DETR 1999a). In 2000, a finalized version of PPG3 stated that
"the national target is that by 2008, 60% of additional housing should be provided
on previously-developed land and through the conversion of existing buildings"
(DETR 2000a, para 23). This national 60% target was relatively easily transferred
into regional and local levels of planning. PPG3: Housing was directly binding for
decisions on planning applications, and thus the 60% target was widely considered
in practice. More recently this target has been adjusted for each of the English
Regions, and regional targets vary between 50% and 100% depending on local
circumstances.

To operationalize the 60% target, and encourage PDL development in general,
the Government focused attention on brownfield development in its Planning for
the Communities of the Future document, which set out a comprehensive vision
for house building (DETR 1998a). The ideas were developed further in the Urban
White Paper of 2000, entitled Our Towns and Cities: the Future, Delivering an
Urban Renaissance (DETR 2000b). The White Paper contains a section entitled
Bringing Brownfield Sites and Empty Property Back into Use. It talks of bringing
previously developed land and buildings into "beneficial economic or social use,
so that they can contribute to, rather than detract from, the urban fabric" (ibid: 42).
The paper urges that new development in towns and cities be built in a sustainable
way. In the paper, the Government stated that it aimed to: "accommodate the
new homes we need ... through a strategy that uses available land, including, in
particular brownfield land and existing buildings in urban areas" (ibid: 29). This
paper was significant as it signaled the introduction of the "Urban Renaissance"
movement in the UK. This was a concerted effort, in policy at least, to attract
people back into cities and towns and to provide high-quality urban environments
through well-designed, high-density residential and mixed-use schemes.

Since the late 1990s, in addition to national planning statements on housing, mentioned above, governmental targets and urban compaction policies have been underwritten by a number of other Planning Policy Statements in England (Dixon 2008). These address, for example, sustainable development, transport, biodiversity, regional spatial strategies, local development, and sustainable economic development, and all have been written or re-written in the last decade to support brownfield development. In some policy statements this is an explicit aim, in others it is implied by policies encouraging development in existing built-up areas, ensuring a sequential approach to land development (where the most appropriate urban sites are used first) and raising residential densities. PPS: 3 Housing was last updated in 2006, and this version included a reinforced message to LAs to increase development on brownfield land, and introduced further safeguards to prevent developers from concentrating only on greenfield sites (DCLG 2006a).

In the early 2000s, however, housing completions were falling and housing demand was still high. Hence, in 2003, the Government launched a prescriptive and detailed housing strategy entitled "Sustainable Communities: building for the future" (ODPM 2003). This plan, known as the Sustainable Communities Programme (SCP), with a £38 billion spending plan attached to it, introduced a two-pronged strategy for housing growth. It consisted of developing large numbers of dwellings in targeted "Growth Areas," mainly in the South East of England, specifically in the Thames Gateway and regenerating declining areas to stimulate demand, via nine "Market Renewal Areas" in Northern cities. In both the Growth Areas and the Market Renewal Areas whole new communities were planned. The vast majority of development delivered under the SCP was to be on brownfield sites, although there was also some development on undeveloped land (Williams 2007).

The SCP also introduced the challenge of developing a new National Brownfield Strategy. This task has been undertaken by English Partnerships (the national regeneration agency), and the Department for Communities and Local Government (DCLG). In 2007, English Partnerships reported a number of key recommendations, and the government published its response in 2008 (EP 2007; DCLG 2008a). This strategy seeks to learn from stakeholders involved in brownfield development to devise recommendations on policy and good practice in four distinct areas: identifying, assessing and preparing brownfield land for reuse; safeguarding the environment and ensuring appropriate levels of regulatory control to ensure efficient reuse of land; enhancing communities through the removal of blight and ensuring long term maintenance of PDL; and improving accreditation and skills by ensuring qualified and experienced practitioners (EP 2007). This Strategy takes the DPL issue forward nationally, and tackles some of the more recently identified barriers to brownfield development.

Are English Brownfield Policies Successful in Reducing the Total Stock of Brownfield Land and Preventing Development on Greenfield Sites?

In order to find out if the policies, strategies and targets outlined above have been effective in quantitative terms, several policy objectives are considered:

- Is the total amount of brownfield land in England being reduced?
- Are all types of brownfield land being recycled, or are there successes and failures in land recycling activity?
- Is the rate of recycling similar across England, or are there spatial successes and failures?
- Is the target of 60% of new housing to be built on PDL being met?
- Is the development of PDL leading to a reduction in development on greenfield sites?

Before answering these questions, it is useful to set out the data available for assessing these quantitative aspects of PDL reuse in England.

Quantitative Brownfield Data Available in England

It is possible to answer questions about the extent of contemporary brownfield development relatively accurately due to recent improvements in land use data collection. Data on PDL in England are collected in two national data sets. The National Land Use Database (NLUD, see NLUD 2009) was established in 1998 to provide a comprehensive record of PDL (although some data are only available from 2001/2). The survey collects data from local authorities (LAs) in England, the latest being collected in 2007. Data on land use change are also collected nationally, and published annually, in the Land Use Change Statistics for England (LUCS, See LUCS 2009). The latest data in this set are provisional estimates for 2008, with the last robust data for 2004–7. These data are useful in tracking the nature of DPL redevelopment, particularly in terms of house-building activity; but data on rural areas are not updated as frequently as urban. In addition to these two data sets, researchers and government also use Ordnance Survey data which are collected as part of periodic map updates.

Is the Total Amount of PDL in England being Reduced?

Overall, the total amount of brownfield land in England has reduced year by year since the introduction of brownfield targets (and NLUD) in the late 1990s (EP 2008). The NLUD shows that in 2007, there were an estimated 62,130 hectares of PDL in England, representing a 6% drop from 66,000 hectares since 2000/1 (when comparative data are available). Hence in terms of absolute stocks, the amount of PDL is being reduced, although this figure does not reveal any information about the 'churn' of land (i.e. how much is being redeveloped compared with how much new PDL is coming on stream).

Are All Types of PDL being Recycled, or are there Successes and Failures in Land Recycling Activity?

The NLUD collects information on the nature of PDL. Specifically it categorizes it into two main types: "vacant and derelict land" and "land that is still in use but with the potential for redevelopment." This latter category is land that is being used, but is allocated in a local plan for redevelopment, or has planning permission for any use. It also includes land with "known redevelopment potential" but no planning allocation or permission. Latest data show that an estimated 33,600 ha of the 62,130 ha total of PDL were "vacant or derelict land or buildings" (54%). The remaining 28,520 hectares were "in use with redevelopment potential" (46%) (Figure 6.1).

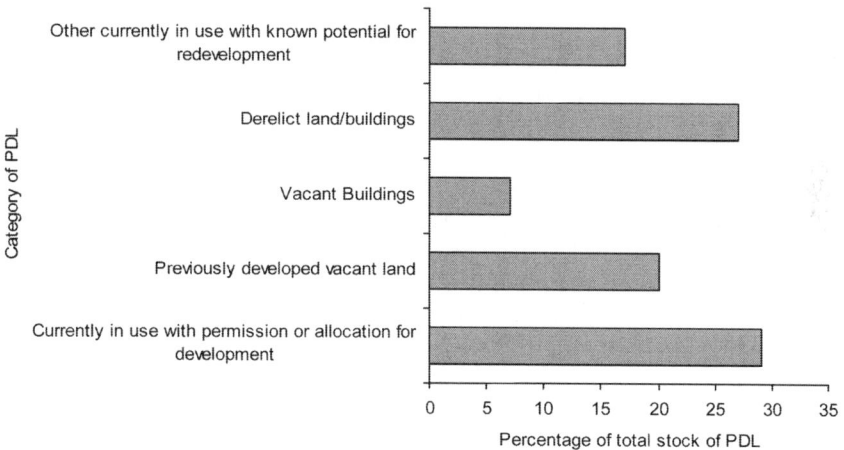

Figure 6.1   Previously developed land by type in England

The monitoring surveys conducted for NLUD show a marked reduction in the extent of derelict and vacant land and buildings since 2001 (Figure 6.2). Year on year these have reduced from 41,000 ha in 2001 to the 33,600 ha in 2007. Research conducted in 2005 (by DTZ Pieda, cited in EP 2007) indicated that even long- and medium-term derelict sites are being returned to beneficial use. Between 2002 and 2005, the stock of these sites had fallen by 29% (in terms of hectares) and 38% by number of sites. What was particularly important was a 37% reduction in the number of long-term derelict sites (EP 2007).

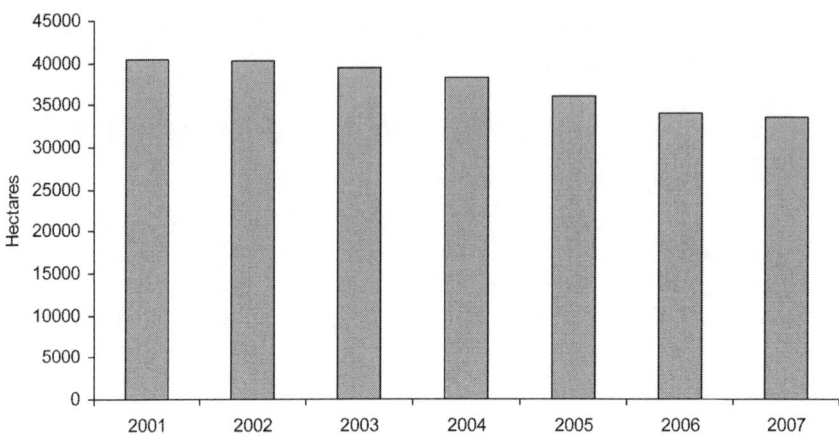

Figure 6.2   Derelict and/or vacant land and buildings in England, annual change in hectares, 2001–2007

The study showed that 23% of the reused land had been developed for housing, 19% for employment, and 17% for mixed uses. Around 10% was used for open space and 26% had reverted to an acceptable natural state and was no longer classed as PDL. These changes have been heralded by government as a marked level of policy success (ibid.: 5).

However, a different picture emerges for land that is "in use but with redevelopment potential" (Figure 6.3). This category of land has shown an annual increase since 2001 from 24,000 ha in 2001 to 28,520 ha in 2007. English Partnerships suggest that this increase may reflect the fact that local authorities are getting better at identifying land for redevelopment, but what is also interesting is that more than 40% of this land has been on the database since its inception in 1998. This means that even though the land has planning permission, and is ready to be developed, it is not coming forward for redevelopment (EP 2007). This type of land is often earning valuable returns for its owners, and may be providing services to local communities, so there is little incentive for it to be released for development. Such sites may also be constrained because of flood risk or ecological issues (EP 2007).

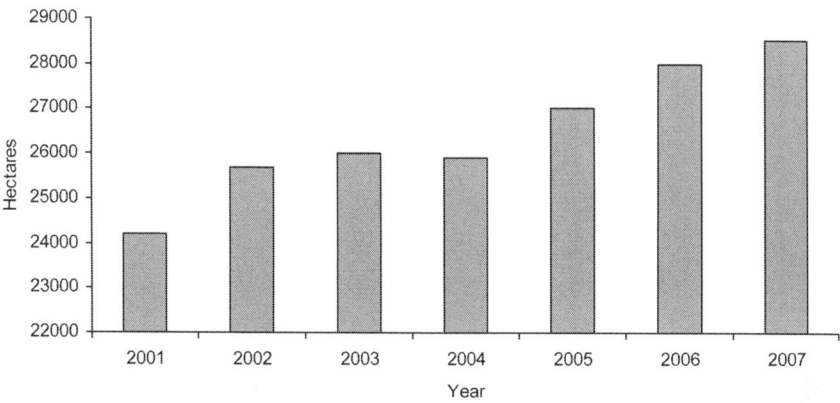

Figure 6.3   Land with planning permission or identified for redevelopment in England, annual change in hectares, 2001–2007

Is the Rate of Recycling Similar across England, or are there Spatial Successes and Failures?

A key feature of the land reuse picture in England is the variation across Regions (Figure 6.4). The greatest concentrations of derelict and vacant sites are in the industrial towns and cities in the Midlands and Northern Regions of England (EP 2007). These sites are often hard to remediate because of physical problems such as contamination, lack of infrastructure, and existing site conditions, but also because they are not economical to develop. In the areas where, until recently, the property markets have been more buoyant, physical restrictions have been more likely to limit redevelopment (ibid.). Land that is still in use but that has planning permission or redevelopment potential is more evenly distributed throughout England than vacant and derelict sites. In places like London and the South East there are more "in use" sites than derelict and vacant sites, and it is these "in use" sites that are often not coming forward for redevelopment.

Turning to the Regional picture over time, the total amount of PDL in two regions has increased since 2002. In Yorkshire and Humber the figure has increased by 14% and in London by 12%. The largest decreases are in the South East (down by 18%) and the North East (16%). There were larger changes in "vacant and derelict land," where every region saw a decrease, with the South East seeing a reduction of 32% and the North East of 30%. The amount of land with development potential increased in most regions, however. The largest increases were 78% in Yorkshire and the Humber, 43% in the East Midlands and 40% in the North East (again perhaps because of Local Authorities becoming more adept at identifying sites). In contrast, the East of England saw a drop of 10% and the South East of 8% (DCLG 2008b).

Reclaiming Brownfields

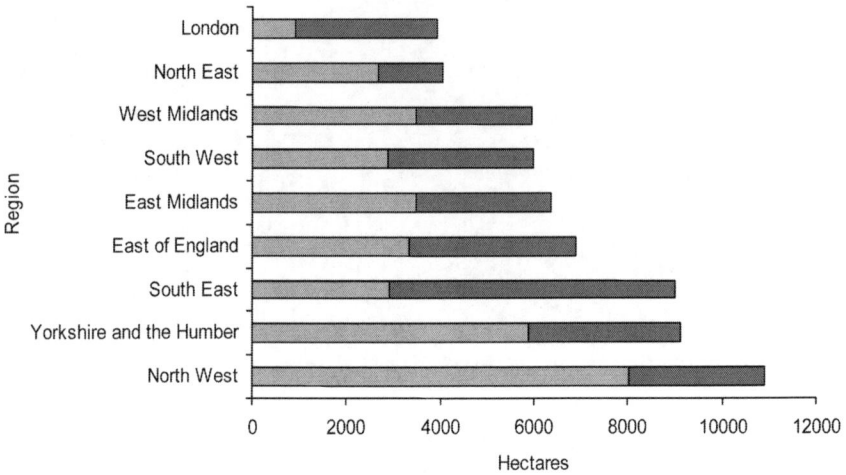

Figure 6.4   Amounts of PDL by region and by type

Is the Target for 60% of New Housing to be Built on PDL Being Met?

The latest LUCS data show that in 2008 (on a provisional estimate) 78% of dwellings (including conversions) were built on PDL, 18% above the target. Since 1998, the proportion of dwellings built on PDL in England has increased by 22% (see Figure 6.5).

Compared with 2003, every region except London showed an increase in the proportion of dwellings built on PDL (See Figure 6.6). This is perhaps unsurprising as London already has the highest proportion of dwellings built on brownfield land (96%), compared with, for example, the East Midlands, which has the smallest proportion (63%).

In fact, the 60% target was achieved nationally in England for the first time in 2000, when 61% of new housing was built on brownfield sites. The achievement of the 60% target, in all Regions and ahead of the original 2008 target date, has led the government to report a major policy success. Some commentators have questioned the value of the 60% target, given the well-documented regional differences, and have viewed regional targets as more valuable, however (Ganser and Williams 2005).

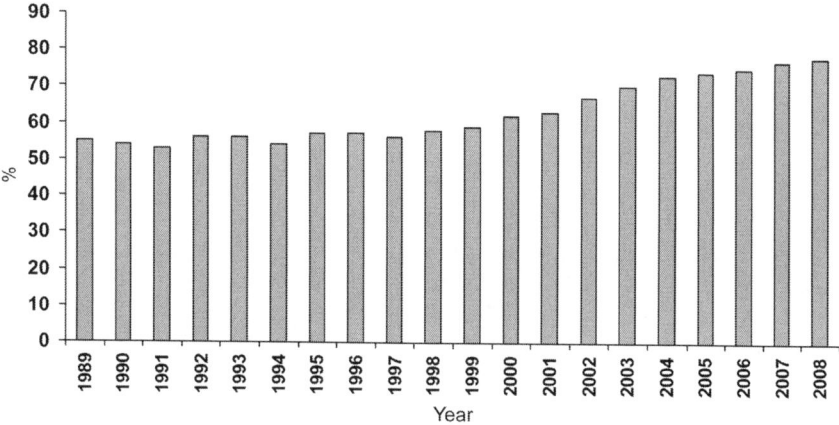

Figure 6.5   Percentage of new dwellings built on PDL, 1989–2008

Others still have questioned if the target itself was too low and suggested an element of political positioning by setting levels that could be reached easily. Specifically critics have questioned the value of divorcing the target (which is proportional) from the number of completed units and the area of land developed (see Adams 2004 and Ganser and Williams 2005 for a fuller discussion of the 60% target).

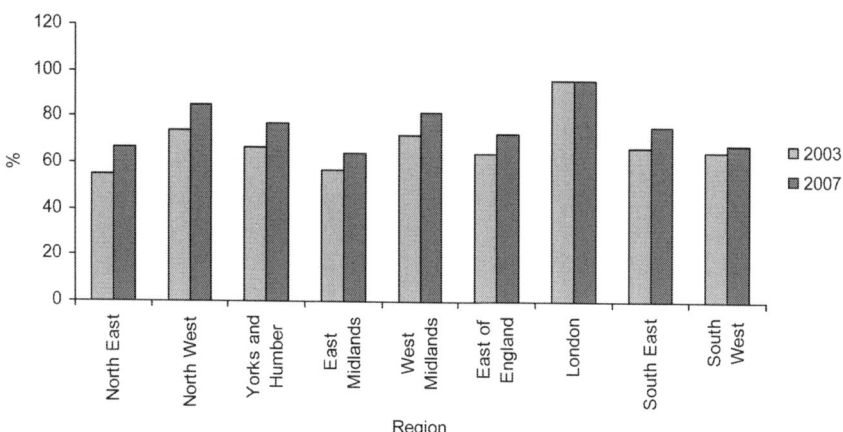

Figure 6.6   Proportion of dwellings on PDL by region 2003 and 2007

Is the Development of Brownfield Land Leading to a Reduction in Development on Greenfield Sites?

The policy ambitions of redeveloping urban PDL are directly linked with goals to stave off development on greenfield sites. So, it is important to ask if the reductions of PDL are commensurate with a reduction in greenfield development to test if the displacement policy is working in practice. Accurate and up-to-date data on the extent of greenfield land development are hard to obtain however. The LUCS record the amount of hectares that change from non-previously developed land to developed land. The latest data available are from 1996–98 (DEFRA 2009) because data from rural areas are updated infrequently.

As no precise data exist, other measures are often used to report house building activity, and to suggest reductions in greenfield development. However, these data are unsatisfactory for a number of reasons. Proportional data on the extent of PDL being used for housing are often presented as an indication of displacement from greenfield sites, but can not be translated directly into information about land take. For example, in 2005, 73% of new dwellings were built on PDL, but only 62% of land for new housing was previously developed. This is because, on average, urban houses are built at higher densities than those on greenfield sites. It is also possible to report on the proportion of land area changing to residential use that was previously undeveloped. This has increased from 46% in 1994 to 55% in 2005, but this is still limited in what it can reveal about the total amount of greenfield land developed. These data are also problematic as they are restricted to housing (and do not include other uses), and do not capture the PDL returned to a natural state.

Perhaps surprisingly, the government currently uses figures from Ordinance Survey map updates (DEFRA 2009). These use cartographic (rather than planning) land classifications. They show that the amounts of forest and woodland, grassland and set aside land increased marginally over the period 1998 to 2005, with crops, grazing, and other agricultural land decreasing slightly. The total amount of "urban and all other land" increased by around 5%. Hence, from these data, it seems that while there have been small gains and losses in different types of "greenfield" land, there has also been a total increase in urban and other developed uses.

Summary of the Quantitative Findings

In answering the questions posed earlier, a mixed picture emerges. There has been an overall reduction in the amount of PDL since the introduction of a concerted target and accompanying policies in the mid-late 1990s. Although the amount of vacant and derelict land has gone down, the amount of "in use" land has increased. The extent to which this is a "real" increase, or one related to LA designations, is difficult to quantify. There are also regional differences in land recycling successes, although all Regions saw a decrease in derelict and vacant land.

In terms of the housing targets, the 60% goal has been achieved and exceeded nationally, but again regional differences emerge, with London building almost all of its new dwellings on PDL, while the East Midlands is building just over 60% on brownfield sites. There has certainly been a "rebalancing" of brownfield and greenfield development since the 1980s, yet significant amounts of greenfield development is still taking place.

The 'Qualitative' Impacts of Brownfield Policies:
Are They Contributing to Sustainable Development and Urban Regeneration?

As set out above, brownfield policies lie at the heart of the English Government's sustainable housing growth and regeneration agendas. The Labour government years saw a series of policies that sought to redevelop brownfield sites as part of a strategic ambition to make existing cities more compact and to regenerate large areas of the country through housing-led development schemes. Alongside the quantitative performance of these policy measures it is important to question the "qualitative" impacts of the brownfield development agenda, specifically the extent to which brownfield development has contributed to sustainable development and urban regeneration.

Qualitative Brownfield Data Available in England

Unlike the quantitative targets for brownfield development that can, notwithstanding some data problems, be assessed relatively easily, the qualitative impacts of the brownfield agenda are far harder to evaluate. Brownfield policies, strategies and targets have evolved and become more multi-faceted over the last 15 years, and complex, wide-ranging benefits are now expected from developing on PDL (Raco and Henderson 2006; Dixon 2008). Brownfield redevelopment is seen as a central component in regeneration, seeking to bring about social and economic development. It also has a major part to play in The Urban Renaissance, with its socio-cultural and urban design aspirations. PDL is also a conduit of environmental improvement, offering the chance to remediate contaminated land and provide new green space. Yet, urban brownfield sites are also the major focus for housing supply, where they are required to increase urban densities and reduce sprawl. Because of this complexity and diversity, compiling evidence to determine whether policy aims have been met is a challenging task. The approach taken here is to use existing research on a number of aspects of PDL policy to assess successes or failures.

The Contribution of Brownfield Reuse to
"Sustainable Development" and "Holistic Regeneration"

The ambitions for brownfield development to contribute to sustainability and regeneration are closely related. They are often dealt with quite separately in practice, however, and hence the analysis here offers an assessment of the extent to which brownfield development has played its part in policy successes within each of the two discourses. Starting with an assessment of the contribution of brownfield development to sustainable development, there are two ways in which this problem can be viewed. The first is to ask if the developments taking place on PDL can be described as "sustainable" schemes in their own right (i.e. do they incorporate good practice in sustainable design, end use etc.), and the second is to assess if the PDL developments, in combination, are leading to sustainable patterns of spatial development, e.g. are they contributing to urban compaction and reducing dispersal?

Are New Developments on Brownfield Sites "Sustainable"?

This analysis deals with sites that have been built on, not those returned to natural uses. This is because the vast majority of redeveloped urban sites are used for hard end purposes. Several researchers have attempted to assess the sustainability of schemes taking place on brownfield sites in England. Most use a broad definition of sustainability that embraces social, economic, and environmental concerns, considered in terms of inter- and intra-generational equity. Dair and Williams (2004) undertook case study research of completed small-scale housing and mixed-use brownfield developments and found that much new development remains unaffected by consideration of the aims of sustainability policy guidance or good practice. For example, they found that the majority of schemes were meeting only limited sustainability standards (ibid.). A particular shortcoming was the omission of environmental sustainability features. Other sustainability requirements such as social housing were being delivered, mainly because planning policies and national guidance were prescriptive about what should be provided. In another study, Dixon (2008) investigated developers' attitudes towards sustainable development on brownfield sites. He found a general lack of awareness of some of the key sustainability objectives and a reluctance to incorporate sustainability aspects into brownfield schemes because of perceptions of increased cost. He also found a particular lack of performance on environmental standards in residential and commercial schemes.

Another source of evidence on the sustainability of completed schemes comes from a number of reports undertaken by Government Commissions, academics, and professional bodies, assessing the major projects built under some of the key government initiatives outlined above (see, for example, Power 2004; Cowans 2006; DCLG 2006b; SDC 2007; Williams and Lindsay 2007). Much of this research evaluates the large-scale new developments in the South East of England, built on PDL in the 2000s. Overall, this research reports both sustainability failures and successes in these projects. On the whole, they are not delivering the flagship sustainable places many would have hoped for. They are not performing well on energy efficiencies, via homes or transport, and are struggling to phase the delivery of housing and public services. There are also difficulties in delivering sustainable transport solutions, with developers tending to put forward "road only" solutions. This said, compared to the developer-led housing schemes of the 1980s and 1990s, these new developments are of higher urban design quality, and offer some improvements in terms of sustainable travel choices and community development.

Is Development on Brownfield Sites Contributing to Sustainable Patterns of Development?

As seen above, brownfield policies and targets have been successful in steering a large proportion of housing development onto PDL, mainly in urban areas. They have also raised average densities and can therefore be seen to be contributing to the compact city agenda. The extent to which these development patterns can be argued to be sustainable depends on how the perceived strategic benefits (e.g. greenfield protection, sustainable transport, options and urban regeneration) are weighed against both local benefits (see below) and potentially unsustainable localized impacts (e.g. overcrowding, pressure on local services, poor urban environmental conditions, and increased traffic congestion).

In England, there is certainly evidence of such localized pressures in urban areas, but determining the extent to which they are linked to cumulative development on PDL is difficult. For example, it is clear that compaction policies, supported by prescriptive density standards, mean that England is now building some of the smallest homes in Europe and that 70% of the population thinks that Britain is too crowded (Williams 2009). Most large English cities also suffer from traffic congestion, and developing brownfield sites has inevitably added to vehicle concentrations in some places.

In addition, the larger scale schemes (delivered in the Growth Areas via the SCP) have also been criticized for their lack of contribution to sustainable spatial development. Many argue that the SCP, in continuing a "predict and provide" approach to growth, is contributing to overcrowding in the South East at the expense of more balanced development in other Regions. In some of the Growth Areas, developments have taken place outside of existing towns and cities and have created new "dormitory" settlements, without the required services or transport connections.

However, these negative issues have to be balanced against alternative locations for growth. There is evidence that, overall, urban PDL development has had some benefits. It has led to more sustainable growth patterns with respect to greenfield sites, and in some areas has certainly produced higher densities to support facilities and services in urban areas. However, the Regional differences in PDL development can be seen as exacerbating "overheating" in some regions and economic and social malaise in others.

Is Brownfield Development Contributing to Urban Regeneration?

Over the last decade, large swathes of inner urban PDL have been redeveloped and new neighborhoods of flats, shops, and restaurants have been delivered in most cities in the UK. The re-population of city centers and inner suburbs has also contributed to the beginnings of an "Urban Renaissance," reversing long-term counter-urbanization patterns in many places. The fact that over 70% of housing is now taking place on mostly urban PDL has had a major impact in some cities, bringing much-needed social and economic activity back into central areas. England now has numerous examples of successful brownfield regeneration schemes, on reused docklands, industrial sites, housing estates, and underused or decommissioned public land. Many such projects have revitalized urban areas with in-fill housing, mixed use schemes, and employment and educational facilities. Many of these schemes showcase high-quality urban design, and add considerably to their host cities' attractiveness.

However, while there are undoubtedly some real successes in the brownfield regeneration story, there are also some problems, and much to learn. In particular some analysts argue that too much is expected from the PDL program. Raco and Henderson (2006) argue that brownfield sites are treated too homogeneously in policy, and that they need to be understood better and set within a wider and more comprehensive set of development projects and policy agendas. Echoing some of the findings about the SCP above, they argue that much recent brownfield development has not benefited local communities, and has not been set in a broader environmental planning context. They argue that brownfield sites can become commodified, with their development taking place "separately" from local communities. This criticism has also been leveled at some projects that have taken place on brownfield sites under the auspices of The Urban Renaissance agenda. Observers have argued that some schemes have gentrified neighborhoods and excluded existing communities. The benefits of regeneration via PDL development have also not been evenly spread across the country. The schemes that have come forward have been attractive to the market (some only because of publicly-funded incentives), yet in many deprived areas brownfield sites remain unattractive to developers and blight neighborhoods. It is here that the issue of "in use" sites earmarked for development but not coming forward needs to be tackled in a more concerted manner.

Conclusions

This Chapter assesses the extent to which brownfield policies in England have reduced the amount of brownfield land and contributed to sustainable development and urban regeneration. It has drawn on data and research to give an assessment of these quantitative and qualitative policy objectives; and has found a mixed picture. Quantitatively, there has undoubtedly been a significant change in the use or reuse of PDL. The last decade has been characterized by visible changes in urban landscapes across the country. Every Region has seen a reduction in total stocks of vacant and derelict land, although many have seen increases in land "in use," but ready for development. There has been a turnaround in spatial patterns of housing delivery, and in the types of dwellings now provided, with a shift away from larger family homes on greenfield sites towards high-density flats and smaller units in urban locations.

In qualitative terms (sustainability and regeneration) development on PDL is providing some benefits. There are examples of urban brownfield schemes that are sustainable in their own right and have enhanced and regenerated urban neighborhoods. There are some exemplars of urban design, community development, and environmental improvements. There are also schemes which have been developed without due consideration of long-term benefits, and which do not contribute to wider sustainability objectives. There are problems with some of the large scale projects, which have failed to deliver in the key area of sustainable growth, and there is also the continuing debate about the extent to which brownfield land is being used to continue the "predict and provide" culture within national planning policy, which is delivering massive growth in the South and South East while less is invested in the Northern Regions.

From the review presented above, it is clear that brownfield development lies at the heart of some of the biggest planning challenges in England today. In the future, the debate is likely to become even more contentious as increasing (and different) pressures shape the land use agenda. For example, it has already been estimated that England will not be able to meet its housing needs through a concentration on PDL. Even if current high-density housing levels are achieved on urban PDL in the future, the country will run out of brownfield sites and need a substantial greenfield development program within the next 50 years (Barker 2006). However, house-building levels are now at their lowest for decades, and it is very likely that there will be a real slowing in development on all types of land, but particularly on more costly PDL. As a result of the economic downturn, more PDL may come on to the registers as businesses close and current uses become uneconomical. Other pressures may also force a reconsideration of the best end uses for PDL. For example, climate change may require a whole new focus on PDL to provide very different elements of urban landscapes, linked perhaps to urban cooling or greening. It will be interesting to see how English brownfield policies (supported by the new National Brownfield Strategy) adapt to both ongoing and new challenges.

References

Adams, D. 2004. "The Changing Regulatory Environment for Speculative House Building and the Construction of Core Competencies for Brownfield Development." *Environment and Planning A* 36(4): 601–24.

Adams, D. and C. Watkins. 2002. *Greenfields, Brownfields and Housing Development*. Oxford: Blackwell Science.

Barker, K. 2006. *Review of Land Use Planning*. HM Treasury, London.

Breheny, M. 1992. "The Contradictions of the Compact City." In *Sustainable Development and Urban Form*, edited by M. Breheny. London: Pion.

Breheny, M., and R. Rookwood. 1993. "Planning the Sustainable City Region." In *Planning for a Sustainable Environment: A Report by the Town and Country Planning Association*, edited by A. Blowers. Earthscan, London: Earthscan.

CABERNET (Concerted Action on Brownfield and Economic Regeneration Network), 2004, *State of the Art Country Profiles*. Available at www. CABERNET.org/publications/reports/php (accessed: February 14, 2010).

CEC (Commission of the European Communities). 1990. *Green Paper on the Urban Environment Commission of the European Communities*. Office for Official Publications of the European Communities, Luxembourg.

Cowans, J. 2006. "Cities and Regions of Sustainable Communities: New Strategies." *Town and Country Planning Tomorrow Series Paper 4*. Town and Country Planning Association.

CPRE (Campaign for the Protection of Rural England). 2001. *Sprawl Patrol: First Year Report*. London: CPRE.

Dair, C. and K. Williams. 2006. "Sustainable Land Re-use: The Role of Different Stakeholders in Achieving Sustainable Brownfield Developments in England." *Environment and Planning A* 38: 1345–66.

DCLG (Department of Communities and Local Government). 2006a. *Planning Policy Statement 3: Housing*. Norwich: TSO.

——. 2006b. *Three Years on from the Sustainable Communities Plan*. Available at www.communities.gov.uk (accessed: February 14, 2010).

——. 2008a. *Securing the Future Supply of Brownfield Land: Government's Response to English Partnerships' Recommendations on the National Brownfield Strategy*. London: DCLG.

——. 2008b. *Previously Developed Land That May be Available for Development: England 2007. Results from the National Land Use Database of Previously Developed Land*. West Yorkshire: Stationery Office/CLG Publications.

——. 2009. "Land Use Change Statistics (England) 2008: Provisional Estimates (May 2009)." *Planning Statistical Release No.1*. London: DCLG.

DEFRA (Department for Environment, Food and Rural Affairs). 2009. *Environmenta; Protection: E-digest Statistics about Land Use and Land Cover*. Available at http://www.defra.gov.uk/environment/statistics/land/ldurban.htm (accessed: February 14, 2010).

DETR (Department of the Environment, Transport and the Regions). 1998a. *Planning for Communities of the Future*. London: HMSO.

——. 1998b. *Planning for Sustainable Development: Towards Better Practice*. London: The Stationery Office.

——. 1999a. *Draft Planning Policy Guidance Note 3 (Revised): Housing*. London: The Stationery Office.

——. 2000a. *Planning Policy Guidance Note 3 (Revised): Housing*. London: The Stationery Office.

——. 2000b. *Our Towns and Cities: The Future, Delivering the Urban Renaissance*. London: The Stationery Office.

——. 2000c. *Millennium Villages and Sustainable Development*. London: DETR.

Dixon, T. 2008. "The Property Development Industry and Sustainable Urban Brownfield Regeneration in England: An Analysis of Case Studies in Thames Gateway and Greater Manchester." *Urban Studies* 44(12): 2379–400.

DoE (Department of the Environment). 1996. *Household Growth: Where Shall We Live?* London: Stationery Office.

DTLR and CABE (Department of Transport, Local Government and the Regions and the Commission for Architecture and the Built Environment). 2001. *Better Places to Live: By Design: A Companion Guide to PPG3*. London: Thomas Telford.

EP (English Partnerships: The National Regeneration Agency). 2006. *The Brownfield Guide: A Practitioner's Guide to Land Reuse in England*. London: EP.

——. 2007. *National Brownfield Strategy: Recommendations to Government*. London: EP.

Ganser, R. and K. Williams. 2005. "Brownfield Development: Are We Using the Right Targets? Evidence from England and Germany." *European Planning Studies* 15(5): 603–22.

Grimski, D. and U. Ferber. 2001. "Urban Brownfields in Europe." *Land Contamination and Reclamation* 9(1): 143–8.

Jenks, M., E. Burton and K. Williams (eds). 1996. *The Compact City: A Sustainable Urban Form?* London: E & FN Spon.

LUCS (Land Use Change Statistics). 2009. Available at http://www.communities.gov. uk/planningandbuilding/planningandbuilding/planningstatistics/landusechange (accessed: February 14, 2010).

NLUD (National Land Use Database). 2009. Available at http://www.nlud.org.uk (accessed: February 14, 2010).

ODPM (Office of the Deputy Prime Minister). 2003. *Sustainable Communities: Building for the Future*. London: ODPM.

Power, A. 2004. *Sustainable Communities and Sustainable Development: A Review of the Sustainable Communities Plan*. London: Sustainable Development Commission.

Power, A. and K. Mumford. 1999. *The Slow Death of Great Cities? Urban Abandonment or Urban Renaissance*. York: Joseph Rowntree Foundation.

Raco, M. and S. Henderson. 2006. "Sustainable Urban Planning and the Brownfield Development Process in the United Kingdom: Lessons from the Thames Gateway." *Local Environment* 11(5): 499–513.

Rudlin, D. and N. Falk. 1999. *Building the 21st Century Home: The Sustainable Urban Neighbourhood*. Oxford: Architectural Press.

SDC (Sustainable Development Commission). 2007. *Building Houses or Creating Communities? A Review of Government Progress on Sustainable Communities*. London: SDC.

TCPA (Town and Country Planning Association). 2004. *Homes for Britain*. Available at http:/www.tcpa.org.uk (accessed: February 14, 2010).

Williams, K. 1999. "Urban Intensification Policies in England: Problems and Contradictions." *Land Use Policy* 16: 167–78.

——. 2007. *"New and Sustainable Communities in the UK."* In *A Tale of Two Cities, China-UK Comparative Study on Housing Provisions for Low-income Urban Residents*. Beijing: Cultural and Educational Section of the British Embassy.

——. 2009. "Space per Person in the UK: A Review of Densities, Trends, Experiences and Optimum Levels." *Land Use Policy* 26: 83–92.

Williams, K., E. Burton and M. Jenks (eds). 2000. *Achieving Sustainable Urban Form*. London: E & FN Spon.

Williams, K. and M. Lindsay. 2007. "The Extent and Nature of Sustainable Building in England: An Analysis of Progress." *Planning Theory and Practice* 8(1): 27–45.

Chapter 7

New Urban Communities: Building on Brownfields for America's Next Generation

Mark L. Gillem and Jill A. Schreifer

Instead of being a place people avoid, a brownfield revitalized for housing and mixed uses can become a place people call home.

And that may be the biggest transformation of all.

(EPA 2009d)

Dynamite. Steel. Concrete. Paint. These are products of America's industrial age. Their making not only led to the country's global reach but also to scores of contaminated sites across the nation. As the industrial age in America, which is now being eclipsed by countries like China, India, and Mexico, comes to a close, the nation is left with the scars of industrial production. The sites where a host of industrial products were made have been abandoned. Yet many layers of pollution and contamination remain a legacy of industrial capitalism and act as barriers to the reuse of these sites. Remediation is costly and time-consuming. Development at these sites is risky and complicated. As a result, many of these "brownfields" have been bypassed and development has been pushed out into "greenfields" across America.

Despite these challenges, redeveloping brownfields can significantly enhance environmental, economic, and social sustainability. This may be one reason why there is high interest and strong support for greener, more sustainable brownfields projects (EPA 2009c). Redeveloped brownfields can support a range of sustainable uses including agriculture and food systems, arts and culture, housing and mixed uses, and other community and civic uses such as greenspace, schools, and health care facilities (EPA 2009d).

This chapter focuses on converting brownfield sites so that they can support housing and mixed uses. Such conversions may help accommodate the housing needs of America's next generation of residents while minimizing the costs of greenfield development. But simply replacing contaminated soils with isolated subdivisions of single-family homes is not the smart move. Rather, the intent is to analyze how principles of what the EPA has called Smart Growth have been applied in theory and application to brownfield redevelopments. Smart Growth, a comprehensive strategy for redevelopment that balances remediation with reconstruction, is one method for converting brownfields into new urban communities. The theory and case studies presented in this chapter demonstrate that the application of Smart Growth principles to the redevelopment of contaminated sites is a strategy that can be used to sustainably leverage the potential resource of brownfields.

In Part I of this chapter, we anchor the discussion within the context of brownfield redevelopment with a particular focus on EPA initiatives and perspectives. The federal government has had a significant role in the process through legislation that unintentionally discouraged redevelopment and then, eventually, encouraged redevelopment. In Part II, we discuss Smart Growth principles and the related planning theory of New Urbanism. Finally, in Part III, we focus on four recent projects built on reclaimed brownfield sites that have used variations of Smart Growth principles. We conclude this chapter with a discussion of the implications of implementing Smart Growth or New Urbanist principles on brownfield sites and argue that, whatever the title, these principles can be used to transform contaminated sites into compelling places to live and work.

Part I: The Importance of Brownfield Redevelopment

Since the 1950s, American developers have largely focused their efforts on greenfield development. The result has been America's unique contribution to planning, known as suburban sprawl—a low-density, auto-oriented development pattern that consumes undeveloped land at metropolitan edges at the expense of infilling within the developed core. Until recently, developers have largely ignored brownfield sites, which are commonly known as properties where the presence of contamination may complicate redevelopment (see 42 U.S.C. § 9601; EPA 2009a). The Government Accountability Office estimates that there may be up to 450,000 brownfields sites in the United States (Link-Wills 2007; EPA 2009a). As an EPA study notes, it is "hard to miss the graffiti-laced walls, the broken windows, the caved-in roofs. It is equally hard to dismiss the unknown environmental contaminants and health hazards brownfields can pose" (EPA 2009d).

The importance of government intervention in brownfields cannot be underestimated—both in initially hindering development and then in providing avenues to overcome associated challenges and encourage revitalization. In 1980, Congress enacted the Comprehensive Environmental Response, Compensation, and Liability Act (CERCLA). The act, often referred to as the "Superfund Act," imposed a tax on the petroleum and chemical industries that would fund remediation of contaminated sites listed on the National Priorities List. The Act also assigned liability to persons responsible for releases of hazardous substances which, perhaps unintentionally, created a risk sufficient to hamper development of contaminated sites for over two decades. The Clinton Administration recognized these concerns and made some adjustments to the rules implementing the Superfund Act; in 1996, the U.S. Environmental Protection Agency removed 27,000 sites from the list of potential Superfund sites and also relaxed the rules on liability for cleanup (Lee 1996).

These efforts, however, did not spur significant redevelopment. But in the early morning hours of December 20, 2002, Congress passed the Small Business Liability Relief and Brownfields Revitalization Act, and previously insurmountable challenges now became opportunities. President Bush signed the Act into law at a former steel mill in Pennsylvania on January 11, 2003, providing "liability relief for innocent developers and landowners of brownfield properties, money to do assessments and cleanups, and money to enhance state voluntary cleanup programs." The bill also provides liability relief for small businesses that were caught in the liability provisions of the original Superfund Law (Sheahan and Coley 2002).

In 1996, EPA Administrator Carol Browner recognized the important physical and symbolic role many of these contaminated sites played in communities. "We have seen so many cases where the shutdown of a plant has taken the heart out of a community. What we're trying to do is give places like that new hope" (Lee 1996). These brownfield sites were, in fact, at the physical heart of many communities. They were sited along prominent waterfronts, at significant crossroads, and in other prime locations, which is why they are in demand for redevelopment today. With their liability concerns addressed by the 2002 legislation, developers began to redevelop brownfields at a rapid pace. They became shopping malls, subdivisions, and parks; some sites were even returned to industrial use.

This redevelopment has also been supported by the EPA's Brownfields Program, which was established in 1993 with the purpose of helping communities return "abandoned or underutilized properties to productive use, clean up the environment, create jobs, and strengthen the social fabric of communities" (EPA 2006a). This program began with one $200,000 grant to Cuyahoga County, Ohio, to assess and remediate a seven-acre site. According to EPA Assistant Administrator Mathy Stanislaus, "141 jobs were created, and two sites were created for healthy new businesses. [The project] also sparked a movement to clean up and redevelop idled, underused, abandoned, and vacant properties throughout the country" (EPA 2009d).

Since that small beginning in 1993, the brownfields program has grown substantially. As of October 2009, according to Stanislaus, "the program has provided more than 2,500 grants totaling more than $600 million in direct funding to communities, which leveraged an additional $12 billion from other sources to assess, clean up, and reuse brownfields. This investment has yielded more than 54,000 jobs—many in disadvantaged communities. While these statistics are impressive, there is also a broad range of additional community-wide benefits that can result from the redevelopment and reuse of brownfield properties" (EPA 2009d).

While the economic benefits of brownfield redevelopment are impressive, more recently the EPA has also recognized the environmental and community sustainability benefits of brownfield redevelopment. According to Stanislaus, "Cleaning up contamination is vitally important to the physical health of America's communities, but putting clean land back into productive use brings with it a range of social and economic benefits that will strengthen those communities for years to come" (EPA 2009a). The EPA also argues, "Cleaning up and reusing contaminated properties can protect the environment, reinvigorate communities, jump-start local economies, preserve greenspace, and prevent sprawl. Revitalized land can be reused in ways that offer the greatest local benefit—from creating public parks and restoring local ecosystems to commercial and residential redevelopment projects" (EPA 2009a).

> Cleaning up and revitalizing brownfields inherently enhances sustainability. Through brownfields revitalization, property that was once contaminated is cleaned up. Property that was previously underutilized due to the perception or existence of contamination is restored to a higher and better use. And greenfields that may otherwise have been developed are left untouched. There are also approaches that can be integrated into brownfields revitalization to improve sustainability (EPA 2009d).

The approaches referenced above refer to sustainable development strategies promulgated by advocates of Smart Growth and New Urbanism. In fact, in its resource guides, the EPA refers readers to websites for smart growth associations as well as the Form-Based Code Institute, which is closely aligned with the Congress of New Urbanism. Furthermore, to enhance sustainability, the EPA suggests that communities "adopt principles of smart growth development" in order to revitalize neighborhoods, protect working lands and open space, keep housing affordable, and provide more transportation choices (EPA 2006a). The EPA knows that many brownfield redevelopment projects have sustainable elements, and "opportunities exist to make even greater strides in sustainable brownfield redevelopment" (EPA 2006a). Greater strides can lead to an increased ripple effect of the benefits of brownfield redevelopment.

As communities use brownfield funding to assess and cleanup sites, jobs are created, contamination is reduced, and the overall community aesthetic is increased. Additional environmental, economic, and social benefits to the community can then be realized.

Environmental Benefits of Brownfield Redevelopment

According to EPA studies, nearly 400 properties have been cleaned up using EPA brownfields funding as of 2009. Removing contaminants from former industrial sites directly impacts and improves the health of the surrounding communities and environment, as contaminants can no longer leach into surrounding air, soil, and watersheds. Developing a brownfield in an area surrounded by existing development can also increase neighborhood walkability, reducing vehicle dependence and thus improving air quality through reduced vehicle emissions (EPA 2009d). Brownfield development also prevents overuse of greenfield sites, maintaining undeveloped land around cities and urban centers, which "serve as a carbon sink, offsetting more than 10% of our nation's greenhouse gas emissions" (EPA 2009d).

Economic Benefits of Brownfield Redevelopment

Brownfields cleanup and redevelopment also attracts business and services to a once-blighted area. The EPA notes how successful brownfield development has been in revitalizing stagnant communities:

> not just at individual properties, but block-by-block and beyond. There are hundreds of examples where the clearing of environmental concerns at one distressed property paved the way for the property to return to productive reuse. We see dozens of examples where blight is reversed with regeneration—where one property's reuse spurs community-wide revitalization. Sidewalks and streets are improved. Trees and flowers are planted. New lighting is installed. A community center gets refurbished. Businesses and residents return to the area. The ripple effects can spread through the community—fear and crime rates fall, access to services and healthcare improves, property values increase, a tax base is restored (EPA 2009d).

Social Capital Benefits of Brownfield Redevelopment

Physical improvements to a redeveloped brownfield property can help redefine a neighborhood and re-establish a sense of place, in essence, breathing new life into an entire neighborhood. Brownfield redevelopment usually spurs the addition of greenspace and public open spaces which provide places for residents to gather and interact. In addition, "simple landscape and building improvements beautify a neighborhood, generate resident pride and make it a more attractive destination for activity and entertainment" (EPA 2009d).

> Increasing a sense of neighborhood cohesion also reduces crime. Community safety is improved as residents, with a greater sense of social connection, look out for one another and take pride in protecting their community. In addition, brownfield development eliminates abandoned buildings and desolate, vacant lots—thereby reducing the attractiveness of such areas for crime (EPA 2009d).

Investment in brownfield redevelopment impacts a community far beyond initial economic, environmental, and social benefits. As Jim Jones of Valley City, Alabama notes, "The momentum from our brownfields project has spilled over into the rest of the community. Our Brownfields Sustainability Pilot has opened the door for us to establish partnerships to redevelop our community and create the type of jobs our citizens need. If we can create these jobs, we can give people hope for their future here" (cited in EPA 2009d).

Moreover, neighborhoods designed to align with the principles of Smart Growth or New Urbanism can directly impact social capital. Neighbors living and working within walking distance of each together have the opportunity to form bonds and watch out for one another. Stronger relationships increase emotional health by providing a strong support network and facilitating recognition of early warning signs of depression or other factors leading to violence or suicide. The idea to create communities with strong social bonds where friends and neighbors can help one another in times of need, where there is strong social trust, and where residents feel a sense of belonging to the community. Research has shown that in communities where social capital is high:

- Children have lower levels of misbehavior and achieve better grades (Szapocznik et al. 2006)
- Marital burnout, childhood injury risk, and violence are reduced (Mannon and Brooks 2006)
- Crime rates are lower (Samson and Bartusch 1998)
- Child maltreatment rates are lower (Coulton et al. 1995)

How can the physical environment contribute to a greater sense of community? The most important task is getting people out of their cars and onto a connected network of sidewalks using sustainable densities. Empirical research has found that a 1% increase in the proportion of neighbors who drive to work is associated with a 73% decrease in the chance that any individual will report having a social tie to a neighbor (Freeman 2001). Similarly, every 10 minutes in daily commuting time reduces involvement in community affairs by 10% (Putnam 2001). Conversely, walking increases the chance that one may have spontaneous encounters with friends and neighbors while on their travel journey, resulting in more social ties.

Research has found that neighborhoods that are walkable and that include a mix of uses in sustainable density patterns have greater neighborhood cohesion and social capital (Leyden 2003). Leyden found, for example, that residents living in highly walkable neighborhoods were likely to score higher on all measures of social capital (2003). Many of the features that make neighborhoods sustainable also add psychological benefits (Biglan and Hinds 2009). Simply put, places designed to facilitate social interaction through walkability have higher levels of social support (Brown et al. 2008). In walkable neighborhoods, research has found that residents feel more connected to their community, are more likely to know their neighbors, are more likely to have faith in others, and are more likely to walk to work (Brown et al. 2008).

Part II: The Role of Housing in Brownfield Development

While brownfield sites can support a wide variety of uses, from urban parks to community gardens, the focus of this chapter is on their conversion to housing and mixed-use neighborhoods. We focus on this aspect for several reasons. First, according to the EPA, "brownfields can provide ideal locations to integrate housing options close to other services, which helps reduce vacancies, improve health, and strengthen neighborhoods" (2009d). Second, "redevelopment also provides housing opportunities for those providing local goods or services such as teachers, police officers or nurses" (EPA 2009d). Third, the EPA has found that "brownfields revitalization can help address our housing challenges because many brownfields are located in historic, older or historically low-income neighborhoods. Located near existing services and infrastructure, including transit, brownfields may offer prime locations for residential construction, higher density housing, transit-oriented development and mixed uses" (2009d). These are compelling reasons that communities use to justify converting brownfields to developments that support residential and mixed-use building types.

Another more timely justification revolves around basic economics. Given their locations within walking distance of many urban amenities, brownfield sites can offer an economic advantage over greenfield development. Supporting and enhancing walkability is a key principle of Smart Growth, of New Urbanism, and even of the emerging Leadership in Energy and Environmental Design (LEED) Neighborhood Development (ND) criteria produced by the U.S. Green Building Council. Walkability matters when it comes to household economics. A study published in August 2009 that looked at the sales of 90,000 homes in 15 markets to determine how much value was associated with walkability (as measured by a neighborhood's Walk Score) found that more walkable neighborhoods had higher property values—as much as a $30,000 relative increase in Charlotte, N.C., Chicago, and Sacramento (Darlin 2010). Similarly, during the gasoline price spike of 2005, researchers found that "distant suburbs had the largest declines in home values, while prices in "close-in" neighborhoods, typically those that were the most walkable, held up or, in a few cases, increased" (Darlin 2010).

In spite of their disadvantages, distant suburbs, however, have been the norm for residential development in the US since the 1950s. The complications of reusing contaminated industrial sites have pushed developers to cleaner and greener sites. State and federal highway policies have subsidized sprawl and encouraged low densities. Federally-backed mortgages for World War II veterans spurred suburban development. And discriminatory banking practices "red-lined" entire neighborhoods and contributed to the "white flight" from inner cities associated with the 1960s. Demographics also mattered—the Baby Boom of the 1950s and 1960s created families that desired larger yards and bigger homes that could not easily be found in the compact cities of pre-World War II America. These factors, coupled together, pushed people away from the urban cores and pulled them into suburbia (Jackson 1987; Hayden 2004; Archer 2005). And as the EPA notes, this type of residential arrangement was supported by a variety of push and pull forces:

> The Federal Housing Administration was created and its mortgage insurance programs incentivized single-family home construction in suburban areas. The post World War II economic boom spurred a dramatic increase in automobile production and large suburban developments. Rising incomes fueled the "American Dream" of a bigger home in the suburbs requiring a car for travel. Zoning was created to separate incompatible land uses. The national highway system was constructed, opening up new areas for development (EPA 2009d).

The end result has been a pattern that has privileged single-family homes in isolated subdivisions. In the US, almost 60% of housing stock is composed of single-family detached homes (EPA 2009d). The costs of this approach have been steep and are only now becoming widely known. The environmental impact of a three-car garage lifestyle is one of the significant contributors to global warming (Rome 2001; Flint 2006). The fiscal impact of extending infrastructure into the hinterlands to support suburban sprawl has been a heavy burden on many municipalities (Duan 2000; Burchell 2002). The health impact of forcing Americans into cars and off sidewalks is contributing to astonishing rates of obesity, diabetes, and asthma (Frumkin, Frank and Jackson 2004). And the social impact of isolating individuals and families in their segregated pod-like developments has diminished the social capital required to build diverse and supportive communities (Oldenburg 1997; Putnam 2000).

But the need for single-family homes in isolated subdivisions is questionable. Nationwide, household size is decreasing and more people are seeking alternatives to the traditional suburban lifestyle. In addition, as the general population grows in environmental awareness, green homes are increasingly popular with home buyers and renters (EPA 2009d). Changing demographics are intensifying the demand for more compact developments. In an America with 300 million people, Ward Cleaver's family is the minority. Married couples make up less than half of the US population. Families with children make up less than one-fourth of the population. Private yards are less relevant for many Americans today. By 2030, when the US population exceeds 400 million, the demand for more diverse living arrangements will be even greater than it is currently.

Many communities are now responding to this changing demand and are attempting to create developments in their urban cores with greater residential densities, improved transit and pedestrian access to employment centers, and increased mixtures of compatible uses. Two-hour commutes, the economic costs of owning and operating multiple cars, and the emotional toll of isolation are now pushing many people away from the suburbs. And the benefits of walkable communities that support a mixture of uses are also pulling people into new development patterns (Breen and Rigby 2005).

One substantial hurdle to such developments, however, is the limited availability of developable land with access to the amenities and opportunities of urban life. This has created an opportunity for redevelopment of brownfield sites across America. Although initially built in the 19th and 20th centuries in the far-flung outreaches of metropolitan areas, many brownfields are now located within an expanded urban core and offer easy access to places of employment and recreation. The barrier to their redevelopment, however, has been the contamination left by previous users.

Today, redeveloped brownfields support a wide variety of housing types, from apartments to high-end condominiums and rowhouses to single-family bungalows. This diversity strengthens communities by allowing people of all different backgrounds, educational levels, and income levels to live near one another. More housing options, according to the EPA, "provide opportunities for people to maintain their community ties as they move through different phases of life. Without housing choice, people are pushed into leaving a neighborhood in which they have close social ties" (2009d). The EPA (2009d) has also identified numerous other benefits to converting brownfields to residential and mixed-use developments:

- Redevelopment projects can enhance the stock of affordable housing and address the needs of special-needs residents, the elderly, and low-income residents.
- Brownfield sites, which are often quite large, can support mixed uses and infill development, which encourages more sustainable communities and minimizes environmental impacts.
- Redevelopment sites can accommodate new housing, which can replace substandard housing. The National Center for Healthy Housing estimates that 5.7 million families currently live in substandard housing, and a range of public health problems, including lead poisoning and asthma, has been linked to older housing in poor condition.
- Brownfield revitalization can improve property values—both on-site and in the adjacent neighborhoods and communities. Increased property values support a stronger tax base, which "provides additional incentives and resource opportunities for local government, public and private property owners to improve their property conditions, leveraging economic and public health benefits for owners, residents and neighbors alike" (EPA 2009d).

Social Equity and Environmental Justice

The "Brownfields Law" is one of the few environmental statutes to explicitly and directly address the concept of environmental justice. This legislation mandates that the EPA evaluate "the extent to which the grant would facilitate the identification and reduction of threats to the health or welfare of children, pregnant women, minority or low-income communities, or other sensitive populations" during the grant application process (EPA 2009b). While brownfield development may create new opportunities for home ownership and expand neighborhoods and communities, the developers must also assume responsibility for evaluating (and mitigating against) how revitalization may adversely affect residents who currently live in the area. The statute ensures that "low- and moderate-income families are not displaced following redevelopment and that communities historically plagued with blighted properties and environmental contamination reap the benefits of environmental cleanup" (EPA 2009b).

A concern over potential unforeseen consequences led the EPA, partnering with the National Environmental Justice Advisory Council (NEJAC) Waste and Facility Siting Subcommittee, to publish a report entitled *Unintended Impacts of Redevelopment and Revitalization Efforts in Five Environmental Justice Communities* in 1996. Demographic analyses of the five communities in this report showed these communities "had higher poverty rates and minority populations, and lower incomes than the national average" (EPA 2009b).

In order to promote social justice, the EPA partners with communities throughout the grant evaluation process to provide education on the critical tenets of "integrating principles of equitable development into the cleanup and redevelopment of brownfields, including the creation of affordable housing, working with minority- and women-owned businesses and environmental contracting firms, creating first source hiring ordinances, ensuring jobs with living wages, partnering with local land trusts, creating commercial linkage strategies, redeveloping brownfields into nonprofit purposes such as clinics and parks, and developing resident shareholding models" (EPA 2009b).

Through this process, communities nationwide can attest to the value of creating a more "socially, economic, and environmentally sustainable future" by "incorporating principles of equitable development and preserving critical aspects of American heritage and culturally diverse neighborhoods" (EPA 2009b).

Smart Growth and New Urbanism

Who could argue against Smart Growth? Is Dumb Growth the opposite? While the title is a bit simplistic, the theory is quite sound and recognizes that growth happens in America. Why not do it smarter? Sprawl, which consists of low-density, auto-oriented, single-use development, is cast as Dumb Growth, which is not surprising given the well-documented environmental, economic, and social costs of sprawl (Duany et al. 2000).

Smart Growth, which consists of compact, walkable neighborhoods with a mix of uses that includes housing and retail shops, emerged as a key planning theory in the 1980s and 1990s and borrowed heavily from the work of two young architects, Andres Duany and Elizabeth Plater-Zyberk. In 1978, they convinced the developer of Seaside Florida, Robert Davis, to build a beachfront community unlike any seen in recent memory along the Florida coast. Rather than build the typical beachfront condominium towers with their backs facing strip malls along the congested arterial highway, the architects had a different idea for the 80-acre site adjacent to the white sand beaches along the Gulf of Mexico. They took Davis on a tour of Florida small towns and found that these places had some common attributes worth emulating in new development. They included a complementary mix of uses, including religious buildings, offices, and a variety of housing types, from bungalows to apartments over retail shops. They incorporated an interconnected network of streets that resulted in small, more pedestrian-scaled blocks. They were centered on public open spaces like village greens and town squares. And they accommodated a greater density of people than traditional suburban development. These attributes became the guiding principles employed in the design of Seaside, which now looks more like a traditional small town than a resort community. The project was widely published in the architectural press and launched the development theory now known as New Urbanism.

However, Seaside specifically and New Urbanism in general is not without its critics (Hall 1998). Perhaps most damning is the fact that it is a very expensive resort community that offers little socio-economic diversity. The heavy-handed design and development rules that are designed to foster architectural compatibility have been the basis for critiques from the academic world as well as the entertainment world. Perhaps the most memorable Hollywood critique was a movie starring Jim Carrey as Truman, a happily married worker unwittingly trapped in a make-believe world controlled by his adoptive father, Christo, who, unbeknownst to Truman was also the director of a reality TV show called *The Truman Show* (which is also the title of the movie). It takes Truman years to find out the hoax—that he is surrounded by actors and actresses that all too convincingly play the role of his family and friends who do not want him to be burdened with the truth of his existence, which would mark the end of Christo's hugely profitable reality show. Ever since *The Truman Show*, New Urbanist developments have been greeted with suspicion.

Even The Walt Disney Company could not hide from the critics of New Urbanism. A few years after Robert Davis showed how profitable New Urbanism could be, Disney hired architect Robert A.M. Stern to plan its New Urbanist town of Celebration. This replica of the classic American small town is awkwardly placed within the sprawling borders of Orlando, Florida. Like Seaside, this new town had plenty of critics. The overt control (residents must use an approved real estate sign and even take their Christmas decorations down on an appointed date) was the subject of several stinging ethnographic studies. Census data even reveals that Celebration has largely failed to live up to its models—the American small town replete with rich and poor, black and white. While Seaside and Celebration may resemble the built environment of small towns, they do not have the socio-economic diversity that is a characteristic of many towns across the US. In fact, Celebration's median family income per the 2000 US Census was $74,231. The US average was $41,994. Similarly, these affluent developments attract a limited range of household types. In Celebration, for example, according to the 2000 US Census, 93.6% of residents were white; in comparison, the national average is 75.1%. In part, household transportation costs associated with living at the metropolitan fringe in New Urbanist developments act as a gatekeeper to socioeconomic diversity.

It is within this context of critique that advocates for Smart Growth followed a slightly different path. Rather than focus on appearances, which required the harsh codes of early New Urbanism, they would focus on underlying morphological structures that supported environmental sustainability.

Regardless of the theoretical terminology employed, the basic development idea of New Urbanism and Smart Growth is fairly straightforward. Rather than build isolated subdivisions that force residents to drive to congested commercial strip arterials, developers should build mixed-use communities that have the amenities of urban life where shops, parks, and schools are within a five-minute walk of one's home (Duany et al. 2000). Traditional suburban development separates land-use by function, often creating distinct zones for commercial, residential, and retail facilities. The consequence of this type of design is suburban sprawl—everyone needs a car or access to public transportation to move between the home, work, and recreation "zones." The concept of Smart Growth focuses on creating a sustainable form over segregating even compatible functions—mixed-use facilities that collocate places to live, work, and play, focused on pedestrians instead of vehicles. Of course, this approach to development is not new—the patterns are, in fact, centuries old. But the auto-age drove developers towards a different model beginning in the 1950s and continuing into the 21st century.

In the early days of the production of New Urbanist communities, the sites used were mostly greenfields outside the core of metropolitan areas. Seaside was certainly a greenfield. And in Orlando, designer Stern took unused Disney property on the outskirts of the city and converted it into the thriving and highly profitable new town of Celebration following the tenets of New Urbanism. In California, architect Peter Calthorpe designed Laguna West on the edge of Sacramento. Critics noted that the focus on developing greenfields was at odds with the environmental rhetoric used by the advocates of New Urbanism, who have boasted about the environmental benefits of walkable communities and greater residential densities (Ellis 2000; MacCannell 2004). But the reality is that families living in these communities must largely rely on their cars to get around. According to the 2000 US Census, in Celebration, 24.3% of families had three or more cars. By comparison, 17.1% of all American families had three or more cars.

Locating Smart Growth or New Urbanist-type developments on brownfields offers an alternative to the sprawling suburbs that are commonplace across the

American landscape and to the greenfield development that for which New Urbanism is known. While the principles for New Urbanist developments on greenfields and brownfields are remarkably similar, the contextual setting of the latter more easily connects residents to a broader array of urban amenities. New Urbanist developments on brownfield sites can marry the internal benefits of a connected, walkable, and denser neighborhood to the external benefits and locational advantage of these former industrial sites, many of which are conveniently located in the heart of major metropolitan areas.

Part III: Case Studies in Sustainable Brownfield Redevelopment

In this section we present four cases studies that have applied principles of Smart Growth or New Urbanism to the redevelopment of brownfield sites. The first two cases are examples of developments built on former industrial sites following the federal legislation in 2002 that reduced liability of developers working on brownfield sites. Built between 2002 and 2006, Glenwood Park in Atlanta, Georgia and Hercules in San Francisco, California's Bay Area show how brownfields can accommodate single-family homes, multi-family apartments, appropriately scaled retail stores, and other commercial uses. They also are examples of the public and private partnerships required to cleanup the sites and codify the new planning model. The next two cases (Atlantic Station in Atlanta, Georgia, and Bay Street in Emeryville, California) demonstrate how brownfield sites can function as alternatives to the suburban shopping mall. In the latest phase of retailing one-upmanship, shops are relocating from enclosed shopping malls to what have been described as lifestyle centers—places devoid of the traditional anchor store and the enclosed pedestrian mall. These new shopping extravaganzas are open-air venues with on-street parking, wide sidewalks, and occasional street trees. Many even have housing above the shops in a manner reminiscent of the traditional and much-loved main streets of the 19th and 20th centuries.

Glenwood Park, Atlanta, Georgia

Glenwood Park in Atlanta, Georgia opened in 2005 and replaced a defunct cement plant in midtown Atlanta. During construction of the $165 million project to create a mixed-use neighborhood, developers recycled 800,000 lbs of granite rubble blocks and 259 million lbs of concrete (60,000 cubic yards)—enough to cover one square acre of land with concrete 36-feet deep. In addition, 250,000 pounds of metal was removed from the site, and 30 million pounds (or 41,500 cubic yards, equal to one football field 30 feet deep) of wood chips were converted into energy in a waste-to-energy plant, producing enough electrical energy to power 900 average size homes for an entire year. The vehicular emission reductions are also impressive—Glenwood Park residents drive over 1.6 million miles less than those in regionally-comparable driving patterns. This reduction equates to removing over 100 cars per annum from Atlanta roads and saves over 54,250 hours per year (glenwoodpark.com).

Glenwood Park's developer, Charles Brewer, envisioned a pedestrian-centric, walkable community exemplifying the latest environmentally-sustainable design and construction techniques. At a density nearly four times greater than the typical suburban ratios, Glenwood Park is a place where residents can both live and work. On the 28-acre site of the abandoned concrete plant two miles east of downtown Atlanta, Glenwood Park has been rebuilt as a traditional town, with several parks, 50,000 square feet of retail space, 20,000 square feet for offices, and 350 single-family homes, apartments, and condominiums. Glenwood Park applied many of the recommendations of the EPA's Brownfields Program, which has led to numerous recognitions. Techniques used included sustainable site selection, water management, planning and design, preservation landscaping, community involvement, and green building (Benfield 2009).

Glenwood Park is designed around a classic main street, lined with street trees and on-street parallel parking spaces, which focuses on a town square flanked by multi-story buildings.

Garages are located off alleys and the buildings define comfortable outdoor rooms.

Figure 7.1   Glenwood Park, Atlanta. Main street design

Figure 7.2   Glenwood Park, Atlanta. New urbanist alley design

The namesake Glenwood Park serves as a storm water retention area and popular playground. Unlike typical suburbs, with their predominance of single-family homes, Glenwood Park's housing variety supports a broad diversity of residents of varied ages and income levels. Glenwood Park also benefits from easy access to MARTA and a planned light rail line to minimize auto-dependency and create a truly vibrant, pedestrian-oriented environment.

Bringing the Glenwood Park vision to fruition has not been without its challenges, however. Developers argued with city officials about street widths and corner radii, as they felt "narrower streets and tighter corners were crucial to the plan's success." Ultimately, it was the partnership with elected officials which led to the passing of a new city ordinance, allowing exemptions in set street dimensions for "qualifying traditional neighborhood developments" (Staff Report 2005). Developers also struggled to obtain approval to fix sewers along a drainage ditch on site, due to confusion over water jurisdiction, and wrestled with the Georgia Department of Transportation (GDOT) about the planned main street. It quickly became clear that GDOT was not in favor of the main street design, which created a central corridor flanked with on-street parking and canopied with street trees. Green Street had to lobby both GDOT and the City of Atlanta for a jurisdictional transfer, which, after a lengthy delay, was granted.

Despite some of these unforeseen challenges, Brewer credits the support of the local community, a strong relationship with permitting authorities, and an insider financing approach for enabling the development to proceed without the need to make a lot of compromises in the vision. Early on, developers reached out to members of the surrounding community and sought their input in the planning process, particularly around the idea of bringing more retail to the area. The positive relationships led to great political support, especially during the permitting process. In fact, developers note, "the cordial relationship with neighbors has been one of the most gratifying outcomes of the development experience" (Staff Report 2005). The developer adds that prioritizing green design and construction wasn't always easy—"It's hard to make (New Urbanism) a reality on the ground" (Woods 2004).

Hercules, California

Hercules, California is a community of nearly 23,000 residents, located approximately 16 miles north of Oakland on the bay side of Interstate-80. Once the home of the greatest production of TNT in the world, and one of the largest-growing suburbs of San Francisco in the 1980s, Hercules is a premier example of Smart Growth applied to a massive brownfield site. Hercules is named after its roots as the site of the once-largest plant of California Powder Works (CPW). In 1917, CPW produced over 1.7 million pounds of its name-brand "Hercules" black dynamite powder. As the need for dynamite decreased, the Hercules plant was converted to a 1,300-acre fertilizer factory. Methanol, formaldehyde, and urea formaldehyde were manufactured in Hercules until overseas competition forced the plant to close its doors in 1967.

Over time, the outlying industrial property (which had effectively served as a buffer zone during the heyday of dynamite production) was sold to developers eager to meet the growing demand for residential housing in the rapidly-expanding Bay Area, and Hercules became a true "bedroom community." During the work day, the town emptied as commuters traveled into San Francisco for work. The 167 acres of land on which the plant's main operations once stood, however, were a contaminated, undeveloped brownfield along the industrial waterfront area. The anticipated cost of remediation left the land to lay fallow for decades. Hercules Property, Inc. purchased the land in 1976 but did not make any improvements until 1985, when the Department of Toxic Substances Control issued a determination of Imminent and Substantial Endangerment, and ordered the previous owners "to remove machinery, debris, and materials from the site" (Brownfield Revitalization Success Story 2005). Twelve years and a $12 million remediation effort later, over 60,000 tons of soil contaminated with lead, zinc, nickel, chromium and petroleum hydrocarbons were removed. Developers were left with a 156-acre parcel that met environmental and health standards for residential construction, and an 11-acre parcel suitable for commercial and industrial use.

Hercules had a unique opportunity. As David Sargent, of Sargent Town Planning remarked, "How many cities are there that have an existing, solid population base; a location close to a thriving urban metropolis, and a 400+ acre blank spot in the middle?" (Hercules Handout 2006). Fortunately, not only did Hercules have what Sargent referred to as an opportunistic "accident of topography and history," but also had a team of city planners and council members committed to creating a community vision for development.

The city knew that "without a true center to anchor it and without a vision to guide its development, the commercial heart of Hercules would gradually become just another anonymous convenience retail node along I-80" (Plan for Central Hercules 2000). In June of 2000, the city of Hercules hired Dover-Kohl and Partners to conduct a collaborative, 10-day design charrette to create a plan for the development of the entire 426-acre site. The developers and city planners were committed to reflecting the desires of the community's residents and, in fact, over 400 participated in the design process. Mike Sakamoto, Hercules' Director of Community Development, shared anecdotally that, post-charrette, "people came up to [me] in the grocery store to discuss the plan." Ultimately, "residents want the city to dictate to developers and not vice-versa." (Plan for Central Hercules 2000). The cycle of collaborative feedback generated enormous public and political goodwill—not surprisingly, the Planning Commission and City Council unanimously passed the resulting Plan for Central Hercules with tremendous support and a deep community commitment.

In a perhaps unprecedented move, the cost of the charrette and plan code documents, totaling nearly $300,000, were split between the city and the developers, Catellus Development and Bixby Land Company. This "spirit of cooperation" and collaboration has pervaded the project since the onset (Smith 2007). Through the charrette, five key design goals emerged to guide development: residents explained that they wanted a coherent, unified vision for central Hercules development; creation of neighborhood centers; and interconnected living, shopping, and employment. In addition, the community committed to maximizing the natural resources of the local area, and creating a pedestrian-oriented, walkable development.

The *Plan for Central Hercules* identifies four areas for development— Waterfront, Central Quarter, Hill Town, and Civic Center/Hospitality Corridor. The Waterfront district consists of a mix of varied-density living units, including single-family homes as well as apartments, all within walking distance of the Historic New Town Center.

The town center is the core of the plan, and provides 160 units of living and working space, along with retail, entertainment, services, and restaurants.

Figure 7.3   Hercules, CA. Waterfront district

In order to support the planned development, the city council has changed the building code and development regulations from "proscriptive (describing what you can't do in page after page of dense text) to prescriptive (describing what buildings fit the plan's intentions)" (Temple 2002a). This type of design code is called Form-Based Code, and it enables people actually to envision the built environment prescribed by the code. Daniel Parolek of Opticos Design, Inc. works for the firm hired by Hercules to oversee and coordinate all of the final design work. "You need to define the character and quality of the public spaces and streets," Parolek said, "and then allow uses to adapt and grow within that framework." Ultimately, according to Parolek, the Form-Based Code pays off by creating "consistently high-quality design of buildings and neighborhoods with day-to-day amenities within walking distance of a variety of residences" (Smith 2005).

Steve Lawton, a longtime Hercules resident, a city planner since 1996, and the Director of Economic and Community Development since 2001, has consistently championed principles of New Urbanism. A member of the Congress on New Urbanism, Lawton was instrumental in bringing design elements such as "a network of human-scaled, walkable streets, forming a pattern of blocks; streets that are framed by buildings, creating well-defined public spaces, faced by doors and windows; and a mix of uses, vertically and horizontally, with everyday needs within a few minutes' walk from home" (*Plan for Central Hercules* 2000) to Hercules, making the town the first in California to adopt a design code based on New Urbanism (Temple 2002a). Like many New Urbanist developments, garages are located off alleys.

This is the view from a Capitol Corridor train window that first attracted Patrick and Lita Tang to Hercules. The cluster of homes they saw that day led to a visit to Hercules, which eventually led to them selling their home in the Oakland Hills so that they could move into one of the new neighborhoods in Hercules. Lita mentions the architectural details of the homes as a key attraction, and was excited not to be in a "typical suburban tract, filled with white stucco homes with front-loaded garages." The views of San Pablo Bay and the "promise of urban convenience" also appealed to the Tangs. "In the near future," Lita notes, "we'll be able to walk literally across the street to a convenience store, restaurants, and transportation connecting to other parts of the Bay Area." Her endorsement could not be more enthusiastic. "It couldn't have been better," she said. "We were instantly sold when we drove up to the development" (Smith 2005). Hercules predicts its vision will bring additional residents like the Tangs—nearly 10,000 are anticipated over the next decade (Adamick 2006).

Figure 7.4   Hercules, CA. New urbanist alley design

From Brownfields to Lifestyle Centers

While Glenwood Park and Hercules show how once-contaminated sites can be successfully converted into medium-density New Urbanist-type developments, brownfield sites can also function as high-density alternatives to the low-density suburban shopping mall. These "lifestyle centers" mimic many aspects of the traditional American Main Street (Southworth 2005). The International Council of Shopping Centers (ICSC) defines a lifestyle center as an open-air shopping mall with at least 50,000 square feet of upscale/specialty retail (Scholl and Williams 2005). Numbers from the ICSC show that these centers are fast replacing the traditional enclosed shopping mall. In 2006, developers built one enclosed mall. By contrast, developers built over 60 lifestyle centers in 2005 and 2006 (Herrick 2006).

Michael Southworth (2005: 152), a professor of Landscape Architecture and Environmental Planning at the University of California, Berkeley argues that "people are experiencing mall saturation or 'mall fatigue' and have become bored with the inwardly-focused, disconnected and placeless suburban shopping center. Instead, people want to be in a town center or a place that has street life, a sense of place and community. In short, people are rediscovering the idea of Main Street." Southworth (2005: 155) suggests that, "Developers have come to realize there is something about main street that people want and if malls are going to survive they are going to have to have some of the features of main street." Despite this desire for a Main Street, the earliest lifestyle centers were no more than high-end strip malls designed to look like main streets, with up to 400,000 square feet of "unanchored" retail and a vast parking lot in front (Ritter 2004).

Architecture critic Margaret Crawford (1992: 26) argues that, "In this overcrowded marketplace, imagery has become increasingly critical as a way of attracting particular shops and facilitating acts of consumption." The image of Main Street sells, and developers are using it to tap into $2.01 trillion that Americans spend each year at shopping centers (Scholl and Williams 2005). In considering these places, the notion of "lifestyle" is important. The "lifestyle" these centers are responding to, according to lifestyle center developer Yaromir Steiner, is one with more pressures, less time, multi-tasking, and the desire for greater convenience, which results in a demand for integrated experiences that collocate shopping, dining, entertainment, and even housing (Steiner 2005).

For the purposes of this chapter, we consider lifestyle centers open-air shopping venues that mimic the cherished main street aesthetic. Like yesterday's main street, lifestyle centers under this definition include housing above ground-floor shops, a variety of restaurants, and even multi-screen movie theaters. Their "main street" has parallel parking, street trees, and wide sidewalks. While the new main street has an emotional pull, it also has an economic advantage over traditional shopping malls. For instance, Common Area Maintenance (CAM) charges at Bayer Property's lifestyle centers average less than half the norm for enclosed malls ($6 per square foot vs. $15 per square foot) (Hazel 2003). Construction costs are also lower with the open-air model. After all, paving a street with asphalt is less expensive than paving a pedestrian mall with ceramic tile. Moreover, sales per square foot are typically higher at lifestyle centers ($298 vs. $242) and the most popular centers, like Easton Town Center outside of Columbus, Ohio, have sales that can easily top $550 per square foot (McLinden 2006).

Perhaps the most striking difference between lifestyle centers and older shopping malls is the new trend of including housing above the shops. Even Walt Disney longed for housing along his main street. "Disneyland's towns were not enough to satisfy Walt's longing," according to Joe Flower (1996: 43). "When he opened the gates to Disneyland and its Main Street in 1955 he knew that it had a flaw: it wasn't real. Nobody actually lived there. He wanted to build a place where people lived." At least Walt himself could live in his namesake theme park—his favorite place to watch the activity on Disneyland's "public square" was his own apartment above the firehouse on the Main Street U.S.A. (Francaviglia 2006).

Placing housing above shops is nothing new. Jeannine Stein (2001) of the *Los Angeles Times* reminds us that, "Living above retail or office space is an old idea that's recently been dusted off, recycled, and redesigned. People have been residing on top of bakeries, butcher shops and tailor's studios for centuries, forging communities in small neighborhoods." Because lifestyle centers do not need the traditional department store anchors associated with suburban shopping malls, they can be built in a wider variety of areas. For example, developers can place lifestyle centers in much smaller areas than traditional regional malls (Scholl and William 2005), making them more adaptable to urban brownfields. Atlantic Station and Bay Street are examples of this approach and they demonstrate how major commercial development can occur on in-town brownfield sites rather than on greenfield sites in the suburbs.

Atlantic Station, Atlanta

Atlantic Station, located in Midtown Atlanta, Georgia, employed smart growth principles to create a live-work-play development on the site of a 138-acre former steel mill. Its pedestrian-friendly neighborhoods and mix of residential, retail, office, and recreational uses allow residents to walk to restaurants and entertainment and use mass transit or walk to work (EPA 2006a).

Atlanta has a storied history as a genteel city, made famous by both its Civil War involvement and Atlanta-resident Margaret Mitchell's grand portrayal of Southern life in *Gone with the Wind*. As the third-largest city in the United States, Atlanta is desperately trying to shed the perception of merely being one of the "great cities of the South," and cement its reputation as one of the most vibrant, progressive urban metropolitan areas in the United States. In some ways, it is hard to understand why this has been such a struggle—Atlanta boasts professional teams in baseball, hockey, football, and basketball; has Hartsfield-Jackson International, the busiest airport in the world; and is home to the world's largest indoor aquarium, the renowned Georgia Aquarium. The third-largest concentration of Fortune 500 company headquarters in the United States, including Coca-Cola, Delta Airlines, CNN, UPS, and The Home Depot, also call Atlanta home.

Despite all of these advantages, however, Atlanta is also known for crippling traffic on the perimeter, Highway 285, which circles the city and connects to the north/south thoroughfares of Interstates 75 and 85. Atlanta traffic can be blamed on its development as a suburb community (historically, housing was built in the communities ringing Atlanta, as opposed to developing more urban housing options in the downtown core), as well as a less-than-adequate public transportation system, the MARTA, which does little to ease commutes.

Atlantic Station, just north of Atlanta's downtown, was built in part as a response to the sprawl, congestion, and lack of a sense of place that is so much a part of metropolitan Atlanta. It is part New Urbanist development and part lifestyle center with the typical high-end franchised shops.

At over $2 billion, the project was, at the time development commenced, the largest brownfield cleanup in the United States. With support from the Environmental Protection Agency, the U.S. Army Corps of Engineers, and the State of Georgia, contractors removed 165,000 tons of contaminated soil that required about 11,000 dump-truck loads (Link-Wills 2007). Upon completion, the 138-acre project will have 11 acres of parks, 2 million square feet of retail, 6 million square feet of offices, and up to 5,500 homes, ranging from single-family bungalows to small studios. Average residential density will be roughly 50 units per acre. Much of this development is on top of a 30-acre two-storey underground parking garage that caps the contaminated site.

Like nearly all brownfield redevelopments, Atlantic Station is an example of the public and private partnerships required to make the deals work. Demonstrating infrastructure subsidies are not only phenomena of the suburbs, the Atlanta City Council issued a $75 million Tax Allocation Bond in 2001 to pay for the first phase of infrastructure development. The 17th Street Bridge was completed in 2004, providing a much-needed connection to Midtown Atlanta and leading the way for subsequent development of the site. Atlantic Station shows us that infrastructure subsidies are not only phenomena of the suburbs.

Atlantic Station broke ground in 1999, as a joint venture between AIG Global Real Estate Investment Corporation and Jacoby Development. It took one year to take the existing buildings down, one year to clean up the site, and one year to put in $300 million in infrastructure (Link-Wills 2007).

Figure 7.5   Atlantic Station, Atlanta. New urbanism combined with
              lifestyle center

The project consists of a "live-work-play" environment subdivided into three main areas—The District, The Commons, and The Village. The District is the heart of the mixed-use community of Atlantic Station, with office, retail, and residential space. Atlantic Station boasts over six million square feet of Class A office space. The first office tower constructed was the first LEED Silver-certified high-rise office building for the core and shell category in the United States. Most of Atlantic Station's two million square feet of retail and entertainment space is located in The District's core, and shops and restaurants were designed as multi-story mixed-use buildings—with space for residential lofts above ground-floor retail that framed pedestrian-scaled main streets.

Restaurants line a "central park"—part of the 11 acres of public parks and greenspace throughout the development. The Commons is the primary residential area in Atlantic Station, and is a mix of single-family duplexes, detached homes, townhomes, condominiums, apartments, hotel rooms, and lofts arranged around a two-acre lake and connected with linear parks. The Village consists of apartments and lofts located on the far west side of the development, adjacent to a smaller retail center.

In September of 1999, Atlantic Station's developers signed a project implementation agreement with the Environmental Protection Agency (EPA). In this agreement, a series of commitments were made around aggressive environmental performance targets, and Atlantic Station has exceeded nearly every one. Atlantic Station's goals of creating a high-density, pedestrian and transit-oriented design are on track. Development is on pace to accommodate over 12,000 residents and employees (currently at 6,500) at an average greater than 180 persons per net acre in the quarter-mile radius around transit stops. Over 33% of blocks in the final site design were to be mixed-use but the number is currently over 50% (Project XL Report 2008). Survey results demonstrate that Atlantic Station is exceeding targets for mode splits and daily vehicle miles traveled (VMT) per resident.

In addition to the environmental benefits of decreasing VMTs, studies are also underway to determine the effects of the built environment in terms of changing behavior around physical activity. Collaboration between Atlantic Station, the Centers for Disease Control and Prevention, and Emory University's Rollins School of Public Health began in 2007 to study the effects of mixed-use, pedestrian-oriented neighborhoods on quality of life among residents. Ultimately, researchers are trying to determine whether people will be more physically active and drive less after moving to a mixed-use residential neighborhood.

Figure 7.6   Atlantic Station, Atlanta. Commercial with upper apartment
            lofts

This would be supported by Atlantic Station's free trolley shuttle, which runs every five minutes and transports passengers from retail to office to residential areas. The free trolley carries 850,000 people a year in its own dedicated lane (Link-Wills 2007). Bike lanes line the streets, and a network of interconnected sidewalks provides pleasant, safe, comfortable walks from one part of the community to another. The research will determine whether or not residents take advantage of these options and adopt a healthier, less car-centric lifestyle, given the access to viable alternatives. Research participants will submit data about their physical activity and transportation choice habits before moving into Atlantic Station, and will document lifestyle changes after living in the community for a year.

For all its innovation and success, however, Atlantic Station is not without its critiques. Due to the site development challenges regarding contamination, developers created a 30-acre, two-storey parking garage and effectively built a new "city" on top.

Figure 7.7   Atlantic Station, Atlanta. 30-acre parking garage

Andres Duany, an architect and urbanist with Duany Plater-Zybeck & Company (DPZ), is a fan: "the way the parking is handled in the massive underground platform is exceedingly clever, as it coincides with the street grid above and thereby retains wayfinding and simplifies the structural systems. It is a great way to handle ultra-high density development, as the parking is integral with the infrastructure, as it should be" (Miller 2006). Georgia Tech University's Ellen Dunham-Jones is not so sure. "It's bizarre. You have a walkable, mixed-use, urban place, but it's two stories above the ground on three sides. The edges of the deck have not been closed in, so right now from a certain views it appears a bit like a city on an aircraft carrier" (Miller 2006).

Aesthetics aside, this is where most of the criticism of Atlantic Station stems—in the combination of form and function, in the struggle to apply theoretically sound ideas when faced with the messy challenges of reality. Duany comments on the "banal architecture" of the large commercial buildings and the "abysmally designed" townhouses, which have more of a suburban vs. urban look and feel. He also notes that the human scale is off, that most of the "streets and open spaces are too large, and the major roads are too speedy." Dunham Jones criticizes the connections: "All edges of the project and the interconnections are still undercooked, and I wince at many of the architectural and urban streetscape details" (Miller 2006).

Developers have admittedly struggled with connections, both internal and external. In early stages of the project, an idea to have the Georgia Department of Transportation build a bridge across Interstates 75 and 85 at 17th Street to link Atlantic Station to Midtown Atlanta was vetoed by the EPA due to a lack of city compliance with air quality laws (Link-Wills 2007). The developers hired DPZ to improve the plan from a sustainability perspective and attempt to build the case that Atlantic Station should qualify as a transportation mitigation effort by bringing 10,000 people to live and work adjacent to Midtown. The bridge was finally completed and, although successful in providing a direct transportation corridor to the heart of Midtown, it is still vehicle-oriented. As Jason Miller describes, "the bridge itself is a problem, since it enters the property 40 feet in the air, and visually discourages pedestrian traffic (even with its dedicated bike lanes and sidewalks on both sides). The 130-foot-wide traffic-mover crosses 12 lanes of traffic and is a daunting structure for people on foot" (Miller 2006).

Brian Leary, Vice President of Design and Development for Atlantic Station, was working on a masters' thesis at the Georgia Institute of Technology when the original ideas for Atlantic Station's development were in their infancy. He was able to get an audience with the developers, who had had "much success building conventional suburban big-box projects," and whose initial plan, Leary says "was rather forgettable" (Miller 2006). Impressed by Leary's foresight, the developers hired him to see the project through to fruition. "We started focusing on creating places for people. We traveled all over looking for examples of good urbanism, trying to unravel the DNA of good places—why they work," said Leary. He overcame the Georgia Department of Transportation's (GDOT) initial objections to the concept by embarking on a joint venture with the Environmental Protection Agency (EPA) and DPZ to hold a three-day charrette and create more opportunities for innovative, sustainable design within the plan (Miller 2006).

One explicit goal of mixed-use developments and "good urbanism" is the idea of creating a sense of community and increasing neighborhood connections. According to John McIlwain, a senior resident fellow for the Urban Land Institute, "Moving to a mixed-use development, a small town, or seeking an urban experience are all elements of the same thing: It's a community where you get to know each other. You're walking around, and you get to know your neighbors, you get to know the shopkeepers, because you meet them on the street" (Greene 2006).

Troy and Mary Kolly were one of the first couples to buy a condo at Atlantic Station but are not convinced that it is a place for building community. Troy notes, "It doesn't really feel like ... a neighborhood. I mean, I don't even know if I thought that was going to be the concept. I don't know my neighbors really well. And do I know the retailers? Well, I guess as much as one would know a local GAP clerk." Their condo is on a double-loaded corridor in a maze of a building. Residents frequently guide lost visitors around hallways to find the correct elevator to access various apartment units, which follow a confusing numbering system. It is hardly an environment conducive to interaction and socialization—in fact, Mary mentioned that the common areas are hardly used, and commented "it feels more like living on top of a shopping mall than in a real downtown."

A sense of community and neighborhood cohesion may be difficult to achieve in Atlantic Station, given both the scale of the community and its ties to suburbia with the big-box retail model. Many townhome and single-family units are located along busy ingress streets and are not designed to capture a neighborhood feel. Shopping at Atlantic Station's two-storey, 366,000 square foot IKEA (the first one built in the Southeast United States) can also be unpleasant—the crowded, overwhelming, and capacity-overflowing shopping experience is definitely not the idyllic model of a mixed-use development creating ideal places for people.

The jury is still out on Atlantic Station. Developers have, through significant efforts, overcome a myriad of political and economic challenges to transform an abandoned, contaminated steel mill into a model mixed-use community and lifestyle center. Thus far, Atlantic Station has far exceeded environmental sustainability goals, increased connections to the city, and provided viable housing options to revitalize Midtown Atlanta. It remains to be seen, however, whether Atlantic Station will eventually be seen as a successful, new urban, brownfield redevelopment—or a place where form never really met function.

Bay Street, Emeryville, California

Like Atlantic Station, Bay Street in Emeryville, California is an example of a mixed-use lifestyle center built on top of a contaminated site. To build Bay Street, the city spent over $36 million to acquire the site's five parcels (which in some cases required eminent domain) and clean up the site (Newman 2004). The city cleaned the land even before any developers had shown interest in the property. According to Ignacio Dayrit, a redevelopment specialist for the city, the ground was full of arsenic, pesticides, and petroleum products (Associated Press 2001). The site was, at various times, home to makers of paints, insecticides, and sulfur. After winning judgments against the former property owners, the city recouped much of the money it invested in cleanup and decontamination. Emeryville's economic development director, Pat O'Keefe, anticipates that $1.3 million in property taxes and $900,000 in sales taxes will flow into the city's treasury every year (Freed 2002).

According to O'Keefe, the city could not advance the retail idea without getting control of the site and cleaning it (Newman 2004). While the city's investments caused little controversy, the potential development of the site itself generated enormous discussion and, in some cases, opposition. As is the case with many urban sites, this one has had multiple users with a vested interest in future development.

Long before the paint plant, the site was home to one of the 10 most sacred burial sites for California's Ohlone tribe. Over the last century, more than 1,000 remains have been excavated at the site (Artz 2002). Any remaining vestige of the 2,700-year old shell mounds at the site, which were used in part to mark Indian burial sites, had to be removed for construction of the upscale shopping center. Sandy Sher, a member of the Emeryville historical society said, "To build on the site when you know there are still bodies there and to drive piles through the bodies—it's very offensive to the Native American descendants" (DelVecchio 2002). To appease the protestors, the city spent $2 million to commemorate the previous use. This included placing a community room in the project, painting a mural on one of the project's blank walls, adding commemorative landscaping, and producing an interpretive web site. But the mural for mall trade-off has upset many activists. According to Kathy Perez, the Native American monitor of the project, "They think they need to do a mural so they appear like they're working with Native Americans and everything is hunky-dory" (Artz 2002b). In 2005, some of the opponents of Bay Street, including Indian leaders and Buddhist monks, began a three-week Peace Walk in the area to highlight the impact of development on sacred grounds (Lochner 2005).

Despite the ongoing protests over the shell mounds, the upfront gamble has certainly paid off for the city. After the cleanup, Madison Marquette purchased the property and began developing it into a lifestyle center oriented around a new main street. Placed on top of the once heavily-polluted but now reclaimed industrial land, Bay Street is to Emeryville (California) what Main Street is to Walt Disney: a retail strip designed to encourage consumption.

Bay Street is a 1,200-foot long, 3-block version of Main Street and includes two to four levels of housing on top of ground floor retail. Its developer calls it a "neotraditional main street block" (DelVecchio 2002b), with a narrow namesake street, on-street parking, storefront windows, and heavily detailed facades completing the main street image on the privately controlled side. But its parking garages and vacant walls turn a blank face to the real public realm.

Nearly 2,000 parking spaces are placed either behind the retail buildings or sandwiched between the retail and residential floors.

The 400,000 square foot retail development, along with its 379 residential units, occupies a narrow strip of land between Interstate 80 and a heavily-used railroad corridor. Its nearest neighbor is the 247,000 square foot IKEA. Like other lifestyle centers, Bay Street has the requisite retailers (e.g. Ann Taylor Loft, Banana Republic, Gap, Talbot's, Williams-Sonoma), a movie theater (16 screens), a bookstore (two-level Barnes and Noble), and 11 restaurants (including the staple of many lifestyle centers, P.F. Chang's China Bistro). But in an odd departure from other lifestyle centers, many of the restaurants cluster around a second-level outdoor food court that feels more like a shopping mall than a main street.

Figure 7.8   Bay Street, Emeryville, CA. Neotraditional main street

Figure 7.9   Bay Street, Emeryville, CA. Parking design

The Jerde Partnership designed the multi-level main street, which according to one observer seems "as if the builder had lifted a street and its flanking buildings from another city with a crane and set it down in the middle of Emeryville" (Newman 2004). Jerde is no stranger to the world of faux main streets. Jon Jerde and his firm have designed dozens of main streets, lifestyle centers, and malls around the world, including San Diego's Horton Plaza and Minnesota's Mall of America. According to the firm's promotional material (Jerde Partnership 2009):

> Jerde Placemaking reinvents the authentic urban experience that has often been lost by modern planning. The world's great cities evolved naturally over centuries, their town squares, streets, and public marketplaces serving as commercial and social centers. The organizations and forms of early cities organically grew out of the natural pedestrian paths people used to move into, about and through them, and these patterns informed the cities' distinct characters, particular forms and mix of uses.

In Emeryville, this focus on the pedestrian experience had to be "reinvented" since it was never the precedent in this industrial city. Easy access to all modes of transportation helped make Emeryville one of the longest-operating industrial areas in the Bay Area. But its location across the bay from San Francisco, adjacent to the heart of Oakland, and next door to Berkeley and its University of California campus, has led to a remake of the gritty city. In the 1990s, power centers, strip malls, and enormous destination retail outlets became the new industry. Then, starting in 2002, condominiums and lofts began replacing paint factories. Bay Street's residential units have been especially lucrative investments. In 2006, the 917 to 1,149 square foot one-bedroom, 1½-bath units ranged in price from $650,000 to $730,000 ($635 to $708 per square foot). Larger two-bedroom, 2½-bath units (1,205 to 1,317 square feet) sold for roughly $735,000 to $750,000 ($569 to $609 per square foot). This compares to an average square foot sales price for US homes of just $135 per square foot.

Bay Street's rental units also command a premium. Rents for one-bedroom (730 square feet) units start at $1,750 per month and larger two-bedroom apartments (1,210 square feet) are renting for up to $3,000 per month (Anders 2006).

Reaction among local architects has been mixed. Greg Van Mechelen of Berkeley-based Architects, Designers and Planners for Social Responsibility argues "It is certainly better than big-box development, but it creates a feeling of artificiality because it isn't a real street that grows organically over time" (Artz 2002a).

Ironically, the place has grown somewhat organically. It has taken nearly four years to complete. The initial housing developer backed out and left the upper levels incomplete for several years. A new developer recently finished the for-sale housing and rental units—five years after the initial opening. In the words of P. Eric Hohmann, Madison Marquette's vice president for acquisitions and development, "This is the antimall. We would like to think that Bay Street is currently the next thing in the evolution of the shopping center" (Newman 2004). Hohmann adds, "Our goal is to make it look like a public street. If you think it's a private street, we've failed in our design intent" (DelVecchio 2002b). But if one scans the informational kiosks set along Bay Street, the long list of rules and regulations makes it clear that this is not a true public space. With its curfews, dress codes, and behavior standards, Bay Street represents the new face of "public" space.

Bay Street and Atlantic Station appear to have started as Smart Growth developments, but their locations, bordered by big-box retail, strip malls, and highway corridors, have kept them disconnected from the urban core. Their parking strategies and their inwardly-focused layout reinforce this disconnect from the public realm. This lack of continuity in an urban core creates an "island" effect, similar to the enclosed shopping mall, despite the designs as "Main Street" type lifestyle centers. In order to walk its walkable streets and enjoy dining in one of the numerous street cafes, non-residents of these developments still have to navigate crowded multi-lane arterials and find a space in the multi-storey garages.

Conclusion

The cases presented here demonstrate the potential of brownfield redevelopment within the context of providing alternatives to suburbia. In fact, former brownfields can support the same types of uses found on greenfields across the American landscape. Single-family homes, apartments, and condominiums can be successfully built on brownfields. Shopping centers, offices, and the chain restaurants that populate suburban shopping malls can also find profitable settings on redeveloped brownfields.

While the cases presented here represent improvements to typical suburban development patterns, they also have limitations—there are many practical issues and problems associated with successful brownfield redevelopment. New Urbanist developments may lead to gentrification and lack of affordability. Lifestyle centers cater to a wealthy clientele and turn their backs on the public realm. While the enclosed mall is dead, and replaced by a faux main street, this is not the typical public street—it is a private street with rent-a-cops and closing times. Large retail floorplates, parking ratios sized for the suburbs, and monotonous rows of stacked flats are the new reality on these "main" streets. And how do you connect land-uses that were intentionally disconnected? Unfortunately, the cases here provide little guidance. Atlantic Station relied on a $76 million bridge over an interstate, but its elevated main street is an odd reality. Bay Street simply ignored the larger city fabric and became an inward-looking development. Hercules contends with Interstate 80 as a barrier. Perhaps Glenwood Park offers the best lesson because it was the least disconnected. By simply extending the existing street grid into the new development and building housing to the edge, the new and old flow seamlessly together.

While the national housing depression has not affected metro Atlanta and the Bay Area's more urban settings quite as badly as many other regions, these areas have followed national trends in one important respect: homes in the center of the region and close to job centers have not suffered as much in terms of price declines. But these urban developments are not immune to the housing crisis. In the spring of 2009, at the posh W Hotel in Downtown Atlanta, 40 condos in Atlantic Station's Elements Tower were auctioned off—their original deals had collapsed. The auctions sparked a controversy when existing owners complained that their own property values would be diminished. But the auctioneer, Jon Gollinger, made a good point when he countered that a half-empty building is not good for property values either (Duffy 2009). The auction went well—175 registered bidders showed up, which shows there is a strong demand for in-town living and that prices may be correcting—they went for about two-thirds of the peak price.

Despite these recent price drops, the trend towards living in walkable neighborhoods, close to the amenities of urban life, will only make many brownfield sites more attractive to mixed-use developments, given their typically prime locations within urban areas.

Christopher Leinberger (2008), a professor of urban planning at the University of Michigan, has found that urban housing carries an enormous premium—between 40 and 200% over traditional suburban development. He argues that, "People are being drawn to the convenience and culture of walkable urban neighborhoods across the country—even when those neighborhoods are small. Demographic changes in the United States are also working against conventional suburban growth, and are likely to further weaken preferences for car-based suburban living" (Leinberger 2008). In fact, by 2025 the US will have equivalent numbers of single households and households with children. These households will help drive the demand for a new type of model that can be accommodated on redeveloped brownfields.

Brownfield sites also offer some distinct advantages for developers interested in creating more sustainable development. Given that many of these sites are within the core of metropolitan areas, they are better integrated into regional transportation systems and offer residents more transportation choices. These choices lead to a reduced reliance on automobile use. In fact, in case study sites in Dallas and Baltimore, redevelopment of brownfields has been found to reduce household VMT by 22–55% when compared to typical greenfield development (Paull 2008). Infill development on brownfields also has a positive impact on the preservation of greenfields. A density analysis conducted by researchers at George Washington University found that one-acre of brownfield development saves 4.5 acres of greenfields (Deason 2001).

In 2007, Sustainable Long Island, a nonprofit research association, conducted an extensive public opinion survey regarding brownfield development and received responses from a diverse population including not-for-profits, community leaders, engineers, developers, lawyers, and municipal officials. Respondents believed that brownfield development can be a key component of sustainable development: 89% indicated that brownfield redevelopment was an essential ingredient in supporting sustainable development; 78% believed that pressure to build on greenfields would be eased by brownfield redevelopment. To facilitate redevelopment of brownfields, respondents indicated that liability protection was the most important factor (29.6%). This was followed by grants for local governments to clean up public land (25.9%) and redevelopment tax credits (18.5%) (Staff Report 2008). Fortunately, the 2002 legislation described above addresses this liability concern.

These beliefs in the environmental benefits of brownfield development are being borne out by empirical evidence. As a mixed-use development, Atlantic Station is a prototype for a new kind of community that can lead to significant environmental benefits. In general, these types of mixed-use developments have been experiencing 25% to 30% fewer net external trips on nearby roads than conventional projects at similar densities. Atlantic Station does even better. The average Atlanta-region resident drives 66 miles per day. Employees in Atlantic Station drive an average of 10.7 miles per day and residents drive just 8 miles per day—an 85% reduction (Dunham-Jones and Williamson 2008). Atlantic Station is also serving as a model for the study of the linkages between CO_2 reduction efforts in New Urbanist communities. According to The Clean Air Campaign's Commute Trak website, Atlantic Station "clean commuters" reduced 351,799 VMTs, eliminated 355,669 grams of Volatile Organic Compounds (VOCs), and saved $175,900 in commuting costs in 2008.

In the end, despite their limitations, these cases show how brownfield development can accommodate residential and commercial uses more frequently found in America's suburbs, and how brownfields are a potentially untapped resource for new urban development. Brownfields can become small-scale suburbs like Glenwood Park and Hercules or they can become new urban centers like Atlantic Station and Bay Street. Changing demographics, real estate economics, and a new national focus on sustainable development will direct growth to infill locations like urban brownfields, and there are a lot of lessons that can be learned from these case studies.

As Hercules' Steve Lawton notes, "In the Bay Area, (traditional planning) leads to ever longer commutes and … a competitive disadvantage for the Bay Area due to housing prices." He stresses that as population increases, current design modes of thought are reaching their practical limits. Land that is still available must be thoughtfully used; it must be dense and self-sustaining (Temple 2002b).

That is a prescription for converting brownfields into new urban communities. These conversions can use principles of Smart Growth, New Urbanism, or even LEED-ND and prosper provided that they address issues of social equity and environmental justice. With public and private collaboration, communities that have applied these principles have succeeded in replacing dynamite factories and concrete plants with neighborhoods that respond to the needs and desires of America's next generation.

References

Adamick, M. 2006. "Hercules CA Sets Sight on Transit District." *Contra Costa Times.* June 26.

Anders, C. 2006. "Living above It All: Emeryville Condominium Complex Sits Atop Retail Village." *San Francisco Chronicle*. February 19.

Archer, J. 2005. *Architecture and Suburbia.* Minneapolis: University of Minnesota Press.

Artz, M. 2002a. "From Big Block to Bay Street," *Berkeley Daily Planet*, October 7.

——. 2002b. "The Search for Healing." *Berkeley Daily Planet*. October 12.

Associated Press. 2001. "Blighted Industrial Sites Home of New Development." *Berkeley DailyPlanet*. August 20.

Atlantic Station Project XL Report. 2008.

——. 2007.

——. 2005.

Benfield, K. 2009. "Meet Glenwood Park—Atlanta's New Showcase Neighborhood of Smart Growth and Green Design." Available at http://switchboard.nrdc.org/blogs/kbenfield/meet_glenwood_park_atlantas_ne.html (accessed: July 22, 2009).

Breen, A. and D. Rigby. 2005. *Intown Living: A Different American Dream.* Washington, DC: Island Press.

Brownfield Revitalization Success Story. 2005. Hercules Powder Works. October.

Burchell, R. 2002. *Costs of Sprawl.* Washington, DC: National Academy Press.

Crawford, M. 1992. "The World in a Shopping Mall." *Variations on a Theme Park: The New American City and the End of Public Space*, edited by Michael Sorkin. New York: Hill and Wang.

Darlin, D. 2010. "Street Corners vs Cul de Sacs." *The New York Times.* January 10.

Deason, J., G.W. Sherk and G. Carroll. 2001. *Public Policies and Private Decisions Affecting the Redevelopment of Brownfields: An Analysis of Critical Factors, Relative Weights and Areal Differentials.* Washington, DC: George Washington University, Environmental and Energy Management Program.

DelVecchio, R. 2002a. "Urban Renewal atop Sacred Past: Ohlone Protest Emeryville Project." *San Francisco Chronicle.* November 20.

——. 2002b. "Brownfield to Bay Street." *Shopping Centers Today*. April.

Duany, A., E. Plater-Zyberk and J. Speck. 2000. *Suburban Nation: The Rise of Sprawl and the Decline of the American Dream*. New York: North Point Press.

Duffy, K. 2009. "Element Condo Units Go To Auction." *Atlanta Journal-Constitution.* February 26.

Dunham-Jones, E. and J. Williamson. 2008. "Retrofitting Suburbs: Instant Cities, Instant Architecture, and Incremental Metropolitanism." *Harvard Design Magazine.* Spring/Summer. Number 28.

Ellis, C. 2002. "The New Urbanism: Critiques and Rebuttals," *Journal of Urban Design* 7: 261–91.

Environmental Protection Agency (EPA). 2006a. *Sustainable Reuse of Brownfields.* October. EPA-560-F-06-247.

——. 2009a. *EPA's Brownfields and Land Revitalization Programs: Changing American Land and Lives.* October. EPA-560-F-09-519.

——. 2009b. *Addressing Environmental Justice in EPA Brownfields Communities.* November. EPA-560-F-09-518.

——. 2009c. *Building a Sustainable Future: A Report on the EPA's Brownfields Sustainability Pilots.* October. EPA-560-F-09-500.

——. 2009d. *Building Vibrant Communities: Community Benefits of Land Revitalization.* October. EPA-560-F-09-517.

Flint, A. 2006. *This Land: The Battle over Sprawl and the Future of America.* Baltimore: Johns Hopkins University Press.

Flower, J. 1996. "Downhome Technopia." *New Scientist.* Issue 2013.

Francaviglia, R. 1996. *Main Street Revisited: Time, Space, and Image Building in Small Town America.* Iowa City: University of Iowa Press.

Freed, D. 2002. "Emeryville's Bay Street to Open." *Berkeley Daily Planet*. November 3.

Frumkin, H., L. Frank and R. Jackson. 2004. *Urban Sprawl and Public Health: Designing, Planning, and Building for Healthy Communities*. Washington, DC: Island Press.

Grant, L. 2004. "Shopping in the Great Outdoors: Elegant, Convenient Lifestyle Centers Encroach on Reign of Indoor Malls." *USA Today*. August 4.

Greene, K. 2006. "Encore (A Special Report). Forget Golf Courses, Beaches, & Mountains." *The Wall Street Journal*. October 2.

Hall, D. 1998. "Community in the New Urbanism: Design Vision and Symbolic Crusade." *Traditional Dwellings and Settlements Review*. V9, N2. Spring.

Hayden, D. 2004. *Building Suburbia: Green Fields and Urban Growth, 1820–2000*. New York: Vintage.

Hazel, D. 2003. "Brave New Format: Lifestyle Centers Look Good, But Are They Earning Their Keep?" *Shopping Centers Today*. May.

Hercules Handout. 2006. City of Hercules. February 3.

Herrick, T. 2006. "City Lite: Fake Towns Rise, Offering Urban Life without the Grit." *The Wall Street Journal*, May 31.

Jackson, K. 1987. *Crabgrass Frontier: The Suburbanization of the United States*. New York: Oxford University Press.

Jerde Partnership. 2009. *Jerde Philosophy*. Available at http://www.jerde.com/flash.php (accessed: July 22, 2009).

Lee, Gary. 1996. "Breathing New Life into 'Brownfields;' Incentives Lure Firms to Contaminated Sites." *The Washington Post*. March 11.

Leinberger, C. 2008. "The Next Slum." *The Atlantic*, March.

Link-Wills, K. 2007. "Atlantic Station: Model City." *Georgia Tech Alumni Magazine*. Summer.

Lochner, T. 2005. "Peace Walk to Protest Mall Built on Sacred Site." *Contra Costa Times*. November 6.

MacCannell, D. 2004. "New Urbanism and its Discontents." *The City Cultures Reader Second Edition*, edited by M. Miles and T. Hall with I. Border. London: Routledge, 382–95.

McLinden, S. 2006. "Proving Ground: New Concepts Cut Their Teeth at Easton Town Center." *Shopping Centers Today*. May.

Miller, J. 2006. "Evolution of a Brownfield." Available at Tndtownpaper.com. Summer (accessed: December 11, 2009).

Newman, M. 2004. "A Onetime Industrial Field Now Sprouting Storefronts." *Commercial Real Estate News*. January 7.

Oldenburg, R. 1997. *The Great Good Place*. New York: Marlow and Company.

Paull, Evans. May 2008. "Brownfields, Infill, and Energy." Washington, DC: Northeast-Midwest Institute.

Putnam, R. 2000. *Bowling Alone: The Collapse and Revival of American Community*. New York: Simon and Schuster.

Ritter, I. 2004. "Yaromir Steiner Defines Town Centers." *Globe Street Retail*. December 6.

Rome, A. 2001. *The Bulldozer in the Countryside: Suburban Sprawl and the Rise of American Environmentalism*. Cambridge: Cambridge University Press.

Scholl, D. and R. Williams. 2005. "A Choice of Lifestyles." *Urban Land*, October.

Sheahan, J. and D. Coley. 2002. "Historic Brownfields Bill Becomes Law: Major Victory for Mayors as Lengthy Lobbying Process Pays Off." Washington, DC: U.S. Conference of Mayors. Available at http://preview.usmayors.org/brownfields/history.asp (accessed: July 22, 2009).

Smith, C. 2001. "Better than Burbs? New Urbanists Build a Town, Not Just Subdivisions, in Hercules." Special to the *San Francisco Chronicle*. March 7.

——. 2005. "Everything Old is New Again." Special to the *San Francisco Chronicle*. March 19.

Southworth, M. 2005. "Reinventing Main Street: From Mall to Townscape Mall." *Journal of Urban Design* 10(2): 151–70.

Stein, J. 2001. "In New Urban Villages." *Los Angeles Times*. November 19.

Steiner, Y. 2005. "What's in a Name? Plenty." *Retail Traffic*. May.

Temple, J. 2002a. "Hercules' 'New Urbanism' Mixes Land Uses." *San Francisco Business Times*. October 4.

——. 2002b. "Hercules Tests New Model." *San Francisco Business Times*. October 11.

Staff Report. 2005. "Unsprawl Case Study. Glenwood Park. Atlanta, Georgia." Terrain.org—*A Journal of the Built and Natural Environments* 17 Fall/Winter.

——. 2008. "Long Island's Brownfields: An Urgent Problem." Beth Page, New York: Sustainable Long Island. September.

Woods, W. 2004. "Internet Pioneer a Green Machine as a Developer in Atlanta." *Atlanta Journal-Constitution*, April 29.

Chapter 8

The Inertia of Environmental Regulatory Enforcement in China: Collusion or Abuse of Authority?

Han Hongyun and Zhao Liange

Introduction

As a transitional economy, China is confronted with the dual task of economic development and environmental conservation. To sustain its economic development, the Chinese government has made environmental protection one of its basic national policies. The situation of a deteriorating environment in China is critical. Many observers blame enterprises' poor awareness of polluting activities as the major contributor to the degraded environment, and the central Chinese Government has sought to encourage environmental protection by setting up a market-oriented institutional environment. But, regulatory instruments have been employed due to the imperfect performance of market-oriented approaches, including closing polluting firms, upgrading technologies, intensifying monitoring, providing firms with training, and subsidizing innovative efforts to curb polluting activities.

Unfortunately, these efforts have not been sufficient to keep pace with the environmental pressures and challenges generated by the rapid growth of China's developing economy. Focusing on multiple instruments and the mix of instruments employed is necessary to identify the role states play in environmental protection because incentives can be most effective within a larger regulatory framework. Thus, government regulation is determining the rate of environmental degradation in China (Jiang and Warwick 2002) but "resources devoted to the monitoring of the regulated community and the enforcement of environmental standards are insufficient" (Dion et al. 1998: 6). And, most existing public policy research has focused on the content of policies and not on the design of enforcement mechanisms.

The choice of instruments for policy intervention has received some consideration. Since the seminal work of Tirole (1986), collusion has become the focus of a growing literature; abuse of authority, however, remains under-explored. Vafaï (2005: 385) investigated "an agency relationship with moral hazard where a principal relies on a supervisor to obtain verifiable information about an agent's output". The existing literature on mechanism design in environmental regulation continues to consider adverse selection and moral hazard separately. Taking into consideration both adverse selection and moral hazard, the principal must provide two types of incentives to agents to induce accurate assessments on the part of agents, revealing types of pollution and efforts undertaken for remediation.

With a background of adverse selection and moral hazard, this chapter aims to explore the tradeoff between informational rents and structures of environmental regulation within the context of collusion and abuse of authority. This will provide some preliminary steps in the construction of a formal political economy of environmental regulation. Consistent with the spirit of unofficial activities under asymmetric information developed by Laffont and Martimort (2000)—in contrast with Vafaï's model in 2005 involving only moral hazard—the research focuses on the cause of inertia in environmental regulatory enforcement with both adverse selection and moral hazard. Our model also differs from previous work as the intermediate layer of hierarchy also supervises some efforts.

Based on models developed by Laffont and Martimort (2000) and Vafaï (2005), this chapter proceeds as follows: first, a multi-level hierarchy with principal-supervisor-agent system is developed; second, based on an assumption of principal and supervisor with the same risk attitude and reservation utility, an analysis using the rules of game theory between principal, supervisor, and agent is given; third, optimal contracts under different organizational structures are argued; finally, a conclusion is provided. The core of the work outlines the rules of incentive mechanisms designed for effective supervision of environmental protection. In contrast with the existing literature, we find that collusion does not affect the efficiency of the organization and can be deterred at zero cost by destroying the profit motive between the agent and the supervisor. Abuse of authority is costly to deter, however.

Institutional Arrangement of Environmental Administration in China

After the establishment of the People's Republic of China in 1949, the country lacked an environmental policy. Since the reforms and opening-up in 1978, much advancement in the field of policy has taken place and a range of regulatory and economic instruments have been developed. Among them, the Three Synchronizations Policy proclaimed in 1973 and the Environmental Impact Assessment (EIA) system in 2003 were fundamental instruments strengthening environmental protections in China. The EIA system is the primary measure to combat environmental degradation and in 2006, the Interim Guideline on Public Participation in EIA was released to involve the public in the EIA process.

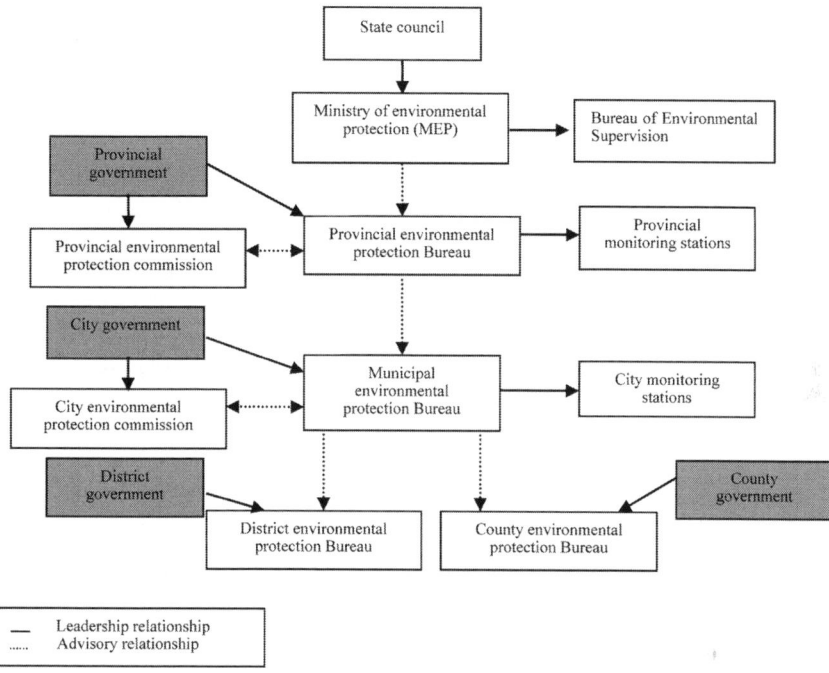

Figure 8.1   The Chinese environmental protection apparatus

In regard to environmental administration, the MEP has overall oversight responsibility in China. As shown in Figure 8.1, the MEP exists at the central level as the chief agency addressing the nation's environmental issues, which fall under the direct leadership of the State Council from which it receives almost all of its funding. MEP is replicated as Environmental Protection Bureaus down through successively lower levels of the administrative hierarchy at the provincial, city, district, county and, in some places, township levels. The chief responsibility of these local environmental units is to enforce laws and implement policies designed by MEP and to assist in drafting local regulations to supplement central ones.

Environmental Protection Bureaus in the provinces and in prefecture governments across the country are responsible for implementation, while licensed research institutes or agencies conduct the actual EIA (Henrik et al. 2006). This vertical hierarchy of administrative entities results in little integration and cooperation among them, and has often become a significant barrier to the effective implementation of environmental protection programs. The EIA system suffers from a number of problems, including interference by local governments, weak enforcement, and poor quality of environmental impact statements.

A Principal-Supervisor-Agent Hierarchy in China's Environmental Supervision

Environmental efforts have lacked effectiveness and efficiency, largely as a result of an implementation gap due to the hierarchy of environmental management in China. There are four relevant stakeholder groups with regard to environmental supervision, including firms, governments at different levels, consumers, and environmental organizations. These parties are at different positions in principal-supervisor-agent relationships. First, there is a principal-agent relationship between MEP and the supervisors at monitoring stations. MEP is in charge of environmental protection, but lacks time and knowledge in monitoring all industries, and employs supervisors to collect information about the environmental protection activities of agents. The supervisor is an agent, assisting the principal supervise firms by collecting information and producing reports on environmental protection, based on hard and verifiable information.

Second, there is a principal-agent relationship between MEP and the manager of a firm; the manager is also an agent of environmental protection. Third, there is a principal-agent relationship between the firm manager and consumers of the environment; the manager is the agent, consumers are principals. Actually, there is no agency problem between the supervisor and the manager since they are all agents in environmental protection.

Environmental organizations and consumers are the most vulnerable groups in society and lack the means to advocate for environmental protection due to a grave imbalance in power between those who own or operate firms and consumers. Consumers lacking self-organization are inferior to managers in the process of bargaining. Therefore, we focus on the relationships among the MEP, the manager, and the supervisors, and ignore the role of consumers and environmental organizations.

Standard agency theory tells us that optimal incentive schemes make use of all available information related to the agent's performance. How to resolve the conflict between efficiency and incentive and the consequence of informational rents with asymmetric information are fundamental issues in the discussion of organizational design (Laffont and Martimort 2000). With an assumption of risk neutrality of the supervisor and the agent, Vafaï (2005: 387) investigates the agency relationship with moral hazard where a principal relies on a supervisor to achieve verifiable information about the agent's effort. He concludes that "abuse of authority is harmful only when supervision technology is sufficiently inefficient" and abuse of authority is more harmful than collusion.

Based on the work of Vafaï (2005), we intend to focus on the institutional sources of the inertia of government supervision in Chinese environmental protection. We further demonstrate efficient results of unofficial activities, collusion and abuse of authority, in organizations with both supervisor's adverse selection and agent's moral hazard. Our model differs from Vafaï's (2005) in the following aspects: the supervisor, as the intermediate layer of our hierarchy does not produce, but has to contribute to producing the report of output of the agent's effort; colluding agents are both risk averse and asymmetrically informed; the collusive offers are made by the agent and the supervisor individually to pursue their own economic benefit rather than collective benefit for coalitions; and the supervisor can only detect the outcome of the agent rather than the effort put forth.

In China, because the manager and the supervisor are two agents selected by public agencies of high rank, in contrast to Vafaï (2005), we believe that the proposition of a supervisor with no effort is unrealistic under the transitional economy. To simplify the analysis, we further assume that the supervisor and the manager have the same level of disutility of effort $\gamma (\gamma \rangle 0)$) and reservation utility $\underline{U}$. The principal relies on the supervisor, whose role it is to make a verifiable report on the agent's output. Evidence on the outcome of the agent's effort is private information for the supervisor, but, once revealed, is verifiable. The principal offers payoffs to the manager and the supervisor based on the report provided by the supervisor.

The Basic Model: The Behavior of Participants

In the model of principal-supervisor-agent hierarchy, MEP is at the top. The bottom of the hierarchy, the manager of the firm, is in charge of implementation of environmental protection. Since the principal cannot observe the agent's effort, the supervisor must produce a report on agent outcome. The principal designs the main contract and offers it to the supervisor and the agent; the principal is risk neutral and the supervisor and agent are risk averse. The supervisor, who has an increasing, differentiable, and strictly concave Von Neumann-Morgenstern utility function, receives a wage from the principal and may also accept a side-payment from the firm.

The firm has the incentive to offer a side-payment to the supervisor in exchange for misinforming the government about his efforts at environmental protection. The supervisor also has an incentive to ask for payoffs from the firm. Hence, the challenge to the principal is to design a set of contracts that result not only in the welfare-improving investment made by the firm, but also truthful reporting provided by the supervisor. In our hierarchical model, the agents decide "to work or to shirk." To maximize the social welfare of the whole society, the principal has to design an incentive mechanism to elicit neutral and professional effort from supervisor and environmental investment of the firm.

Without loss of generality, we assume that the manager has the choice between two effort levels of $e \in \{0,1\}$; $e = 1$ indicates that he just invests in environmental protection; otherwise $e = 0$. By virtue of nature, the firm can only achieve a high outcome of environmental protection X_h with probability $\pi (0 \langle \pi \langle 1)$, with probability $1 - \pi$ to achieve a low level of outcome of environmental protection. If he exerts no effort, he can only achieve a low level of environmental protection, say $X_l = 0$. The output is publicly observable but only verifiable to the supervisor.

In the absence of an incentive contract, the manager would shirk and claim that the low outcome of environmental protection is due to the low productivity type as opposed to his low level of effort.

Supervision Technology

To obtain information about the agent's effort, the supervisor must exert an unobservable effort. He has the choice between two supervisory effort levels, $m \in \{0,1\}$; at the zero supervisory level he observes nothing. We assume that the supervision technology is not totally efficient; when a supervisor chooses $m = 1$, he verifies the agent's effort level only with probability p. The supervisor's report belongs to $I = \{X_h, X_\varphi, X_l\}$. The principal cannot tell whether the failure of exact reporting on a firm's environmental protection comes from the imperfect character of supervision technology or the shirking of the supervisor or both. With a probability $1 - p$, the supervisor will report X_φ; if a supervisor exerts no effort, then a selfish manager will exert zero effort, hence, the environmental protection investment is 0.

The information provided by the supervisor can be concealed but not forged. It is impossible for the informed supervisor to misreport low effort as high effort or vice versa. The supervisor's discretion gives him an opportunity to engage in unofficial activities. When the supervisor verifies that the agent shirks, the agent might pay him a bribe to induce the supervisor to conceal the truth from the principal; in this case, the supervisor colludes with the agent. The bribe is the payment actively provided by the agent to the supervisor to conceal bad information from the principal. When the supervisor threatens the agent that he will conceal favorable information about remediation efforts unless the agent offers an amount of tribute, this is abuse of authority; tribute is the payment, which is actively asked by the supervisor and is more detrimental than a bribe to society.

The Contracts

Information on the outcome of environmental investment is visible and not open to manipulation. Hence, the only way the supervisor can "doctor" information is to conceal it by reporting X_φ. The supervisor may conceal information when he verifies $X_l = 0$, and provide a report X_φ; however, he cannot provide a report $X_h \rangle 0$ when he verifies $X_l = 0$. In this case, he colludes with the agent. The supervisor can only threaten the agent with concealing information from the principal when he observes $X_h \rangle 0$. We name this form of unofficial activity "abuse of authority" or "collusion."

Based on supervisor reports, the principal offers contracts to both supervisor and agent. When the supervisor provides a report $\{X_l, X_\varphi, X_h\}$, the principal offers a contract $\{w_l, w_\varphi, w_h\}$ to the manager and $\{s_l, s_\varphi, s_h\}$ to the supervisor simultaneously.

When the supervisor exerts effort, he can verify the outcome of environmental investment of the firm as X_l or X_h with probability p, and then the supervisor gets the payoffs s_k or s_h. Due to the inefficiency of supervision technology and uncertainty, however, even if the supervisor works diligently he may not be able to verify anything, so the supervisor can get payoff s_φ with probability $1 - p$. Hence, a supervisor may provide a report X_φ under the following conditions: he exerts zero effort; he exerts some effort, but he cannot verify anything; he exerts effort and verifies X_l, but he conceals the real information. Contracts are publicly observable and all wages must be non-negative due to the limited liability of the supervisor and the agent; that is, $w_l \geq 0, w_\varphi \geq 0, w_h \geq 0$ and $s_l \geq 0, s_\varphi \geq 0, s_h \geq 0$.

The Unofficial Activities

Unofficial activities consist of unregistered activities aimed at deriving benefits in either monetary or natural forms, which create new values or cause redistribution effects. Actual information on environmental investment is concrete and not forgeable; information can be concealed but not forged. Hence, the only way the supervisor can manipulate information is to conceal it by reporting X_φ. The supervisor may conceal information when he observes $X_l = 0$; if the manager offers the bribe B, and he provides a report X_φ, the manager actively colludes with the supervisor to get the payoff w_φ, and the supervisor gets the extra payoff B. He cannot provide report $X_h \rangle 0$ when he observes $X_l = 0$; however, in the case of collusion with the agent, the supervisor can only threaten the agent with concealing his/her effort from the principal when he observes $X_h \rangle 0$. We refer to this form of unofficial activity as abuse of authority. If the supervisor has verified that the level of environmental investment is X_h and he threatens that he will provide a report X_φ unless he gets the side-payment F, at this moment the supervisor actively colludes with the manager to get the tribute F.

Technical Efficiency of Bribes and Tributes: Who Will Police the Police?

Agency relationships create a demand for monitoring. Because many agency relationships rely on intermediate agents to seek agent-related information, ways to overcome the possibility of collusion between the agent and the supervisor have attracted more and more attention. There is a trade-off between regulatory flexibility and the creation of counter powers (Laffont and Tirole 1990). Collusion is not frictionless: "the lack of enforceability of the side-contract generates endogenously some transaction costs" (Martimort 1999). Because of transaction costs, the regulator can only get a fraction less than one of the agents' informational rent. Although the manager is willing to provide a bribe or tribute, the transaction costs will affect the effectiveness of the payoff transfer. We assume that the transfer efficiency of a bribe is $k_B \in [0,1)$, and the transfer efficiency of a tribute is $k_F \in [0,1)$. So, even if the manager has provided the bribe B or tribute F, the supervisor can only get the payoffs $k_B B$ or $k_F F$.

The Payoffs

The principal-agent relationship we analyze can be described as follows. A risk-neutral principal wants to hire an agent to perform a certain task, but he/she cannot monitor the agent's actions. The principal receives the residual profits of the vertical structure and pays the other members of the hierarchy. His/her objective is to maximize the expected utility of the whole society. The supervisor and the agent receive their monetary payments from the principal and try to maximize their expected monetary income. The reservation utility of the manager and the supervisor is $\underline{U}$. The utility function of the manager is $EU(W - g(e)) \geq \underline{U}$. For the sake of simplicity, we suppose that the utility function is $U^A(w,e) = w - \gamma e$. The utility function of a supervisor is $EU(s - g(m)) \geq \underline{U}$, $U^s(s,m) = s - \gamma m$; w and s are modeled as monetary payments of the manager and supervisor from the principal, respectively.

With regard to the allocation of benefits generated by side contracts, "the issue of how the supervisor and the agent split the surplus generated by their side contract is a matter of bargaining power" (Trole 1986: 192). An allocation is "incentive efficient" if it has no objection from the larger coalition; the set of enforceable contracts is the set of incentive compatible contracts (Vohra 1999). Both trust and reciprocity sustain the credibility of commitments between the supervisor and the manager (Laffont and Martimort 1998; Faure-Grimaud et al. 2002; Vafaï 2002). If and when all agreements are concluded, a coalition structure forms. Each coalition in this structure is now required to allocate its worth among its members as dictated by the proposals to which they were signatories.

Under participation and incentive constraints, the principal maximizes the social welfare function. The purpose of the principal is to design an incentive contract to elicit the maximum output at the least cost. The utility function of the principal is $U^P(w,s,m) = \pi X_h + (1 - \pi)X_l - c(w,s,m)$; here, $X_l = 0$, so the utility function of the principal can be summarized as

$$U^P(w,s,m) = \pi X_h - c(w,s,m), \text{ here}$$

$$c(w,s,m) = p[\pi(w_h + s_h) + (1 - \pi)(w_l + s_l)] + (1 - p)(w_\phi + s_\phi) \ .$$

To summarize, we present the steps in the contract process:

1. The principal offers the contract $\{s_l, s_\varphi, s_h\}$ to the supervisor, and at the same time, offers the contract $\{w_l, w_\varphi, w_h\}$ to the manager, which depends on the report of the supervised firm's environmental protection provided by the supervisor;

2. The supervisor and the manager decide to accept or refuse the contracts; if two agents reject the contract, then this is the end of the game; if instead contracts are accepted, the game continues as follows;
3. The supervisor and manager take actions and the agent decides whether to work or to shirk;
4. The supervisor and manager make a decision with collusion or abuse of authority;
5. The supervisor provides a report about the effort of the manager to the principal; and
6. Payoffs are realized.

The Optimal Contract Design under Different Organization Structures

Optimal Contracts without Unofficial Activities

As a benchmark, let us consider the case where unofficial activities are impossible, that is, $k_B = k_F = 0$.

(1) The optimal contract for the manager
If unofficial activities are impossible, then the agent's incentive compatibility constraint is $p[\pi w_h + (1-\pi)w_l] + (1-p)w_\varphi - \gamma \geq pw_l + (1-p)w_\varphi$.

The agent's contract must also satisfy his/her participation constraint, $p[\pi w_h + (1-\pi)w_l] + (1-p)w_\varphi - \gamma \geq \underline{U}$ and limited liability constraints $w_l \geq 0, w_\varphi \geq 0, w_h \geq 0$.

(2) The optimal contract for the supervisor
If unofficial activities are impossible, then the supervisor's incentive compatibility constraint is $p[\pi s_h + (1-\pi)s_l] + (1-p)s_\varphi - \gamma \geq ps_l + (1-p)s_\varphi$.

The supervisor's contract must also satisfy his/her participation constraint, $p[\pi s_h + (1-\pi)s_l] + (1-p)s_\varphi - \gamma \geq \underline{U}$.

Optimizing yields proposition 1:

Proposition 1 (see Appendix 8.1):

1. The optimal contract for the manager in the absence of unofficial activities is $w_\varphi^0 \in [0, \underline{U}/(1-p)]$, $w_l^0 \in [0, \underline{U}/p - (1-p)w_\varphi/p]$, $w_h^0 = (\gamma + \underline{U})/(p\pi) - (1-p)w_\varphi/(p\pi) - (1-\pi)w_l/\pi$;

2. the optimal contract for the supervisor is

$$s_{\varphi}^0 \in [0, \underline{U} / (1-p)], \ s_l^0 \in [0, \underline{U} / p - (1-p)s_{\varphi} / p],$$

$$s_h^0 = (\gamma + \underline{U}) / (p\pi) - (1-p)s_{\varphi} / (p\pi) - (1-\pi)s_l / \pi;$$

3. The expected production cost and supervision costs are $C^0 = 2(\gamma + \underline{U})$;

4. The utility function of the principal is $U^{p0}(w, e, m) = \pi X_h - 2(\gamma + \underline{U})$.

The Optimal Contracts with Unofficial Activities

If unofficial activities are possible, that implies the combination of $k_B \neq 0, k_F = 0$, $k_B = 0, k_F \neq 0$, or $k_B \neq 0, k_F \neq 0$, and we discuss organizational design under unofficial activities, respectively.

Optimal Contracts in the Presence of Collusion

(1) The Optimal Contract for the Manager
In the presence of collusion ($k_B \neq 0, k_F = 0$), the agent's incentive compatibility constraint is

$$p(w_{\varphi} - B) + (1-p)w_{\varphi} \geq pw_l + (1-p)w_{\varphi}.$$

The agent's contract must also satisfy his/her participation constraints $p(w_{\varphi} - B) + (1-p)w_{\varphi} \geq \underline{U}$.

(2) Optimal contract for the supervisor
The supervisor's incentive compatibility constraint is
$$p(s_{\varphi} + k_B B) + (1-p)s_{\varphi} - \gamma \geq ps_l + (1-p)s_{\varphi} - \gamma.$$
The supervisor's contract must also satisfy his/her participation constraint
$p(s_{\varphi} + k_B B) + (1-p)s_{\varphi} - \gamma \geq \underline{U}$.
Optimizing yields proposition 2:

Proposition 2 (see Appendix 8.2):

If only collusion is possible, that means $k_B \neq 0, k_F = 0$: then the optimal contract to the manager is $w_{\varphi}^C = w_l^C = \underline{U}$: the optimal contract to the supervisor is $s_{\varphi}^C = s_l^C = \gamma + \underline{U}$. The expected production and supervision costs are $C^C = \gamma + 2\underline{U}$; the utility function of the principal is $U^{pC}(w, e, m) = -c(w, s, m) = -(\gamma + 2\underline{U})$.

Optimal Contracts in the Presence of Abuse of Authority

(1) The optimal contract for the manager
The agent's incentive compatibility constraint is
$$p[\pi(w_h - F) + (1 - \pi)w_l] + (1 - p)w_\varphi - \gamma \geq p[\pi w_\varphi + (1 - \pi)w_l] + (1 - p)w_\varphi - \gamma.$$

The agent's contract must also satisfy his/her participation constraint of abuse of authority $p[\pi(w_h - F) + (1 - \pi)w_l] + (1 - p)w_\varphi - \gamma \geq \underline{U}$.

(2) The optimal contract for the supervisor
From the viewpoint of the supervisor, if the manager agrees to provide a tribute F, the payment to the supervisor is $s_h + k_F F$; otherwise, the monetary payment to the supervisor is s_φ.

The supervisor's incentive compatibility constraint is
$$p[\pi(s_h + k_F F) + (1 - \pi)s_l] + (1 - p)s_\varphi - \gamma \geq p[\pi s_\varphi + (1 - \pi)s_l] + (1 - p)s_\varphi - \gamma.$$

The supervisor's contract must also satisfy his/her participation constraint of abuse of authority
$$p[\pi(s_h + k_F F) + (1 - \pi)s_l] + (1 - p)s_\varphi - \gamma \geq \underline{U}.$$
Optimizing yields proposition 3:

Proposition 3 (see Appendix 8.3)

If only the abuse of authority is possible, that is, $k_B = 0, k_F \neq 0$ let us denote $\tilde{p} = \gamma / (\gamma + \pi U)$.
If 1) $p \leq \tilde{p}$, the optimal contract to the manager is $w_l^{A1} = 0$, $w_\varphi^{A1} = w_h^{A1} = \gamma / p\pi$; the optimal contract to the supervisor is $s_l^{A1} = 0$, $s_\varphi^{A1} = s_h^{A1} = \gamma / p\pi$; the expected production and supervision costs are $C^{A1} = 2(1 - p + p\pi)\gamma / p\pi$; the utility function of the principal is $U^{PA1} = \pi X_h - 2(1 - p + p\pi)\gamma / p\pi$.
If 2) $p\rangle\tilde{p}$, the optimal contract to the manager is $w_l^{A2} \in [0, [p\pi\underline{U} - (1 - p)\gamma] / (p,$ $w_\varphi^{A2} = w_h^{A2} = [(\gamma + \underline{U}) - p(1 - \pi)w_l^{A2}] / (1 - p(1 - \pi))]$, the optimal contract to the supervisor is $s_l^{A2} \in [0, [p\pi\underline{U} - (1 - p)\gamma] / (p\pi)]$, $s_\varphi^{A2} = s_h^{A2} = [(\gamma + \underline{U}) - p(1 - \pi)w_l^{A2}] / (1 - p(1 - \pi))]$; the expected production and supervision costs are $C^{A2} = 2(\gamma + \underline{U})$; the utility function of the principal $U^{pA2}(w, e, m) = \pi X_h - 2(\gamma + \underline{U})$.

Optimal Contracts in the Presence of Collusion and Abuse of Authority

(1) The optimal contract for the manager
In the presence of collusion and abuse of authority ($k_B \neq 0, k_F \neq 0$), the agent's incentive compatibility constraint is
$$p[\pi(w_h - F) + (1 - \pi)(w_\varphi - B)] + (1 - p)w_\varphi - \gamma \geq pw_l + (1 - p)w_\varphi$$

The agent's contract must also satisfy his/her participation constraint
$$p[\pi(w_h - F) + (1 - \pi)(w_\varphi - B)] + (1 - p)w_\varphi - \gamma \geq \underline{U}$$
And, it has to satisfy the constraint of limited liability $w_l \geq 0, w_\varphi \geq 0, w_h \geq 0$.

(2) The optimal contract for the supervisor
The supervisor's incentive compatibility constraint is
$$p[\pi(s_h + k_F F) + (1 - \pi)(s_\varphi + k_B B)] + (1 - p)s_\varphi - \gamma \geq ps_l + (1 - p)s_\varphi .$$
The supervisor's contract must also satisfy his/her participation constraint
$$p[\pi(s_h + k_F F) + (1 - \pi)(s_\varphi + k_B B)] + (1 - p)s_\varphi - \gamma \geq \underline{U} .$$
Optimizing yields proposition 4:

Proposition 4 (see Appendix 8.4)

If collusion and abuse of authority coexist, then $k_B \neq 0, k_F \neq 0$ let us denote
$\tilde{p} = \gamma / (\gamma + \pi U)$.

If 1) $p \leq \tilde{p}$, the optimal contract to the manager is $w_l^{CA1} = 0$,
$w_\varphi^{CA1} = w_h^{CA1} = \gamma / p\pi$, the optimal contract to the supervisor is $s_l^{CA1} = 0$,
$s_\varphi^{CA1} = s_h^{CA1} = \gamma / p\pi$, the expected production and supervision costs are
$\hat{C}^{CA1} = 2(1 - p + p\pi)\gamma / p\pi$; the utility function of the principal is
$U^{PCA1} = \pi X_h - 2(1 - p + p\pi)\gamma / p\pi$.

If 2) $p \rangle \tilde{p}$, the optimal contract to the manager is
$w_l^{CA2} = [p\pi\underline{U} - (1 - p)\gamma] / (p\pi)$; $w_\varphi^{CA2} = w_h^{CA2} = (\gamma + \pi\underline{U}) / \pi$ the
optimal contract to the supervisor is $s_l^{CA2} = [p\pi\underline{U} - (1 - p)\gamma] / (p\pi)$;
$s_\varphi^{CA2} = s_h^{CA2} = (\gamma + \pi\underline{U}) / \pi$; the expected production and supervision costs are
$\hat{C}^{CA2} = 2(\gamma + \underline{U})$; the utility function of the principal $U^{PCA2} = \pi X_h - 2(\gamma + \underline{U})$.

The Inertia of Environmental Supervision in China: Collusion or Abuse of Authority?

If Unofficial Activities are Impossible, the Principal Can Design an Incentive Mechanism to Elicit the Efforts of the Supervisor

If the supervisor can verify and truthfully report the outcome of environmental protection activities, then the principal can offer exact payoffs to the manager and the supervisor. The proposition states that, in the absence of unofficial activities, the principal does not need to provide rents to obtain information. Unfortunately, this is unrealistic when the regulator uses possible discretionary power to pursue personal benefits by colluding with the regulated agent (Tirole 1986). Particularly in China, local environmental protection bureaus at all levels have incomplete vertical relations with the national MEP. While the national MEP leads the professional work and policy decisions of the local bureaus, local party committees and governments allocate their personnel and administrative funds.

China's pollution levy system is one of the most extensive in the world. While the central government sets up the level and structure of the pollution levy, local environmental authorities are responsible for collecting levies from industrial facilities and determining how much of the calculated charges to collect from each facility (Wang et al. 2002). The local environmental bureau is part of the local government. It is difficult for a bureau to enforce environmental regulations on polluting industries or projects sponsored by local governments; more importantly, local environmental bureaus can only obtain minimal financial support from the government treasury. Most need to generate financial resources by collecting pollution fees or by providing paid consulting services to the industries they regulate.

This situation creates severe conflicts of interest in regulatory enforcement. In addition, high tax rates, onerous official regulations, predatory behavior by government officials, criminal gangs, and the inadequacy of the institutional environment are external factors resulting in collusion (Johnson et al. 2000). The performance objectives of local leaders, the pressures to raise revenues locally to finance un-funded mandates, and the limited accountability to local populations have generally meant that economic priorities have overridden environmental concerns and the effectiveness of environmental regulation in China is limited.

If Only One Unofficial Activity is Possible, Then the Principal Can Preclude it through Cutting off the Source of Materials

When collusion is the only possible way for unofficial activity, it can be prevented at no cost by setting $w_\varphi^C = w_l^C = \underline{U}$ and $s_\varphi^C = s_l^C = \gamma + \underline{U}$. This indicates that collusion can be prevented by creating incentive payments for the supervisor. However, to elicit high levels of effort from the supervisor, the principal has to offer higher payments to the supervisor.

When abuse of authority is the only unofficial activity, then optimal contracts will vary with the efficiency of supervision technology. If technology is inefficient, abuse of authority could be prevented through two means: if the principal destroys its stake by setting a payoff of $w_\varphi^{A1} = w_h^{A1} = \gamma / p\pi$ to the manager, or if the principal sets the contract to the supervisor to eliminate the motive for abuse of authority, $s_\varphi^{A1} = s_h^{A1} = \gamma / p\pi$.

If the supervision technology is efficient ($p\rangle\tilde{p}$), the principal depends on collective information and makes no discrimination between two levels of agent efforts X_φ and X_h while differentiating two levels of X_φ and X_l. The principal will select an option between tolerating and deterring abuse of authority. It is clearly optimal for the principal to deter abuse of authority by setting $w_\varphi^A = w_h^A = \gamma / p\pi$, that is, by destroying its stake: rather than tolerating it, by instead setting $w_h^{A2}\rangle w_h^{A1}$.

*If Abuse of Authority is the Basis for Collusion, It Will Benefit the Manager
Rather than the Supervisor*

If $p \leq \tilde{p}$, the principal must set the contract to the manager and the supervisor $w_l^{CA1} = 0$ and $s_l^{CA1} = 0$, respectively. If $p \rangle \tilde{p}$, the principal can set the contracts to the manager and the supervisor depending on the report from the supervisor. To eliminate the stake of abuse of authority, the principal does not differentiate between the two levels of the agent effort, X_φ and X_h, and offers the same payment to the manager. Replacing $w_l^{A2} \leq [p\pi\underline{U} - (1-p)\gamma] / (p\pi)]$ with $w_\varphi^{A2} = w_h^{A2}$, we can get $w_\varphi^{A2} = w_h^{A2} \rangle \gamma / \pi + \underline{U}$, or equivalently, $w_\varphi^{A2} = w_h^{A2} \rangle w_\varphi^{CA2} = w_h^{CA2} = (\gamma / \pi + \underline{U})$. Collusion is a mutually advantageous agreement between the supervisor and the agent to the detriment of the principal, but abuse of authority benefits the manager rather than the supervisor. The condition for the manager to refuse the demand for abuse of authority is $w_\varphi - w_l \geq \gamma / p\pi$, while the sufficient precondition for collusion is $w_\varphi \geq w_l$.

Once the supervisor has offered abuse of authority and received tribute from the manager, then a coalition is formed, and the abuse of authority provides the basis for collusion. "Regulatory capture is not enforced through an explicit side-contract since no court of justice can verify such an illegal contract. Instead, collusion must be self-enforcing" (Martimort 1999: 932); "one-sided favors call for reciprocated ones" (Tirole 1986: 185). Long-term cooperation between the supervisor and the agent makes regulation inefficient and private income becomes more important. Environmental policy has little control over these social relations (Muldavin 2000). Social power and influence relations are important determinants of the ensuing political-economic equilibrium (Zusman and Rausser 1990).

*The Improvement of Supervision Technology Can Reduce the Expected Costs of
Production and Supervision*

If $p \leq \tilde{p}$, the expected cost of production and supervision is $C^{A1} = C^{CA1} \rangle C^0 \rangle C^C$, and consequently, $U^{PC} \langle U^{P0} \leq U^{PA1} = U^{PCA1}$, social welfare, is lowest when collusion is the only form of unofficial activity. That is, if the supervision technology is inefficient, it will have the same cost of production and supervision when only with collusion as with two forms of unofficial activities. This means the principal can deter the collusion at no cost; if unofficial activities are impossible, the expected cost of production and supervision lies between abuse of authority and collusion. Abuse of authority just increases the cost of production and supervision. In the presence of collusion, the expected cost of production and supervision is low. The sharing of information between the supervisor and the agent can reduce the cost of production and supervision, but it comes at the cost of decreased social welfare.

If $p \rangle \tilde{p}$, $C^0 = C^{A2} = C^{CA2} \rangle C^C$, $U^{PC} \langle U^{P0} = U^{PA2} = U^{PCA2}$, with high efficiency of supervision technology, social welfare with only one form of abuse and with two types of unofficial activities is the same, insuring that the principal can deter collusion at no cost. Collusion is less detrimental for the organization as the transaction costs of side-contracting are greater. So, improving the efficiency of supervision technology is critical to improve the level of social welfare.

The Efficiency of Transfer is Meaningful to Environmental Enforcement

"The private efficiency of a side contract between a regulator and the firm depends only on the underlying structure of their relationship," including tenure of the regulator, frequency of the firm-regulator relationship, and the amount of information they share (Laffont and Martimort 1999: 243). The effect of collusion on the organization is mixed (Villadsen 1995), while discretionary power provides the basis for collusion and abuse of authority (Vafaï 2002). When the supervisor and the agent share information, social welfare is low. Abuse of authority is one of the most widespread and harmful forms of unofficial activity in hierarchies (Vafaï 2005).

The lack of enforceability of the side-contract endogenously generates some transaction costs. Transaction costs will affect the effectiveness of the payoff transfer. If K_F is increased, it will reduce the temptation for the supervisor's abuse of authority; hence, it helps eliminate the possibility of collusion between the agent and the supervisor. The lack of personnel in supervision agencies and public participation in environmental protection will inevitably limit the effect of supervisory pressures (see Table 8.1). Community capacity and willingness to enforce environmental norms determines the effectiveness of enforceability. The abuse of authority of supervisors is the fundamental origin of poor levels of environmental protection in China. The core of policy making is to reduce the level of discretionary power through technological efficiency (Tirole 1986; Laffont and Martimort 1999).

Table 8.1   Background of environmental administration in China

Year	Number of Agencies (Unit)	Scientific and research institutions (Unit)	Monitoring station (Unit)	Supervising and administrative station (Unit)	Total number of staff and workers (Persons)	Monitoring personnel (Persons)	Supervising and administrative personnel (Persons)
1997	9,207	205	2,138	2,142	103,180	36,773	20,449
1998	9,167	197	1,926	2,003	105,932	34,857	22,567
1999	10,811	226	2,203	2,398	121,049	40,105	28,039
2000	11,115	240	2,250	2,552	131,092	40,674	31,228
2001	11,090	246	2,229	2,567	142,766	43,269	37,934
2002	11,798	269	2,356	2,693	154,233	46,515	41,878
2003	11,654	263	2,305	2,795	156,542	45,813	44,250
2004	11,555	266	2,289	2,800	160,246	45,849	47,189
2005	11,528	273	2,289	2,854	166,774	46,984	50,040
2006	11,321	260	2,322	2,803	170,290	47,689	52,845
2007	11,932	243	2,399	2,954	176,988	49,335	57,427

Source: The Ministry of Environmental Protection, a series of bulletins of environmental statistics since 1997. http://www.zhb.gov.cn/plan/hjtj/qghjtjgb/199806/t19980605_84246. htm.

Conclusions and Implications for Policies

China has established a basic regulatory framework for environmental protection. Political and institutional mechanisms for addressing the interrelated environmental, social, and economic problems have been constituted. Implementation of regulations and enforcement is currently a top priority. China needs to strengthen the effectiveness and efficiency of its environmental policies by addressing the incomplete vertical relations between local environmental protection bureaus at all levels and the national MEP, which have been reinforced by the complicated interactions between supervisors and supervised firms.

Although the principal hopes to elicit high effort from agents through incentive schemes, unofficial activities together with uncertainty about the nature of pollution and imperfect technology make incentive mechanism design complex. Discretionary power allows the supervisor to engage in unofficial activities. The supervisor's abuse of authority provides a basis for collusion between the supervisor and the agent. The actual principal-agent relationship between the supervisor and the agent is particularly abused in transitional China; and the supervisor will pursue information rents by concealing bad information about the manager. Abuse of authority benefits the manager, while after the point of collusion, the supervisor gets entangled in the trap of unofficial activities.

The result of an optimal contract varies with the efficiency of supervision technology. When supervision technology is inefficient, selfish supervisors have a strong motivation to pursue abuse of authority. The principal can deter collusion in a cost-free manner but it is costly to deter abuse of authority. With the improvement of supervision technology, the efficiency of deterring two forms of unofficial activities is the same as that of deterring abuse of authority. Due to the fact that the payment to the supervisor depends on the effort of the agent, and the share of information on production, the supervisor can save the expected cost of production and supervision; with that saving, however, comes the cost of reduced social welfare.

If only one form of unofficial activity occurs, the principal can deter it by destroying its benefit. When the supervisor actively pursues abuse of authority, there is no trade-off between destroying the stake and limiting discretionary power; the core of state policy must be to regulate the action of the supervisor.

There is a need for much stronger monitoring, inspection, and enforcement capabilities to establish a better mix of incentives and sanctions. Abuse of authority provides a basis for collusion; we conclude that abuse of authority of the supervisor is the fundamental cause for inertia in environmental protection enforcement in China. It is clear that China still faces many challenges in establishing an effective regulatory system, including strengthening its governing capacity in order to address issues of the rule of law, independent regulation, information disclosure, and public accountability more effectively.

References

Boyer, M. and J.-J. Laffont. 1999. "Toward a Political Theory of Emergence of Environmental Incentive Regulation." *Rand Journal of Economics* 30(1): 137–57.

Dion, C., P. Lanoie and B. Laplante. 1998. "Monitoring of Pollution Regulation: Do Local Conditions Matter?" *Journal of Regulatory Economics* 13: 5–18.

Faure-Grimaud, A., J.-J. Laffont and D. Martimort. 2002. "Collusion, Delegation and Supervision with Soft Information." USC Center for Law, Economics and Organization, Research Paper No.C02-9. Available at http://ssrn.com/abastract_id=279522 (accessed: January 27, 2010).

Frascatore, M. 1998. "Collusion in a Three-Tier Hierarchy: Credible Beliefs and Pure Self-Interest." *Journal of Economic Behavior & Organization* 34: 459–75.

Johnson, S., D. Kaufmann and C. Woodruff. 2000. "Why Do Firms Hide? Bribes and Unofficial Activity after Communism." *Journal of Public Economics* 76: 495–520.

Kofman, F. and J. Lawarrée. 1996. "A Prisoner's Dilemma Model of Collusion Deterrence." *Journal of Public Economics* 59: 117–36.

Laffont, J.-J. and D. Martimort. 1991. "The Politics of Government Decision-making: A Theory of Regulatory Capture." *The Quarterly Journal of Economics* 106(4): 1089–127.

——. 1998. "Collusion and Delegation." *RAND Journal of Economics* 29(2): 280–305.

——. 1999. "Separation of Regulators against Collusive Behavior." *RAND Journal of Economics* 30(2): 232–62.

——. 2000. "Mechanism Design with Collusion and Correlation." *Econometrica* 68(2): 309–42.

Laffont, J.-J. and J. Tirole. 1990. "Politics of Government Decision Making: Regulatory Institutions." *Journal of Law, Economics, & Organization* 6(1): 1–31.

Low, S.H. and N. Maxemchuk. 1998. "A Collusion Problem and Its Solution." *Information and Computation* 1(40): 158–82. Article No. IC972678.

Martimort, D. 1999. "The Life Cycle of Regulatory Agencies: Dynamic Capture and Transaction Costs." *Review of Economic Studies* 66: 929–47.

Maskin, E. and J. Tirole. 1992. "The Principal-Agent Relationship with an Informed Principal, 2: Common Values." *Econometrica* 60(1): 1–42.

Mishra, A. 2002. "Hierarchies, Incentives and Collusion in a Model of Enforcement." *Journal of Economic Behavior and Organization* 47: 165–78.

Muldavin, J. 2000. "The Paradoxes of Environmental Policy and Resource Management in Reform-Era China." *Economic Geography* 76(3): 244–71.

Strauz, R. 2005. "Honest Certification and the Threat of Capture." *International Journal of Industrial Organization* 23: 45–62.

Tingsong. J and W.J. McKibbin. 1986. "Hierarchies and Bureaucracies: The Role of Collusion in Organizations." *Journal of Law, Economics, and Organization* 2(2): 181–214.

——. 1992. "Collusion and the Theory of Organizations." In *Advances in Economic Theory*, Vol. 2, ed. J.-J.Laffont. Cambridge: Cambridge University Press, 151–206.

——. 2002. "Assessment of China's Pollution Levy System: An Equilibrium Pollution Approach." *Environment and Development Economics* 7(1): 75–105.

Vafai, K. 2002. "Preventing Abuse of Authority in Hierarchies." *International Journal of Industrial Organization* 20: 1143–66.

——. 2005. "Abuse of Authority and Collusion in Organizations." *European Journal of Political Economy* 21: 385–405.

Villadsen, B. 1995. "Communication and Delegation in Collusive Agencies," *Journal of Accounting and Economics* 19: 315–44.

Vohra, R. 1999. "Incomplete Information, Incentive Compatibility and the Core." *Journal of Economic Theory* 86: 123–47.

Wang H. and D. Wenhua. 2002. "The Determinants of Government Environmental Performance: An Empirical Analysis of Chinese Townships." Policy Research Working Papers are also posted on the Web. Available at http://econ.worldbank. org (accessed: January 27, 2010).

Wang H., N. Mamingi, B. Laplante and S. Dasgupta. 2002. "Incomplete Enforcement of Pollution Regulation Bargaining Power of Chinese Factories." Policy Research Working Papers are also posted on the Web. Available at http://econ.worldbank.org (accessed: January 27, 2010).

Zusman, P. and G.C. Rausser. 1990. "Endogenous Policy Theory: The Political Structure and Policy Formation." GATT Research Paper 90-GATT 27.

Appendices to Chapter 8

Appendix 8.1: Proof of Proposition 1 –
Optimal Contracts without Unofficial Activities

1. The Optimal Contract for the Manager

If the unofficial activities are impossible, then the agent's incentive compatibility constraint is $p[\pi w_h + (1-\pi)w_l] + (1-p)w_\varphi - \gamma \geq pw_l + (1-p)w_\varphi$

Or equivalently, $w_h - w_l \geq \gamma / p\pi$ (1)

This equation makes the agent prefer to exert effort in equilibrium. The agent's contract must also satisfy his/her participation constraint,

$$p[\pi w_h + (1-\pi)w_l] + (1-p)w_\varphi - \gamma \geq \underline{U} \qquad (2)$$

And limited liability constraints $w_l \geq 0, w_\varphi \geq 0, w_h \geq 0$
Then we can get $w_h \geq (\gamma + \underline{U}) / p\pi - (1-\pi)w_l / \pi - (1-p)w_\varphi / p\pi$

To the principal, the high payoffs to the agent is at extra cost, so he/she will set the payoff

$$w_h = (\gamma + \underline{U}) / p\pi - (1-\pi)w_l / \pi - (1-p)w_\varphi / p\pi \qquad (3)$$

Replacing (1) in the equation (3), it entails $w_l \leq \underline{U} / p - (1-p)w_\varphi / p$, accordingly $w_\varphi \leq \underline{U} / (1-p)$ due to the fact of $w_l \geq 0$.

2. The Optimal Contract for the Supervisor

If unofficial activities are impossible, then the supervisor's incentive compatibility constraint is

$$p[\pi s_h + (1-\pi)s_l] + (1-p)s_\varphi - \gamma \geq ps_l + (1-p)s_\varphi$$

Or equivalently $s_h - s_l \geq \gamma / p\pi$

The supervisor's contract must also satisfy his/her participation constraint

$$p[\pi s_h + (1-\pi)s_l] + (1-p)s_\varphi - \gamma \geq \underline{U}$$

Optimizing yields the following proposition:

Proposition 1: The optimal contract for the manager in the absence of unofficial activities is $w_\varphi^0 \in [0, \underline{U}/(1-p)]$, $w_l^0 \in [0, \underline{U}/p - (1-p)w_\varphi/p]$, $w_h^0 = (\gamma + \underline{U})/(p\pi) - (1-p)w_\varphi/(p\pi) - (1-\pi)w_l/\pi$; the optimal contract for the supervisor is $s_\varphi^0 \in [0, \underline{U}/(1-p)]$, $s_l^0 \in [0, \underline{U}/p - (1-p)s_\varphi/p]$, $s_h^0 = (\gamma + \underline{U})/(p\pi) - (1-p)s_\varphi/(p\pi) - (1-\pi)s_l/\pi$; the expected production cost and supervision costs are $C^0 = 2(\gamma + \underline{U})$, the utility function of the principal is $U^{p0}(w, e, m) = \pi X_h - 2(\gamma + \underline{U})$.

Appendix 8.2: Proof of Proposition 2

1. The Optimal Contract for the Manager

When the supervisor verifies the effort of environmental activity the result is X_l; the manager will provide bribe B to induce the supervisor to provide a report X_φ rather than X_l; if the supervisor accepts the bribe and agrees to cooperate, then the manager can get the payment w_φ, the supervisor can get the extra payment B; otherwise, the manager will get the payment w_l. The agent's incentive compatibility constraint is, then,

$$p(w_\varphi - B) + (1-p)w_\varphi \geq pw_l + (1-p)w_\varphi, \text{ that is, } w_\varphi - B \geq w_l$$

If and only if $w_\varphi - B \geq w_l$, it is profitable for the manager. The upper limit for the bribe is $B^M = w_\varphi - w_l$. To keep the bribe from the manager, the principal can offer payment to the manager $w_\varphi = w_l$. Hence, the necessary condition for collusion is $w_\varphi \geq w_l$.

The agent's contract must also satisfy his/her participation constraint

$$p(w_\varphi - B) + (1-p)w_\varphi \geq \underline{U}$$

Or equivalently, $w_\varphi \geq pB + \underline{U}$. The principal can offer $w_\varphi = pB + \underline{U}$ in case of extra costs.

2. Optimal Contract for the Supervisor

The supervisor's incentive compatibility constraint is

$$p(s_\varphi + k_B B) + (1-p)s_\varphi - \gamma \geq ps_l + (1-p)s_\varphi - \gamma, \text{ that is } s_\varphi + k_B B \geq s_l$$

To deter the likely collusion, the principal can set $s_l \geq s_\varphi + k_B B$, and hence set $s_l = s_\varphi + k_B B$ in the case of extra costs.

The supervisor's contract must also satisfy his/her participation constraint, $p(s_\varphi + k_B B) + (1-p)s_\varphi - \gamma \geq \underline{U}$, which entails $s_\varphi \geq -pk_B B + (\gamma + \underline{U})$.

If the principal is unwilling to offer high wages to the supervisor, then we can conclude that the principal can provide the payment $s_\varphi = -pk_B B + (\gamma + \underline{U})$, $s_l = (1-p)k_B B + (\gamma + \underline{U})$.

Optimizing yields the following proposition:

Proposition 2: If only collusion is present, the optimal contract for the manager is $w_\varphi^C = w_l^C = \underline{U}$; the optimal contract for the supervisor $s_\varphi^C = s_l^C = \gamma + \underline{U}$; the expected production cost and supervision costs are $C^C = \gamma + 2\underline{U}$; the utility function of the principal is $U^{pC}(w, e, m) = -c(w, s, m) = -(\gamma + 2\underline{U})$.

Appendix 8.3: Proof of Proposition 3

1. The Optimal Contract for the Manager

The agent's incentive compatibility constraint is,

$$p[\pi(w_h - F) + (1-\pi)w_l] + (1-p)w_\varphi - \gamma \geq p[\pi w_\varphi + (1-\pi)w_l] + (1-p)w_\varphi - \gamma$$

that is, $w_h - F \geq w_\varphi$ \hfill (4)

if and only if $w_h - F \geq w_\varphi$, the manager can benefit from the collusion. Of course, he can offer the maximum tribute $F^M = w_h - w_\varphi$, so $w_h - w_\varphi \geq 0$ is the necessary condition for the manager to accept the offer of abuse of authority. If the principal can set the payment to the manager $w_h - w_\varphi = 0$, the principal can destroy the value of collusion.

The agent's contract must also satisfy his/her participation constraint $p[\pi(w_\varphi + (1-\pi)w_l] + (1-p)w_\varphi - \gamma \geq pw_l + (1-p)w_\varphi$,

That is $w_\varphi - w_l \geq \gamma / p\pi$ \hfill (5)

The agent's contract must also satisfy his/her participation constraint,

$$p[\pi(w_h - F) + (1-\pi)w_l] + (1-p)w_\varphi - \gamma \geq \underline{U}$$ \hfill (6)

From (6), we can get $w_h \geq (\gamma + \underline{U}) / p\pi + F - (1-\pi)w_l / \pi - (1-p)w_\varphi / p\pi$

Then the principal can set the payment to the manager

$$w_h = (\gamma + \underline{U}) / p\pi + F - (1 - \pi)w_l / \pi - (1 - p)w_\varphi / p\pi \tag{7}$$

From $w_h = w_\varphi$ and (7), we get

$$w_\varphi = w_h = [(\gamma + \underline{U}) - p(1 - \pi)w_l] / [(1 - p(1 - \pi))] \tag{8}$$

From (5) and (8), we can get $w_l \leq [p\pi\underline{U} - (1 - p)\gamma] / p\pi$.

When $p \rangle \gamma / (\gamma + \pi U)$, then $0 \leq w_l \leq [p\pi\underline{U} - (1 - p)\gamma] / p\pi$; when $p \langle \gamma / (\gamma + \pi U)$, due to the limited liability of the manager, the principal can set $w_l = 0$ and $\overline{w}_\varphi = (\gamma + \underline{U}) / [(1 - p(1 - \pi))]$; from (5) and $w_l = 0$, we can get $w_\varphi \geq \gamma / p\pi$.

If $p \leq \gamma / (\gamma + \pi U)$, then $\dfrac{\gamma + \underline{U}}{1 - P + P\pi} \leq \dfrac{\gamma}{P\pi}$. So the principal can set a payment to the manager, $w_\varphi = w_h = \gamma / p\pi$.

2. The Optimal Contract for the Supervisor

From the viewpoint of the supervisor, if the manager agrees to provide tribute F, then the payment to the supervisor is $s_h + k_F F$; otherwise, the monetary payment to the supervisor is s_φ. The supervisor's incentive compatibility constraint is

$$p[\pi(s_h + k_F F) + (1 - \pi)s_l] + (1 - p)s_\varphi - \gamma \geq p[\pi s_\varphi + (1 - \pi)s_l] + (1 - p)s_\varphi - \gamma$$

That is $s_h + k_F F \geq s_\varphi$

The necessary precondition for the supervisor to agree with an abuse of authority is $s_h + k_F F \geq s_\varphi$; if and only if $k_F F \geq 0$, the supervisor will benefit from the abuse of authority. The principal can deter the abuse of authority by setting $s_h + k_F F \leq s_\varphi$. However, the incentive mechanism needs $s_h \geq s_\varphi$, so the principal can only set the payment $s_h = s_\varphi$ to eliminate the motivation of abuse of authority of the supervisor. At the same time, if the manager disagrees with the demand for tribute, the expected profit could not be less than that of shirking. The supervisor's incentive compatibility constraint is

$$p[\pi(s_\varphi + (1 - \pi)s_l] + (1 - p)s_\varphi - \gamma \geq ps_l + (1 - p)s_\varphi$$

That is, the necessary precondition for a supervisor's hard work is $s_\varphi \geq s_l + \gamma / p\pi$.

The supervisor's contract must also satisfy his/her participation constraint, $p[\pi(s_h + k_F F) + (1 - \pi)s_l] + (1 - p)s_\varphi - \gamma \geq \underline{U}$

Optimizing yields the following proposition:

Proposition 3: if only abuse of authority is possible, that is, $k_B = 0, k_F \neq 0$, let us denote $\tilde{p} = \gamma / (\gamma + \pi U)$

If 1) $p \leq \tilde{p}$, the optimal contract to the manager is $w_l^{A1} = 0$, $w_{\varphi}^{A1} = w_h^{A1} = \gamma / p\pi$; the optimal contract to the supervisor is $s_l^{A1} = 0$, $s_{\varphi}^{A1} = s_h^{A1} = \gamma / p\pi$; the expected production and supervision costs are $C^{A1} = 2(1 - p + p\pi)\gamma / p\pi$; the utility function of the principal is $U^{PA1} = \pi X_h - 2(1 - p + p\pi)\gamma / p\pi$.

If 2) $p \rangle \tilde{p}$, the optimal contract to the manager is
$w_l^{A2} \in [0, [p\pi \underline{U} - (1 - p)\gamma] / (p\pi)]$,
$w_{\varphi}^{A2} = w_h^{A2} = [(\gamma + \underline{U}) - p(1 - \pi)w_l^{A2}] / (1 - p(1 - \pi))$;
the optimal contract to the supervisor is $s_l^{A2} \in [0, [p\pi \underline{U} - (1 - p)\gamma] / (p\pi)]$,
$s_{\varphi}^{A2} = s_h^{A2} = [(\gamma + \underline{U}) - p(1 - \pi)s_l^{A2}] / (1 - p(1 - \pi))$; the expected production and supervision costs are $C^{A2} = 2(\gamma + \underline{U})$; the utility function of the principal $U^{pA2}(w, e, m) = \pi X_h - 2(\gamma + \underline{U})$.

Appendix 8.4: Proof of Proposition 4

1. The Optimal Contract for the Manager

In the presence of collusion and abuse of authority ($k_B \neq 0, k_F \neq 0$), the agent's incentive compatibility constraint is

$$p[\pi(w_h - F) + (1 - \pi)(w_{\varphi} - B)] + (1 - p)w_{\varphi} - \gamma \geq pw_l + (1 - p)w_{\varphi} \tag{9}$$

From (9), we can get following condition

$$w_h \leq \gamma / p\pi + F - (1 - \pi)(w_{\varphi} - B) / \pi + w_l / \pi \tag{10}$$

If the principal sets $w_h - w_{\varphi} = 0$ to destroy the stake in abuse of authority, the necessary precondition for the manager to agree with it is $w_{\varphi} - w_l \geq \gamma / p\pi$. If and only if $w_{\varphi} \geq w_l$, collusion between the supervisor and the manager is possible.

The agent's contract must also satisfy his/her participation constraint,

$$p[\pi(w_h - F) + (1 - \pi)(w_{\varphi} - B)] + (1 - p)w_{\varphi} - \gamma \geq \underline{U} \tag{11}$$

And it has to satisfy the constraint of limited liability $w_l \geq 0, w_{\varphi} \geq 0, w_h \geq 0$
From (11), we can get

$$w_h \geq (\gamma + \underline{U}) / p\pi + F - (1 - \pi)(w_{\varphi} - B) / \pi - w_{\varphi}(1 - p) / p\pi$$

The principal can set the payment to the manager

$$w_h = (\gamma + \underline{U}) / p\pi + F - (1 - \pi)(w_\varphi - B) / \pi - w_\varphi(1 - p) / p\pi \qquad (12)$$

From (10) and (12)

$$w_l \geq [\underline{U} - (1 - p)w_\phi] / p \qquad (13)$$

From (13), the necessary condition for the manager to agree with abuse of authority is $w_\varphi - w_l \geq \gamma / p\pi$, then we can get

$$w_\phi \geq (\gamma + \underline{U}) / \pi \qquad (14)$$
$$w_l \geq [(P\pi\underline{U} - (1 - P)\gamma] / P\pi \qquad (15)$$

From (13), we can get

$$w_\varphi \leq \underline{U} / (1 - p) \qquad (16)$$

From (14) and (16), we can know that only if $p\rangle\gamma / (\gamma + \pi\underline{U})$, $(\gamma + \pi\underline{U}) / \pi \leq w_\varphi \leq \underline{U} / (1 - p)$, the principal can set the payment to the manager $w_\varphi = (\gamma + \pi\underline{U}) / \pi$.

If the principal sets the payment to the manager $w_h - w_\varphi = 0$, then he can deter the abuse of authority. Let $F = B = 0$, and replacing $w_h - w_\varphi = 0$ in (12), we can get $w_\varphi = (\gamma + \pi\underline{U}) / \pi)(\gamma + \underline{U})$. To satisfy the incentive constraint and participation compatibility, the principal must offer the payment to the manager, $w_h = w_\varphi = (\gamma + \pi\underline{U}) / \pi$. When the supervisor actively pursues abuse of authority, the manager is likely refuse it only when $w_\varphi - w_l \geq \gamma / p\pi$. To induce the manager to refuse to shirk, the principal has to offer extra money to the manager.

From (15), we can get $w_l = [p\pi\underline{U}(1 - p)\gamma] / p\pi$

If $p \leq \gamma / (\gamma + \pi\underline{U})$, the principal can set the payment $w_l = 0$; to destroy the origin of the abuse of authority, the principal can set $w_h = w_\varphi$. From (5) and $w_l = 0$, we can get $w_\varphi \geq \gamma / p\pi$. If and only if $p \leq \gamma / (\gamma + \pi\underline{U})$, then $\gamma + \underline{U} \leq \gamma / p\pi$, so to satisfy the incentive constraint, the principal must set $w_\varphi = w_h = \gamma / p\pi$.

2. The Optimal Contract for the Supervisor

The supervisor's incentive compatibility constraint is

$$p[\pi(s_h + k_F F) + (1 - \pi)(s_\varphi + k_B B)] + (1 - p)s_\varphi - \gamma \geq ps_l + (1 - p)s_\varphi$$

That is, $s_h \geq \gamma / p\pi - k_F F - (1 - \pi)(s_\varphi + k_B B) / \pi + s_l / \pi$.
To preclude the collusion and abuse of authority, the principal must set $s_h \leq \gamma / p\pi - k_F F - (1 - \pi)(s_\varphi + k_B B) / \pi + s_l / \pi$

The supervisor's contract must also satisfy his/her participation constraint

$$p[\pi(s_h + k_F F) + (1 - \pi)(s_\varphi + k_B B)] + (1 - p)s_\varphi - \gamma \geq \underline{U}$$

$$s_h \geq (\gamma + \underline{U}) / p\pi - k_F F - (1 - \pi)k_B B / \pi - (1 - p)s_\varphi / p\pi$$

Then the principal sets

$$s_h = (\gamma + \underline{U}) / p\pi - k_F F - (1 - \pi)k_B B / \pi - (1 - p)s_\varphi / p\pi$$

If abuse of authority is possible, then $s_\varphi - s_l \geq \gamma / p\pi$. But only when $s_\varphi \geq s_l$, is collusion good for the supervisor, so abuse of authority provides a basis for collusion. Only when $s_\varphi - s_l \geq \gamma / p\pi$ is a tight constraint, from $s_h + k_F F \geq s_\varphi$, we can get that if and only if $k_F F \rangle 0_\varphi$, abuse of authority is good for the supervisor, then the principal can set $s_h = s_\varphi$.

Optimizing yields the following proposition:

Proposition 4: If collusion and abuse of authority coexist, then $k_B \neq 0, k_F \neq 0$, let us denote $\tilde{p} = \gamma / (\gamma + \pi U)$

If 1) $p \leq \tilde{p}$, the optimal contract to the manager is $w_l^{CA1} = 0$, $w_\varphi^{CA1} = w_h^{CA1} = \gamma / p\pi$; the optimal contract to the supervisor is $s_l^{CA1} = 0$, $s_\varphi^{\check{C}A1} = s_h^{\check{C}A1} = \gamma / p\pi$; the expected production and supervision costs are $C^{CA1} = 2(1 - p + p\pi)\gamma / p\pi$; the utility function of the principal is $U^{PCA1} = \pi X_h - 2(1 - p + p\pi)\gamma / p\pi$;

If 2) $p \rangle \tilde{p}$, the optimal contract to the manager is
$$w_l^{CA2} = [p\pi \underline{U} - (1 - p)\gamma] / (p\pi),$$
$$w_\varphi^{CA2} = w_h^{CA2} = (\gamma + \pi \underline{U}) / \pi \text{ the optimal contract to the supervisor is}$$
$s_l^{CA2} = [p\pi \underline{U} - (1 - p)\gamma] / (p\pi)$, $s_\varphi^{CA2} = s_h^{CA2} = (\gamma + \pi \underline{U}) / \pi$; the expected production and supervision costs are $C^{CA2} = 2(\gamma + \underline{U})$; the utility function of the principal $U^{PCA2} = \pi X_h - 2(\gamma + \underline{U})$.

Chapter 9

Strategic Land Management in Germany: One Key for Brownfield Cleanup and Sustainable Development

Detlef Grimski, Fabian Dosch and Herbert Klapperich

Introduction

Soil and groundwater contamination on brownfields was for many years the most dominant aspect of brownfield concern in Germany. Approaches for risk assessment and cleanup technologies have been regarded as key issues for the redevelopment of contaminated sites. The term "contaminated site" was almost used interchangeably with the term "brownfield." Land management and land reuse aspects from the planners' perspective were of minor relevance in the "contaminated land community."

Meanwhile, the German government has developed a codified legal scheme for the management of contamination. Since the beginning of the century strategies for brownfield redevelopment were frequently discussed as part of a broader land management scheme. Brownfield management in this context means a type of land management in the larger scale as well as inner-city planning. The interdisciplinary character leads to an improvement in communications among different stakeholders as well as in research and policy making. It demands "out of the box" thinking and future perspectives that focus on activities and visions. The technical and economic domains related to the brownfield process as well as the ecological ones, are governed by an umbrella-like "stakeholder view" including policy makers, owners, developers, investors, planners, risk-transfer providers, financiers, and marketing.

The management of derelict land is a traditional issue for urban planners in local authorities. Such previously used properties typically arise from land use changes in the process of industrial decline and structural change. Thus, brownfields are quite a common issue in industrialized countries like Germany. For the German planning administration, it is not new to deal with brownfields and to cope with the existence of derelict land. Especially during periods of structural change in German coal and steel mining, local governments have set a specific focus on bringing former land from the coal and steel sector to new uses. After German reunification in 1990, specific brownfield problems also emerged in the new German states as a consequence of industrial decline. Action in the past was locally based and targeted to specific local problems, however.

Germany nowadays—and even more in the future—is affected by far-reaching demographic and economic structural changes. Low birth rates, a dwindling population in a number of old-industrialized regions, an increasing number of smaller households, and an aging population are leading to an even more differentiated demand for land and residential, commercial and industrial buildings. A shift away from unlimited land consumption to policies that foster the redevelopment of existing buildings and land appears imperative.

Since the early 1990s, attention across the globe and in Germany has been given to the revitalization of brownfield properties—with real or perceived contamination. Cleanup strategies and physical methods have been successfully developed. The focus has occurred at all levels of government and in the private sector. Brownfield management today means integrated approaches to remediation, sustainable redevelopment, reuse, and financing, considering social and ecological dimensions (Klapperich et al. 2003; Proceedings 2007).

Reduction of Land Consumption

Since 2002 a new countrywide policy focus has been placed on the development of brownfield land. The German Federal Government's National Sustainability Strategy defines the reduction of greenfield development and the initiation of sustainable land management as a central task, fostering the recycling of derelict land (Perspectives for Germany Our Strategy for Sustainable Development 2002 and progress reports 2004 and 2008, http://www.bundesregierung.de/nn_6516/ Content/EN/StatischeSeiten/Schwerpunkte/Nachhaltigkeit/nachhaltigkeit-2006-07-27-die-nationale-nachhaltigkeitsstrategie.html). One of seven priority fields of action is the reduction of land consumption. In order to achieve sustainable land use, the quality and efficiency of land use must be improved, i.e. through high-quality densification of housing. This also contributes to a significant reduction of CO_2 emissions and the retention of CO_2 sinks. The close relationship and synergies between land use, energy efficiency, and climate protection becomes particularly evident in moderately densified settlement structures.

From a larger perspective, this involves a reorientation focusing on the renewal of old housing stock instead of further expansion of settlements. Commercial premises, housing, and leisure and recreational areas can be built on these old sites, which frequently have good infrastructure, thus reducing pressure on greenfields. The redevelopment of brownfield sites, vacant lots, and empty buildings offers a great opportunity for reducing land consumption. Brownfield redevelopment is identified as one major tool; within the National Sustainability Strategy the government identified a number of indicators and goals to achieve sustainable development for the country. The target for the reduction of land consumption—the use of land for new settlement areas and traffic—is a maximum of 30 ha per day until the year 2020. It is based on a rate of 129 ha per day in the year 2000. The actual rate is still officially 113 ha per day (2004–2007). Nevertheless, during this consumption the portion for new building and related open space alone declined by 50% since the year 2000. Compared to developments abroad, building prices remained moderate in Germany and even declined in shrinking regions. On the other hand, the pressure to redevelop on derelict land remained weak.

Definitions and Data

"Brownfields" are not legally defined in Germany. The term is widely applied to unused or underused land that needs intervention to reintegrate it back into the property market. Thus, the term indicates sites which had previously served commercial, industrial, or military purposes and for which no new use has been found. Given the extent of urban redevelopment activity in many parts of Germany, the term brownfields should also include abandoned housing areas, railway yards, and even abandoned shopping malls and offices. The length of time that the land has remained vacant is also often considered as a decisive factor for classification as a brownfield. This factor is only of minor importance with regard to the regeneration process, however.

In many cases the term brownfield also includes the need to manage contaminated land and groundwater. In fact, a considerable number of brownfields in Germany have pollution problems arising from their previous industrial use. Hence, contamination and associated problems (reuse options, costs, and duration for remediation, liability etc.) are serious obstacles for brownfield redevelopment, to be solved within the regeneration process. Some brownfield-types are regarded as land use potentials for inner-city development together with other land use categories, e.g. gaps between buildings, unplanned interior areas, and vacant buildings.

Germany has no federal statistics on brownfields, their re-use potential, or on-site recycling although there is a periodic survey by the Federal Office for Building and Regional Planning. In 2006, municipalities recognized that around 63,000 ha of brownfields could be reactivated. Generally, more areas are falling into disuse than are being recycled. The German Federal Environment Agency estimates that there are around 150,000 ha of brownfields in the country today (Presse- und Informationsamt 2008). The Federal Statistical Office has calculated that the total amount of derelict land within towns and cities across Germany increased by 12.7 ha per day between 1993 and 2000 (Federal Statistical Office 2003).

Basically, the location of brownfields varies across the country depending on the spatial structure of the area. Regions with a historic industrial structure like the Ruhr area or the Saar region are more affected by industrial and structural change than rural parts of the country. Thus, the density of land falling into disuse is significantly higher in such areas. Therefore, brownfields are providing a considerable potential for urban renewal in city centers and suburbs and substantial conditions for internal development in many places. Especially for inner city development there are excellent opportunities to include brownfields in a broader development strategy which also encompasses other land categories like empty lots, unplanned interior areas, vacant buildings, or other underused properties that are likely to become abandoned in the foreseeable future.

After German reunification in 1990, specific brownfield problems emerged in the new German states. High greenfield consumption promoted by tax incentives met the decline of industry and military conversion, as well as migration from East to West. According to the latest BBR (Federal Office for Building and Regional Planning) survey and estimates, municipalities in former East Germany have an average of more than three times as much derelict land with regeneration potential than their western counterparts. In 2006 the availability of commercial land for reuse was almost 10 times higher than the annual statistical breakdown of demand. Even though this is only a theoretical reflection, one should keep in mind that there is also considerable additional potential of brownfields from military sites, disused railway land, and other similar uses.

Hardcore Brownfield Sites

In many parts of Germany there is a lack of demand for building sites due to a declining population and economic structural change. This is a major hindrance for the successful redevelopment of brownfields in such areas. A significant proportion of brownfield land, specifically in areas with shrinking populations, is not immediately commercially viable enough to be brought back into beneficial use. Without any form of public intervention these sites will remain unused and potentially derelict for the foreseeable future. The consequence is a blight on surrounding areas and communities and the loss of an opportunity to renew the community in a sustainable manner. High costs of reclamation and redevelopment and low market values constitute specific challenges for many cities and regions. The problems associated with these sites particularly relate to the fact that market forces are not the driver for redevelopment. Future use is often limited to unprofitable soft end uses, short-term reuse is unrealistic, and sites are not eligible for the majority of public programs focusing on redevelopment for economically beneficial uses. This status leads to hardcore brownfields that will remain unused or under-used for long periods of time. This extended derelict phase in turn can cause considerable associated urban problems for the economic and social redevelopment of the whole area. One possible option or basic solution for these sites is soft end uses, whether permanently or for interim (or transitional) use (Ferber 2009) or reverse site conversion. The approach of improving the visual impact of the site by the establishment of interim uses is becoming more and more important in other countries, like the United Kingdom (Communities and Local Government Publications 2008: Previously-developed land that may be available for Development: England 2007 Results from the National

Land Use Database of Previously-Developed Land. http://www.communities. gov.uk/-publications/corporate/statistics/previouslydevelopedland2007 [01.07.2009]). The broad majority of brownfields and vacant land, especially in old industrialized regions, is going to be deconstructed, dismantled, and in many cases renaturalized or ecologically upgraded (www.bbsr.bund.de/BBSR/DE/FP/ ExWoSt/Forschungsfelder/StadtquartiereImUmbruch/03__Ergebnisse.html [in German only]).

Responsibilities and Legislation

The Federal government is not directly engaged in the enforcement of land planning or contaminated land remediation. There is no specific brownfield legislation in Germany. The legal basis for brownfield management is laid down in Germany's Environmental and Spatial Planning legislation. It sets the framework for the country. Enforcement, however, is with the Federal States, which have mostly delegated land management and land remediation responsibilities to regional or municipal administrative bodies. The local planning authority is responsible for the approval of land use and buildings; the local environmental authority is responsible for permission and approvals concerning the remediation of contaminated soil and groundwater and for the disposal of waste generated through the redevelopment process. Generally, projects are controlled and guided by one authority—usually the planning authority—within local administration. If other aspects—environment, cultural monuments—are concerned, the guiding authority must coordinate necessary approvals with the other authorities affected.

The Federal Soil Protection Act (Federal Soil Protection Act, http://www.bmu. de/files/pdfs/allgemein/application/pdf/soilprotectionact.pdf) sets the framework for the assessment and remediation of contaminated soil on brownfield land. The Federal Building Code (Baugesetzbuch, BauGB) is the most important law for project planning and building design and defines the best available tools for urban planning. The Federal Building Code emphasizes in its General Provision and in a special section for "*Urban Redevelopment Measures*" the preference for brownfield redevelopment prior to new developments on greenfields. Despite this general principal, greenfield consumption in Germany has increased consistently. To foster inner-city development and the redevelopment of brownfield-sites some amendments were fixed by new federal legislation in the years 2007 and 2008 ("Inner-development law" (Gesetz zur Erleichterung von Planungsvorhaben für die Innenentwicklung der Städte, Bundesgesetzblatt Jahrgang 2006 Teil I Nr. 64, ausgegeben zu Bonn am 27. Dezember 2006); revised version of spatial planning law (Gesetz zur Neufassung des Raumordnungsgesetzes und zur Änderung anderer Vorschriften [GeROG] vom 22. Dezember 2008, Bundesgesetzblatt Jahrgang 2008 Teil I Nr. 65, ausgegeben zu Bonn am 30. Dezember 2008; in particular §§ 2 (2) 6)).

Governmental Action

There is a high level of governmental assistance for brownfield redevelopment related to the national strategy on sustainable development and related efforts to reduce land consumption. The need for integrated work across stakeholder groups and across disciplines encouraged various Ministries to co-operate in this area. The Ministry for the Environment, the Ministry for Transport, Building and Urban Affairs and the Ministry for Education and Research work very closely together to develop strategies and tools for sustainable land use. The overall objective is to integrate brownfield redevelopment into a sustainable land use policy for the country. Goals can be characterized by the following: reorienting from urban sprawl to urban renewal and regeneration; reducting of land consumption by decentralized concentration of urban expansion; fostering inner city development; improving benefits to costs for both developers and local authorities; and preserving natural resources like soil or groundwater.

In this context in 2004, the Ministry for Education and Research launched a special research program "Research for the Reduction of Land Consumption and for Sustainable Land Management" (REFINA), linked to the government's political approach to reducing land consumption in the future. All three Ministries and representatives from Federal State and community governments are steering the research program, which concludes in 2010. The main aspects of REFINA are:

- Model concepts for innovative land management in selected regions with different development conditions;
- Analyses, methods and evaluation methodology for sustainable land management and brownfield redevelopment; and
- Development of new information and communication structures.

Creating new life at old sites is associated with enormous challenges, which are being addressed in a number of recently finished or still ongoing REFINA projects (Federal Ministry of Education and Research 2008). In order to prepare sites in the most inexpensive way and in accordance with demand, it is necessary to identify innovative remediation methods, concepts of planning and use, and new forms of funding so that complex project management tasks can be addressed. The projects regarding recycling of brownfields can be summarized as follows:

Management of Brownfield Projects

Whether in the town or countryside, whether small sites or large areas, the revitalization of derelict land is always a complex management task in which planning, remediation, land preparation, development, financing, and marketing have to be connected. At the same time, the challenge is also to address a wide range of stakeholders stakeholder interests. Typical stakeholders in such processes are: land owners, government authorities, developers, consultants, citizens, and the general public. The site-specific differences of individual cases make it difficult to identify a simple general rule for successful conversion projects. In some of the REFINA research projects special focus is given to the improvement of site data and information about the planning status of land (i.e. reserve building land) using new digital information systems. The development of a private sector funding model for the mobilization of brownfields and its implementation is also important.

Grey Turns to Green

As mentioned above, in many cases, there is little demand for land. In cases where extensive remediation is required it can be difficult to develop derelict land for new commercial purposes at reasonable costs. Ecological upgrading of such unmarketable land may be a meaningful approach—whether as a transitional step (interim uses) or as a permanent solution. Special plants not only provide green cover, they sometimes also have the ability to support necessary remediation processes by taking up and absorbing pollutants. This leads to valuable "second-hand" natural areas that improve residential amenities in conurbations and combat exodus from cities. It is likely that the transition of a site from an abandoned or derelict status to a reserve status could be fairly immediately fulfilled, especially at sites which are already owned by a public authority. For public authorities this approach should be a preferable option due to its cost effectiveness. Hence, the need to explore specific planning and technical approaches for the conversion of brownfields into a reserve status was identified as an important research topic. A few research projects developed advanced strategies and tools for cost-effective utilization and for remediation management of non-competitive (public-driven) sites. New models concerning respective organizational, management, and funding schemes are being developed within several REFINA projects.

Intelligent Remediation Methods

Soil contamination on brownfields is often a significant obstacle for revitalization. Investors cannot be attracted due to long-term liability risks, unpredictable remediation costs, and uncertainties regarding additional remediation requirements that may be detected during the redevelopment process and may affect the intended reuse scheme. The keys are legal and planning security for both investors and public authorities. Research in this field focuses on improved approaches to assessing legal and technical risks and on tools and strategies to cope with them.

Community Co-operation

Regarding land management strategies, the regional cooperation of communities can help to integrate brownfields systematically and comprehensively into land management schemes crossing community borders. There are several options:

one is to have a common regional portfolio of old industrial land and share the development risks among several communities. This would support regional planning and reduce both community competition on the property market and demand for greenfield land in the region. Concerning the properties of large scale private enterprises, the goal is to establish an interface with public authorities, ensuring that private sector portfolio management meets the objectives of public spatial planning on the regional level. Finally, communities can also coordinate their management of shared old industrial landscapes characterized by a high content of hardcore sites by means of green space master planning. For instance, the integration of such areas into regional landscape parks—on occasion associated with brownscape design (the term brownscape design is used for a complex process that integrates cleanup, ecological restoration, landscape design, and land use management involving all affected stakeholders at different steps of the process)—can establish attractive public green belts and set quality standards for different types of land use. Several research projects cover these topics.

Major results of the research program REFINA focus on cost-benefit analysis tools and models, making the follow-up costs of settlement development more transparent (Preuß et al. 2009). In particular, mid- and long-term follow-up costs for the maintenance of technical and social infrastructures have the potential to burden municipal budgets considerably in the future. As such, it appears advisable to promote cost transparency before zoning new land for development by appraising the actual revenues and expenditures this would incur. Comparative follow-up cost analyses of potential building sites aid in strategically defining spatial and temporal priorities for land development in the municipality. This allows analysis and, if necessary, comparison of an array of land development strategies, such as development of building gaps, settlement intensification, settlement expansion, and brownfield revitalization.

General Procedure of the Redevelopment Process

Brownfield redevelopment is a complex process with regard to its underlying legislation, the number of disciplines and stakeholders involved (economists, planners, environmentalists, sociologists), and the variety of management models for the process (Grimski et al. 2001; CLARINET 2002). The management of the redevelopment process can include different organizational and legal institutions which allow for varying intensity of cooperation between partners with specific degrees of liability: informal groups or task forces, planning groups, regional or special-purpose associations, and organizational forms under private law such as associations, limited liability companies, or joint stock companies (see example below).

The complexity of the process is displayed in the flow chart in Figure 9.1 developed by CABERNET (CABERNET: Concerted Action on Brownfield and Economic Regeneration Network, www.cabernet.org.uk).

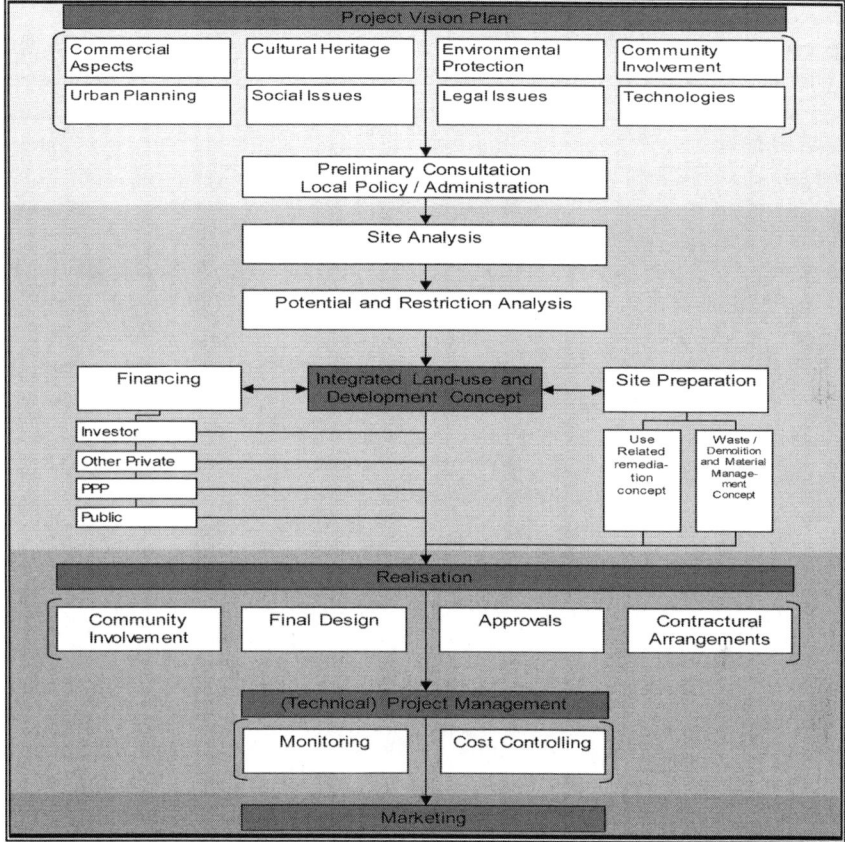

Figure 9.1   CABERNET process

Generally the process depends on a number of factors related to the individual site:

- location,
- type of land use after redevelopment,
- site conditions (contamination, infrastructure, existing buildings etc.),
- preparation needs,
- number of stakeholders to be involved,
- ownership,
- legal status,
- availability of funds, and
- extent of policy involvement.

In some Federal States, like Thuringia or North Rhine Westphalia, the process is promoted and supported by Land Development Agencies. The "Landesentwicklungsgesellschaft Nordrhein-Westfalen," for instance, established a Property Fund in 1982, and thus made the redevelopment of brownfield sites and disused buildings central to its policy of creating an integrated urban development model. The activities of the Property Fund go beyond the establishment of new attractive business parks. They include:

- accumulating wide experience in the economic framework of and prudent dealing with contaminated and derelict sites;
- placing quality targets in urban construction before purely economic considerations, whether it is the architectural plan for the commercial building itself, or surrounding landscaping and recreational areas and facilities;
- preserving industrial architecture which had been abandoned and bears witness to the history of the industrialization which once was important;
- safeguarding monuments such as the coalmine Zollverein XII in Essen or the Landscape Park in North Duisburg, which are now becoming new tourist and cultural attractions within the Ruhr area.

Generally, having a reliable data base of comprehensive site information is the principal requirement for effective land use management and brownfield redevelopment. Due to divergent departmental stipulations, many municipalities currently track derelict land in different administrative sections simultaneously. Environmental agencies record contaminated land or former location registers. Business promotion departments store site information in their trading site records and planning offices trace them through their construction land records. In the future, integrated and proactive action must increasingly replace these sectoral approaches if solid foundations for planning and decision-making are to be created for land recycling and land use management.

Triggers for Kicking off Projects

Since there is no stringent law requiring the redevelopment of brownfields, no typical procedure for the initiation of brownfield redevelopment projects in Germany can be stated. Projects mainly begin on an ad hoc basis, with no over-arching brownfield policy driving them. As a rule, each site is individual and exceptional. Instrumental approaches which sought to reorganize spatial structures by redeveloping vacant land, suitably integrating planning, finance, cooperation, and organizational aspects were never comprehensively discussed. Despite economic growth in the 1990s in both Eastern and Western Germany, the absence of land management concepts was one of the main reasons that the significant potential of derelict land did not stimulate the restructuring and modernizing of established settlement patterns by brownfield redevelopment. Today, in numerous regions the result is the release of new greenfield sites for construction, leading to a derelict land surplus from the 1990s to the present.

There are some aspects that can be considered as almost typical triggers to start the redevelopment process; in practice, however, a reliable trigger is the availability of money, either by funding or other sources. Another starting point for brownfield projects is the purchase of land by private land development companies, which have emerged to buy contaminated land for redevelopment. Their business plan is to clean up the land, make it fit for after-use purposes, and resell it. Typically, these companies are small- and medium-sized and invest mainly in small- and medium-sized real estate.

Furthermore, as other examples of policy-triggered brownfield redevelopment, it is worthwhile to mention two successful International Building Exhibitions in Germany with a special focus on brownfield redevelopment: the current Fürst Pückler Park Building Exhibition ("IBA See") and the more famous International Building Exhibition Emscher Park. Many brownfield redevelopment projects have been part of the International Building Exhibition (IBA) at Emscher Park. The latter was completed in 1999 and aimed at providing overall stimulus for ecological, economic, and social restructuring by establishing new landscape schemes, brownfield redevelopment for technology centers, and innovative housing schemes.

The combination of the factors mentioned above along with soil contamination and the need for action frequently also trigger reclamation of derelict land for reuse. The steps required for the management of contaminated sites are laid down in the Federal Soil Protection Act and the Soil Protection Acts of the individual Federal States. The general procedure for contaminated site management includes identification, assessment, remediation, and monitoring of contamination. The intended reuse of land must be considered during each of the management steps, especially for setting clean up targets by the local authority. Thus, it provides significant contributions for planning security and reliability for the determination of remediation targets.

Drivers, Failure Factors, and Perspectives

The decision making of local administrators is based on the legislation mentioned above, which has supported economic competitiveness, social fairness, and ecological sustainability. The keyword is sustainable development, which closes the circle on Germany's national strategy on sustainability that was mentioned in the introduction. It aims for reduction of land consumption, i.e. preserving green space. It can be argued that this strategy has been a significant driver for a number of initiatives on the Federal State and community levels to put brownfield redevelopment higher on the political agenda. While there are many Federal initiatives for inner-city development and urban sprawl avoidance, the redevelopment of brownfields is one activity within the broader approach of sustainable land management contributing to the protection of natural resources.

Nowadays there are "green potentials under a blue sky" in old industrialized regions. Despite brownfield "lighthouse projects" long-term disused spaces are increasing and the loss of open spaces is continuing. The processes of decline require new strategic regional governance for the treatment of patchwork settlements and open space structure. It should extend beyond the presently segmented area-related governance. The aim is a green, attractive agglomeration area, networked with increasing leisure and recreational areas. Important measures include new interim uses and concepts for the adoption of public space through the greening of disused areas. New approaches for the revitalization of spaces in the Ruhr-area could be based on energy use. The formal basis of all of which could be an integrated policy framework for circular land use management (Dosch et al. 2008).

The planning practice of local authorities needs to do more to meet the specific demands of land revitalization. This applies equally to private developers. Strategies for reactivating brownfields must penetrate the highest planning levels. The high cost of brownfield redevelopment and the reluctance of banks and insurance companies to support efforts (due to real or perceived environmental risks) must be tackled as barriers to private initiatives. In contrast to planning on greenfield sites, the juxtaposition of living and working environments, the existing infrastructure, and the uncertainties connected with potential contamination all require flexible planning approaches. One aspect of this is that municipal planners must reach agreement with investors and future users at an early stage.

It is recommended that the urban planning process for reclassifying an abandoned industrial site should be implemented in two phases:

- Phase one—pre-planning: the municipality sets the general strategy for the future use of an area in consultation with project sponsors. An informal urban development plan should be the outcome of this phase.
- Phase two—binding construction planning: the development plan takes on more precise contours and results in legally binding commitments in the form of a charter and a land use or infrastructure development plan.

According to the latest (2006) BBR survey on derelict land and brownfields redevelopment, municipalities have recognized the growing potential of brownfields as land for business use on greenfields decreases, while brownfields increase. Accordingly the ratio of building activities on green- and brownfields is getting narrower. Looking at intended uses, the perception since the survey of 2000 has not changed significantly. According to the vision of local administrators, two-thirds of derelict land is destined for future commerce, although this is rather unrealistic due to a lack of demand. Demolition of building infrastructure is a common tool of urban revitalization especially in old industrialized regions. According to a survey of the federal program Stadtumbau Ost, "Urban Redevelopment in East Germany," of 223 municipalities in Eastern Germany, 85% of the designated demolition areas are intended for interim-uses or renaturalization, or in other words not for housing or building purposes (Bundestransferstelle Stadtumbau Ost 2007).

Due to the complexity of the process there are still a variety of project specific barriers to success. The most significant are the following:

- competition with greenfields,
- running out of money due to limited funding,
- running out of money due to inappropriate cost calculation,
- delay of approvals/lack of approvals,
- slow decision making and deceleration of the project,
- lack of communication among stakeholders with different interests,
- bad communication on the level of local administration,
- bad process management and deceleration of the project,
- inflexible requirements of the local administration concerning site remediation for the intended reuse option, and
- subsequent requirements of the local administration.

Economic Classification of Brownfield Projects

The classification of brownfields according to project types in terms of cost/yield ratios is a promising approach to land use management and a tool for understanding economics, brownfields and priority setting. Different types of brownfield regeneration projects, in relation to their economic status and funding, are illustrated by the **A-B-C Model**. (CABERNET: Concerted Action on Brownfield and Economic Regeneration Network, www.cabernet.org.uk).

Depending on the cost of regeneration and the value of the land, sites can be classified as:

- Project category A "Self-sustaining land": Projects can finance themselves, e.g. they are optimized by integrating land use and regeneration planning and by profits from planning.
- Project category B "Development land": These projects are characterized as being on the borderline of profitability. Projects are only made possible by public support through start-up financing and/or risk-sharing between private investors or developers and state funders, e.g. through public-private partnerships. Greater risks must be accepted when the cost/profit ratio is close to one.
- Project category C "Reserve land": these projects represent mainly public sector or municipal projects driven by public funding or specific legislative instruments. Short and medium-term regeneration projects cannot be expected to succeed without support. Low land prices, extensive start-up costs, and frequently high spatial concentration of derelict land are the main reasons this type of project is not profitable.

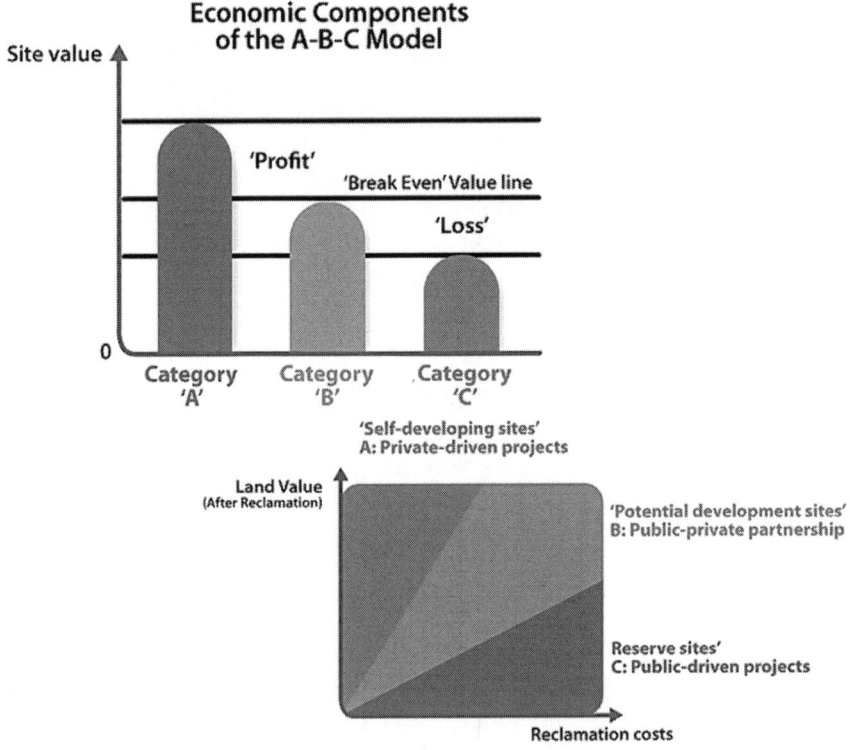

Figure 9.2   Economic components

Projects that are completely private are mainly market driven (A projects). Governmental intervention and steering are possible through the approval process. Projects of category B receive public support. Consequently, various local administrative priority criteria determine the execution of the project, for example:

- additional funding by third parties,
- compatibility with a specific overall land management strategy (i.e. circular land use management, see below),
- urban planning targets (i.e. inner-city development),
- environmental effects (soil and groundwater cleanup),
- effects on social environment (housing etc.),
- economic effects/economic prospects of the region,
- revenues for the public (taxes, employment), and
- social acceptance.

Redevelopment projects of category C sites are rather rare. A precondition for such a project is that there must be strong public interest to justify short term economic loss.

A State wide mapping in Thuringia, in 2002, for instance, concluded that 90% of all brownfields are located in suburban or remote areas. On the Federal level this figure seems to be far too high, but brownfields in suburbia are growing (Ruff et al. 2008). Many of these category C brownfields are not really "on the radar of the local representatives." They present a big challenge for the future because their short term reuse will be impossible and their visual appearance is an additional obstacle to attracting investors. A promising approach for local activity related to such sites is currently under investigation. The approach aims at converting the site into a reserve status by

- demolition as well as clearing in the area;
- simple landscaping, covering and construction of terraces, and planting; and
- where appropriate, construction of recreational paths.

This approach would at least remove the negative appearance of the site (also for investors) and improve the environment. Public action on C sites can also be triggered by legislative tools. For example, the Federal Soil Protection Act requires that soil and groundwater contamination be remediated if there are hazards for individuals or the general public. According to German legislation the owner is responsible for the action. However, if the owner is insolvent or cannot be identified, safeguarding of the site by the public authority through "execution by substitution" is permitted.

Funding

Germany has funding sources for cleanup of contaminated land, for urban renewal, and for structural change on the local, Federal State, and national level. In former East Germany, a special governmental funding program has been set up to reduce investor liability for soil and groundwater contamination caused in the past on former publicly-owned industrial land. The funds are allocated for the remediation of soil and industrial groundwater and are particularly aimed at attracting investors to redevelop former GDR sites.

Besides specific funds for contaminated land there are also funds to support urban renewal and to assist communities coping with the challenges of structural change. Essential national funding sources are the urban development assistance funds. The Federal level provides assistance to towns and cities most affected, within the scope of the program of urban restructuring in the old Laender. The regeneration of derelict urban sites resulting from the process of economic and military structural change can be part of projects within a broader context so that they also function as a trigger for brownfield redevelopment. The same is true for the structural funds of the European Commission. They have a broader focus but enable communities to redevelop brownfields within broader projects.

Brownfields in a Broader Land Use Management Strategy

Redevelopment of brownfields is an important part of any land use management strategy. Significant progress for a systematic approach on brownfield redevelopment is represented by a new, action-oriented model: "Circular land use management" (Preuß et al. 2008). The idea is to ensure that new uses of land reutilize existing areas (land rotation through land recycling) to avoid the development of greenfields. Circular land use management is a key policy and strategic approach for implementing the two-pronged strategy of quantity and quality management necessary to achieve sustainable land use. Circular land use management primarily focuses on systematically exploiting the potentials of existing structures and reusing derelict land. It describes a cyclic process encompassing planning, utilization, termination of use, abandonment, and finally reintegration. The reactivation of brownfields as part of this process is welcome not only from an environmental standpoint, but also because it satisfies economic requirements (by avoiding investment in new infrastructure and optimizing exploitation of existing infrastructure) and social needs (by contributing to functional and social inclusion). If the recycling mentality and efficient use of resources can become standard in urban and regional land management then sustainable urban development will progress from theory to reality.

Circular flow land use management embodies a different philosophy of use, which is expressed by the motto: "avoid—mobilize—revitalize" (http://www.flaeche-im-kreis.de/english_version.phtml). This management approach accepts the development of greenfield sites under specific conditions, but primarily it seeks to utilize the potential of all existing sites including brownfield and greyfield areas.

Circular flow land use management also aims to provide an integrated political and governance approach that includes the whole spectrum of policy areas and fields of activity. It is implemented at both local and regional levels, combining them in an integrated urban and regional land development policy. The cycle relies on the interplay between strategies and instrument, in different fields of activity and with a suitable comprehensive deployment of tools (instrument mix) in these areas, including planning, land information, cooperation, organization and management, investment and support programs, marketing, and legislation (Preuß et al. 2006). The realization of a Circular land use management policy requires on-site analysis, the mapping of potential areas, regulations and agreements on land use, the elaboration of integrated action plans, and the use of a policy mix that pools existing and new tools in planning, information, organization and cooperation, funding and budgeting, marketing, and dispositions.

New tools that provide goal-oriented incentives and are capable of influencing current land use and zoning practices effectively are necessary because existing instruments have only limited potential to achieve these objectives. Shrinking regions and cities require additional specifically tailored instruments to address the consequences of population loss, vacant buildings, functional deficits, and surplus space. Urban redevelopment, funding programs, and brownfield strategies must consider specific regional aspects. Certain urban redevelopment procedures can implement measures for the conversion of developed sites into open spaces. This also requires future programs preventing the land from being used as building sites and allowing temporarily free sites (green interim uses), by rezoning. In the long run such funding and organizational approaches (B-Site Pools, C-Site Funds) can contribute to low-cost consolidation of low-demand markets in shrinking regions. This would, in turn, create better opportunities for the reuse of B Sites. Thus, urban redevelopment concepts play a prominent role in shrinking regions because they are inherently integrative and implementation-oriented. Under certain conditions public and real estate industry activities can also serve as an impetus for the development of brownfields and other empty lots. The main role of public authorities is marketing activities like attracting investors and providing stewardship.

Final Remarks

Basically, environmental and planning legislation in Germany is quite advanced and provides a good foundation to redevelop brownfield sites in a sustainable way. The biggest project-related challenge is coordination of the process including communication with stakeholders. This is true for internal project management as well as at the administrative level. For the project level, good experience has been had with multi-skilled project managers who are sensitive to all aspects of the process and who have the power to become widely accepted. On the administrative level, more leadership, coordination, and integration are favorable for the process. A precondition is the will of the community to tackle the problem of derelict land. Although multidisciplinary project groups within local authorities are increasing, there is still a lot of uncoordinated work. This results in a lack of synergy.

Furthermore, there is still a lack of stakeholder-related project tools. One example is a tool to scan for the most important information to stimulate a project. Such a start-up plan was developed in a funded project of the Ministry for Education and Research as part of the cooperation between Germany and the U.S. EPA (http://www.bilateral-wg.org/). The idea behind it is to provide a holistic project and business plan tailored to a specific vacant space and stakeholder perspectives. It focuses on data relating to information, communication, project planning, and acquisition of funding, which are of primary importance to the target groups concerned.

The start-up plan for brownfields collates site-specific information for stakeholders previously only available from a wide variety of sources. The plan thus facilitates crucial information flows and helps to stimulate interest in brownfields and overcome existing prejudices. This makes it a particularly suitable instrument to kick off category B or C projects, which depend on interventions from public and private stakeholders. This tool neither replaces the general construction and remediation planning required by law, nor the wide array of successful informal planning measures, however.

Finally, in summary, there are several general lessons learned by reviewing the brownfield policy situation in Germany:

- Brownfield projects need a long time for preparation and even longer for execution. If this takes too long, the project fails.
- Without money, there is no brownfield redevelopment—funds are urgently needed to stimulate investment in brownfields.
- For many C site brownfields, short term redevelopment remains unrealistic.
- New approaches like interim uses and reverse site conversion are needed. The greening and renaturalization of abandoned or derelict land and the connection of green belts are major tasks of adaptation to climate change in city regions.
- No brownfield redevelopment will be successful without cooperation across stakeholder groups and disciplines—the project manager must have the skills to integrate needs and requests.
- Information exchange about best practice approaches is essential to encourage brownfield redevelopment.

References

CLARINET. 2002. Brownfields and Redevelopment of Urban Areas: A report from the Contaminated Land Rahabilitation Network for Environmental Technologies, Austria, Federal Environmental Agency. Available at http://www.umweltbundesamt.at/fileadmin/site/umweltthemen/altlasten/clarinet/brownfields.pdf (accessed: December 19, 2009).

Dosch, F. and L. Porsche. 2008: "Green Potentials under a Blue Sky. New Approaches for Area Revitalization and Open Space Development in the Ruhr Area." In *The Ruhr Area in the Focus: Today's Situation and the Road Ahead.* Information zur Raumentwicklung IzR 9/10.2008.

Federal Environmental Agency 2007. "Proceedings 2nd International Conference on Managing Urban Land—Towards More Effective and Sustainable Brownfield Revitalisation Policies." *SAXONIA*, 697.

Ferber, Uwe. 2008. Federal Ministry of Education and Research: Paths to Sustainable Land Management – Topics and Projects in the REFINA Research Programme. Research for the Reduction of Land Consumption and for

Sustainable Land Management, brochure, 24, Berlin. Available at www.refina-info.de/en/index.phtml (accessed: December 19, 2009).

——. 2009. Available at www.refina-info.de → KOSAR (accessed: December 19, 2009).

Grimski, D. and U. Ferber. 2001. "Urban Brownfields in Europe. Land Contamination and Reclamation." 9(1): 143–8. Available at http://epppublications.books.officelive.com/Documents/09-1-14.pdf (accessed: December 19, 2009).

Klapperich, H., V. Franzius, D. Medearis and C.D. Shackelford. 2003. "Proceedings International Conference on Green Brownfields II – From Cleanup to Redevelopment." vol. 1: *Verlag Glückauf, Essen*, 392 S.; vol. 2, CiF e. V. publication 1—2003, 281 S.

Presse- und Informationsamt der Bundesregierung. 2003. Federal Statistical Office: Umwelt. Umweltproduktivität, Bodennutzung, Wasser, Abfall, Wiesbaden.

——. 2008. *Fortschrittsbericht 2008 zur nationalen Nachhaltigkeitsstrategie.* Berlin, p. 145 f.

Preuß, T. and U. Ferber. 2006. "Circular Flow Land Use Management: New Strategic, Planning and Instrumental Approaches for Mobilisation of Brownfields." Difu occasional papers. online Version.

——. *Bundestransferstelle Stadtumbau Ost. 2007. 5 Jahre Stadtumbau Ost—eine Zwischenbilanz.*

——. 2008. "Circular Land Use Management in Cities and Urban Regions – A Policy Mix Utilizing Existing and Newly Conceived Instruments to Implement an Innovative Strategic And Policy Approach," difu-papers. Available at www.difu.de/index.shtml?/publikationen/liste-jahr.phtml (accessed: December 19, 2009).

——. 2009. "Making the Follow-Up Costs of Settlement Development More Transparent. Cost-benefit Analysis Tools and Models." Difu occasional papers. online Version.

Ruff, A. and C. Wittemann. 2008. "Flächenrecycling in suburbanen Räumen." Nordhäuser Hochschultexte. *Schriftenreihe Ingenieurwissenschaften Heft 3.*

PART III
Brownfield Case Studies

As witnessed by the chapters in this volume, there is a wide and ever expanding literature on public efforts to spur the remediation and redevelopment of brownfields. Much of this literature, however, is limited in scope. It tends to be primarily descriptive, often focusing on formal rules and programs that have been implemented across various political jurisdictions. Existing comparative analysis tends to remain at the level of formal structure. The chapters presented in this section try to make the brownfield discussion more concrete by focusing on specific projects and outcomes. As these chapters clearly show, it is critically important that scholars and practitioners have a clear idea of the actual impact of specific projects as well as an understanding of broad program organization and goals.

Exploring the impact of brownfield redevelopment at the project level leads to consideration of what are perhaps the two most common questions asked by citizens about any public policy. The first question is simply whether there is any evidence that a given program "works?" The second question asks under what circumstances the program is most likely to be successful. Before either question about brownfield remediation can be answered, however, there first needs to be a consensus on how success is defined. Perhaps the most useful organizing principles for brownfield policy success are the classic economic ones of effectiveness, efficiency and equity. Effectiveness can be measured in terms of whether incentives are actually a factor in achieving community goals. In the case of brownfield remediation and redevelopment activities that might include qualitative improvements in the state of the land, reduced health threats to residents, and/or increased job or housing opportunities as the result of new uses for brownfield sites. Efficiency measures the costs and benefits of brownfield policies against an alternative set of policies that might be expected to have similar results; cost is often the principal consideration. The principal equity question is how brownfield policies (or potentially lack of policies) affect residents or other proximate land uses and how action or inaction in one community or jurisdiction impacts others.

Too often public policies are not designed or selected with goals or community definitions of "success" in mind. They are selected because they are available, used by other communities, familiar to public officials, seemingly low cost or, most frequently, because they have been used in the past. This leads to a situation where *policies* drive the brownfield redevelopment enterprise rather than a community's *goals*. Ideally, in rational decision processes, goals promote the selection of policies. Thus, policy-makers need to identify the challenges and goals of their particular community, and then assess alternative development policies in light of those goals as well as a consideration of fiscal resources—both public and private—and scientific knowledge about potential risks and benefits of any remediation plans. In short, policy choice should logically be connected to community goals.

The process of determining development goals is complex and can be time consuming. There are a number of alternative mechanisms for gaining input to weigh and determine community goals: discussions with elected and appointed officials, business and citizen surveys and focus groups, analysis of economic profiles and forecasts, planning charrettes, liaison with chambers of commerce and other business groups, meetings with neighborhood associations and so on. There are both indirect costs and indirect benefits to be considered. Although a detailed quantitative analysis is a theoretical ideal, in practice the necessary data are extremely difficult to obtain. When an analysis is done, it is often a mixture of quantitative and qualitative measures. In addition to dollar values for costs and measurable benefits, analysis could include attempts to at least identify the populations that will be affected, even when exact values cannot be assigned to the costs and benefits.

In different ways, each author in this section helps identify existing complexities, ambiguities and sometimes contradictions in forces driving the definition of a successful brownfield program. One tension, which has already been frequently cited, is that between the goal of site remediation and the goal of economic development. A consistent finding in this volume is that although brownfield redevelopment is typically framed as an environmental policy, economic concerns almost always trump environmental ones. Consider, for example, the nature of published evaluations of brownfield programs. Typically economic indicators such as jobs created, wages paid, land values, or changes in economic production dominate.

Berman et al. explore this question of how to define success by reviewing several alternative frameworks which might be utilized to evaluate specific project outcomes. These authors especially consider efforts to link brownfield redevelopment to indicators of public health and sustainability in two neighborhoods in the City of Milwaukee. Berman and his colleagues note that, while goals of public health and sustainability have widespread popular support, efforts to incorporate relevant indicators of these concepts within an evaluation scheme typically involve significant effort and cost. The practical value of such efforts is also brought into question by the fact that, at least in the case of Milwaukee, political authorities have little interest in using identified indicators as measures of success or failure. They argue that the local political process forces decision makers to continue to focus on more narrow economic returns.

Welsh and Jones examine outcomes in 57 brownfield projects that received public funding through the Michigan Department of Environmental Quality. They conclude that one important source of the ambiguity in defining program success is the multiplicity of goals assigned to such projects. For many these goals extend beyond the familiar ones of environmental cleanup and local economic development. For example, community and neighborhood advocates argue that brownfield projects should be evaluated in terms of the project's impact on the entire community in which it is located rather than the site alone.

The range of goals assigned to brownfield redevelopment projects is quite large. Some see such efforts as a tool to limit urban sprawl and promote some variation of "smart growth." The argument is based on the claim that brownfield redevelopment can divert development from green fields to previously developed locations. Note that the likely result of such a diversion would be to channel development to older, often depressed, urban centers. Indeed, some political authorities in Michigan see brownfield redevelopment as the state's major urban revitalization tool. The appeal of this potential indirect management of development is particularly strong in a state like Michigan that has extraordinarily decentralized land-use authority. Advocates of controlled growth have long been frustrated by the fact that the state has almost no power to modify local development decisions and have eagerly embraced what seems to be a politically viable alternative to such regulation.

Although they identify a number of dimensions on which success might be measured, Welsh and Jones's empirical discussion relies on a much more straightforward criterion of whether the proposed construction on a given site was actually completed. In many ways this seems to be a defensible decision in that project completion is likely a necessary if not sufficient condition to generate the externalities that many argue will follow from a successful brownfield program. They identify a number of correlates of program completion. Several of these factors go beyond the specific project, and include evidence that projects have been developed within the context of broader regional planning. There also is evidence that fostering strong public awareness and support for the project predicts project completion. Perhaps not surprisingly, a clear vision of the redevelopment and strong and enthusiastic leadership is also associated with program completion.

Yang extends the analysis of program success by identifying a set of contextual variables that impact outcomes. She examines program outcomes in Pittsburg and Cincinnati. The comparison is interesting in that both cities are of similar size and, more importantly, share a similar history of industrial development and decline. Both also exist in a policy framework created by the city's respective states. Each has a brownfield program that emphasizes relief from potential liability and offers a variety of financial incentives. In spite of such similarities, the actual implementation of specific programs was quite different in the two cities. Relative to projects in Pittsburg, Cincinnati's tend to be designed and implemented in a more "closed" process in which disclosure about details is more limited. As a result the aggregate statistics which might be used to compare the two cities are problematic since many of the Ohio cleanups are thought to be what Yang calls "off the books."

Alexander reviews two brownfield projects from an even narrower prism, exploring the process by which one project succeeded and one failed in a single city (Rochester, New York). Given that both of these projects occurred in the same city, a number of variables often cited to explain differences in outcomes are held constant. At a conceptual level Alexander attempts to build an integrated framework linking policy tools and public management, what he terms "contextual interaction theory." He argues that variations in stakeholder strategies and goals will drive policy tool selection and ultimately outcomes. Specifically, he posits that differences in policy outcomes can occur even within similar environments. For example, he notes that although both Rochester projects sought some measure of liability relief, one project emphasized the comprehensiveness of relief, and the other the speed at which the relief could be obtained. The difference in the policy tool selected was best explained by the goals of local officials and potential developers. Ultimately, however, the more rapid liability relief did not provide sufficient protection to developers to spur redevelopment on the target site.

Taken together, these chapters point to the inherent complexity of any evaluation effort. The most practical, concrete constraint is obtaining sufficient data to estimate project outcomes. Consider, for example, the difficulty of estimating the magnitude of job creation. Few states have the capacity to monitor on-site employment, much less estimate the proportion of that employment generated by site redevelopment efforts. Indeed, it seems clear that any estimate of employment growth will be problematic.

Beyond the important issues of indicator definition and estimation are the critical conceptual issues of how one defines project success. Several chapters in the section, even after noting the multidimensional character of expected outcomes, relied on the relatively minimal criteria of whether the site obtained basic environmental protection standards and if on-site construction occurred. Such a definition, however, fails to capture most of the public justification for brownfield programs. The design and implementation of adequate evaluation schemata will undoubtedly continue to be a major component of the ongoing public review of brownfield redevelopment.

Chapter 10

Fostering Brownfields Development in Rust Belt Cities:
A Comparison of State Approaches and their Impacts in Cincinnati and Pittsburgh

Andrea E. Yang

Brownfields redevelopment, or the cleanup and development of former industrial sites that are or are perceived to be contaminated by pollutants, has been discussed as a means of fostering urban revitalization, an alternative to sprawling development of open land, and promoting spatially efficient living patterns that reduce automobile traffic and air pollution. According to U.S. EPA's website, for every acre of brownfields redeveloped, an estimated 4.5 acres of greenspace is preserved, so that brownfields redevelopment is an "integral component to smart growth." Beyond minimal grant assistance, the federal government's efforts to foster brownfields development has been through limitation of federal CERCLA/ Superfund liability for new land owners taking on cleanup and redevelopment responsibilities (Cohn 2004: 674).[1] State environmental protection and development agencies typically provide technical and financial assistance, loans and tax credits, and limitation of state liability.

Despite these oft-pronounced benefits, metropolitan areas have experienced varying levels of interest and success in brownfield development. Using two rust belt cities—Cincinnati, Ohio and Pittsburgh, Pennsylvania—as examples, this chapter compares state programs and legal frameworks and their success in fostering urban brownfield development. Part I describes the challenges former industrial properties pose to redevelopment of rust belt cities, focusing on Cincinnati and Pittsburgh. Part II briefly describes federal liability issues under CERCLA and Small Business Liability Relief and Brownfields Revitalization Act of 2002 (BRERA). Part III compares the approaches adopted by the Ohio and Pennsylvania legislatures. Part IV examines the relationship between state approaches and local development by comparing Cincinnati and Pittsburgh. In particular, this section will attempt to link achievements in brownfields development in the two cities to

---

1   BRERA authorized $200 million per year for assessment and remediation grants and loans compared to the $878 million to $2.4 billion estimate in lost local tax revenues (Cohn 2004: 674).

differences in state programs and legal frameworks, local economic conditions, local government involvement, and political will.

Brownfields: A Challenge to Revitalizing Rustbelt Cities

From the mid-1800s, heavy manufacturing fueled the development of Cincinnati, Pittsburgh and other Midwestern Rust Belt cities. In 1900, Cincinnati and Pittsburgh were respectively the tenth and eleventh largest cities in the US (Gibson 1998) (see Figure 10.1). In the late 1970s and 1980s, however, US manufacturing and Rust Belt cities experienced a major downturn (Faberman 2002).[2] The sectoral shift from a manufacturing to service economy combined with the geographic shift of residential, commercial, and industrial development to suburban municipalities and counties had a dramatic impact on these urban areas (Hamilton County Regional Planning Commission [HCRPC] 2004: 16-17, 3).[3]

During this period, the City of Pittsburgh lost over 90,000 manufacturing jobs (Rause 1989). From 1950 to 1990, its population fell by nearly half from 676,000 to 369,000 (Gibson 1998). Cincinnati experienced a similar manufacturing downturn with its population falling from 504,000 to 364,000[4] (HCRPC 2004, 16-4). Though the 1990s brought some recovery, this growth was largely outside the urban core (U.S. Census Bureau 1998; HCRPC 2004: 16:11, 8).[5] New infrastructure, housing, and commercial, and industrial development sprawled on greenfields on the metropolitan fringe, while urban industrial properties were abandoned, including the very properties that fed the city's historic growth (HCRPC 2004: 16-4, 9; Ohio EPA 2005: 1).

---

2   Rust belt states averaged 0.7% annual growth, about one-third the national average.

3   A majority of regional growth since the 1980s has occurred outside Hamilton County (Cincinnati). This pattern of central city population and investment loss and concomitant growth in suburban fringes is common to US major metropolitan areas (HCRPC 2004: No. 16-76-8).

4   In the 1960s, manufacturing supplied 35% of Hamilton County jobs, while services made up 17%. By 1987 manufacturing jobs dropped to 20%, while service industry jobs rose to 28%.

5   Cincinnati and Pittsburgh, and other cities have seen population and economic loss in the urban core, which correlate with the growth of outlying areas. According to 1960–2000 census data, Hamilton County's (Cincinnati) population as a percentage of the metropolitan area decreased from around 56% to 43%, while Pittsburgh's population went from around 60% to 55% of the county population. Note that around 1980 the decline of Pittsburgh's county (Allegheny) leveled off as a percentage of the region, while Cincinnati (Hamilton County) continued its rate of decline.

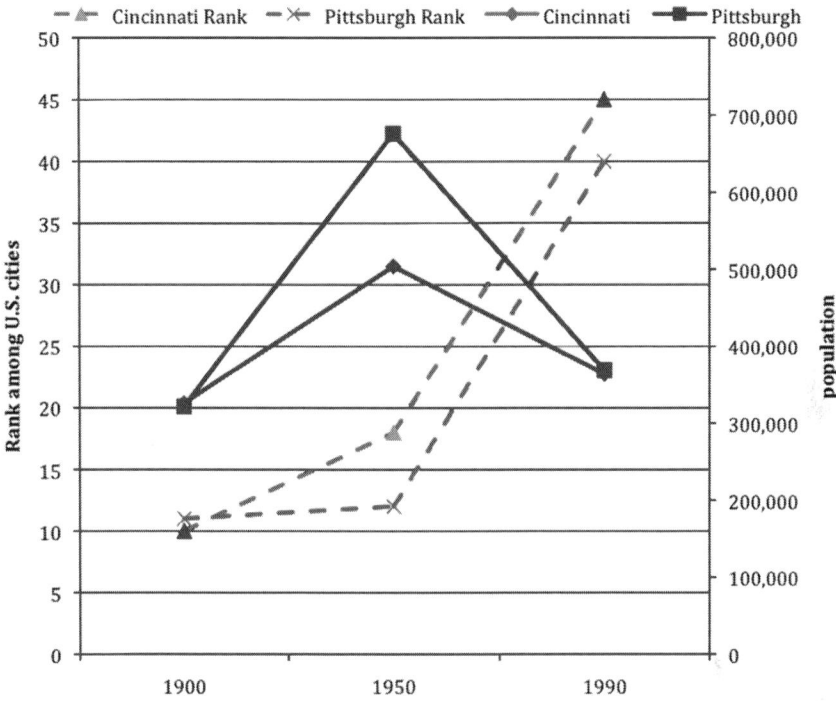

Figure 10.1 City rankings

The federal Comprehensive Environmental Response, Compensation, and Liability Act of 1980 (CERCLA), defined brownfields as "real property, the expansion, redevelopment, or reuse of which may be complicated by the presence or potential presence of a hazardous substance, pollutant, or contaminant." The Urban Land Institute estimates there are over 150,000 acres of abandoned industrial land in the nation's major cities, while over 450,000 brownfields sites are located across the country (Turner 2005; U.S. Economic Development Admin. 2005; U.S. GAO 1996: 4). Ohio alone has an estimated 4,000–6,000 brownfields sites (U.S. GAO 1996: 4). In older industrial cities, 15% of a city's property may be vacant or abandoned, (Davis 2002: 5; HCRPC 2004: 16:11, 9; LeBlanc 2006: 4), presenting an added challenge to revitalizing central cities and urban neighborhoods. For example, Dayton, Ohio is home to about 25 brownfield sites estimated at 250 acres with over 50 acres surrounding and impeding the economic expansion of its downtown (U.S. Conf. of Mayors 2003: 23; Turner 2005: 4). Cleveland, Ohio has an estimated 14,000 acres of brownfields, leading its former Mayor Mike White to name environmental contamination as the number-one obstacle to urban redevelopment (Davis 2002: 9; GAO 2005: 14).

Federal Framework for Brownfields Development

The abandonment of urban industrial sites may be traced to federal and state laws creating strict cleanup liability for property owners and operators (U.S. Conf. of Mayors 2003). In the wake of Love Canal, Congress passed the 1980 Comprehensive Environmental Response, Compensation, and Liability Act (CERCLA) to stop illegal discharge of pollutants. The 1,250 most contaminated industrial sites were listed on the National Priorities List (NPL) making them eligible for federal cleanup funds (Davis 2002: 6). For sites with lower levels of contamination, Congress gave the federal government power to hold "potentially responsible parties" (PRPs)—including past and present owners and operators, arrangers and transporters of waste—strictly, retroactively, and jointly and severally liable for the cost of remediation and damage to natural resources regardless of their role in causing the contamination (Davis 2002: 9; *B.F. Goodrich Co. v. Murtha*, 958 F.2d 1192, 1198 [2d Cir. 1992]). CERCLA liability deterred investment in industrial property because even an unsuspecting purchaser who later discovers contamination could be held liable (Davis 2002: 6–7). Lenders and owners of potentially exposed adjacent properties could also be liable as PRPs. For example in *U.S. v. Fleet Factors Corp.*, 901 F.2d 1550, 1557 (11th Cir. 1990), *cert. denied*, 498 U.S.1046 (1991), the Eleventh Circuit Court of Appeals held that a lender could be liable under CERCLA if it exercised a degree of financial management that indicated a capacity to influence the corporation's treatment of its hazardous waste (Davis 2002: 8). Lenders refused loans for projects with any possibility of environmental liability (Bartsch and Collaton 1997: 15–29; Davis 2002: 8). Property owners, unable to sell or fearful that cleanup might trigger liability, began to abandon or "mothball" their facilities (Davis 2002: 2). Thus, earlier CERCLA versions unintentionally inhibited cleanup and redevelopment of retired industrial sites, creating a preference for undeveloped "greenfields" and an additional drag on revitalization of urban communities (Davis 2002: 8; U.S. Conference of Mayors, 10).

Although the 1986 and 1996 amendments attempted to limit innocent purchaser, lender, and fiduciary liability, only 22 years after CERCLA's enactment did BRERA, the Brownfields Revitalization and Environmental Restoration Act bring comprehensive liability relief (Small Business Liability Relief and Brownfields Revitalization Act 2002, Title II, PL 107–18; Hird 2002: xxxv). BRERA increased EPA's financial and technical assistance for brownfields assessment and cleanup; clarified liability and exceptions for prospective purchasers, innocent and contiguous landowners, and small volume contributors, and expanded CERCLA enforcement limitations for voluntary cleanups under state programs (U.S. EPA 2002).

BRERA's expanded purchaser liability exemptions were heralded as "perhaps the most important" to encourage brownfields redevelopment (Hird 2002: xxxvii). The 1986 amendments (CERCLA §§101(35), 107(b)(3)) included an "innocent purchaser" defense if owners were without knowledge of contaminants at time of purchase, had made "all appropriate inquiries," and exercised due care after learning of contamination. Under BRERA, knowledge of contamination at time of purchase is not a bar to the "bona fide prospective purchaser" defense, as long as the purchaser meets §107(a) requirements—such as making appropriate inquiries as to contamination and agreeing to exercise "appropriate care" to prevent and stop releases of hazardous substances (Hird 2002: xxxvii; Mintz 2002: 410–15). Contiguous landowners who did not cause or consent to the release, did not know or have reason to know about contamination from a contiguous property, take reasonable steps to prevent and stop the release and limit exposure to the hazard, and comply with other requirements may be eligible for liability exemption (CERCLA §107(q)). Although historical fears are difficult to overcome, these protections may encourage some projects, especially where states are active in promoting brownfields redevelopment (Davis 2002: 8; Hird 2002: xxxvi).

BRERA also aims to encourage state governments to address brownfields development by enhancing their role and available resources and providing for greater coordination between state and federal brownfields efforts (BRERA §1281 U.S. EPA 2004: 3). Prior to BRERA, state voluntary cleanups approved for state liability protection were still subject to liability under federal law (Green 2004: 580–81). BRERA allowed a streamlined process for developers to receive protection from both CERCLA and state law liability for state programs meeting federal statutory requirements (Hird 2002: xli).[6]

State and Local Programs: Laboratories for Brownfields Development

Even as BRERA addresses federal impediments to brownfields development, EPA recognizes that state programs should be "at the forefront of brownfields cleanup and redevelopment" (U.S. EPA 2004: 3). With an estimated 95,000 acres of abandoned brownfields in 195 cities (U.S. Conf. of Mayors 2003: 12), state and local governments have a greater stake in and are better positioned to stimulate brownfields development. These idle properties represent loss of local tax revenue, employment, and multi-billions of dollars of wasted investment in abandoned or underused urban roads, transportation, water, and other infrastructure (Bartsch and Collaton 1997: 2).[7] Brownfields hinder waterfront

---

6   Note that the CERCLA enforcement bar for voluntary cleanups under state programs is more limited than the bonafide prospective purchaser, innocent landowner, and contiguous landowner exemptions. A number of exceptions or "reopener" provisions allow the federal government to take enforcement actions, for example, if the "contamination migrates across state lines, or presents an imminent and substantial endangerment to public health, welfare, or the environment."

7   Of the mayors' survey respondents, over 60% stated that if brownfields were redeveloped, they could realize $790 million to $1.9 billion annually in additional tax revenues. Further, 148 cities responded that over 570,000 new jobs could be created on

and downtown revitalization, invite arson, dumping, rodents, and vandalism, and threaten the health, safety, and property values of nearby residents (Bartsch and Collaton 1997: 2–3; Davis 2002: 3; Citizen's Guide to Cincinnati Brownfields, 5).

Early on, many states enacted liability provisions parallel to CERCLA, with similar negative impacts on property cleanup and development (Green 2004: 582). Since the mid-1990s states have attempted to reverse this trend by addressing a variety of redevelopment barriers including ambiguous legal liability, unclear or inconsistent cleanup standards, lack of expertise, high and unpredictable assessment and remediation costs, insufficient financing, inconsistent and inadequate redevelopment policies, public opposition, limited demand for redeveloped sites, and competition from greenfields (Davis 2002: 9). State Voluntary Cleanup programs (VCP) typically encourage any volunteer, including owners and new purchasers, to participate in the cleanup and redevelopment of brownfield properties, by offering incentives, including covenants-not-to-sue under state environmental law for voluntary cleanups that meet state standards of public health and environmental protection (Ohio EPA 2001: 1; Green 2004: 582; Penn DEP 2005: 2). In addition, state programs may offer financial elements (grants and loans, tax incentives, and insurance), technical elements (assessment and remediation standards, methods, and technical assistance), and public notification and participation requirements (U.S. EPA 2004: 3). However, there are differences among these programs. Some programs emphasize coordination between economic development and environmental remediation resources through assistance in marketing brownfields sites to potential developers. Some states have tried to reach more properties by reducing enforcement and oversight to "efficiently" use limited staff resources (Bartsch and Collaton 1997: 11; Green 2004: 582). State VCPs also differ in the degree of agency oversight, sources of program funding, and fees charged for services (U.S. EPA 2004: 3). Finally, some states have developed Memoranda of Agreement (MOA) with U.S. EPA where the VCP standards and processes meet federal standards to allow cleanup volunteers to simultaneously obtain a non-enforcement agreement under state law and a limited EPA enforcement bar under CERCLA (Hird 2002: xxi–xxii). BRERA §128(b) expanded the availability of this federal enforcement limitation such that voluntary cleanups under state programs which meet the minimum standards and processes set out in the newly codified sections of CERCLA would also be eligible for the limited CERCLA enforcement bar, even without an MOA (Hird 2002: xliii–xliv).

The following section compares two programs—Ohio's relatively hands-off private sector approach and Pennsylvania's engaged approach to brownfields redevelopment.

---

brownfields sites." Over half of the 244 responding cities said that they could accommodate 43 million additional residents without additional infrastructure.

Ohio

Ohio's brownfields program features certification of private sector professionals, state cleanup standards, liability limitations and technical and financial assistance. Created in 1994 and fully implemented in 1997, Ohio's Voluntary Action Program (Ohio VAP) (Ohio Revised Code [ORC] § 3746 et. seq.), emphasizes the use of state certified private sector professionals to assess and cleanup an eligible site to Ohio EPA standards—with or without Ohio EPA oversight (Ohio EPA 2001). This approach is thought to harness private sector skills and resources to facilitate cleanups that might otherwise have stalled because limited state oversight resources were committed to the most contaminated properties (Ohio EPA 2001: 1).

Ohio Administrative Code (OAC) § 3745-300 defines the Ohio VAP standards, which vary by the intended land and groundwater use. The numerical soil standards for each of over 40 chemicals vary according to industrial, commercial, or residential use, from least to most stringent (Ohio EPA 2001). Groundwater standards are similarly linked to potable or non-potable use. Some large cities, such as Cleveland, Columbus, and Toledo, have specified certain industrial areas as "urban setting" groundwater areas for which the non-potable standard applies (Davis and Kwasniewski 2002: 813).

Three different cleanup paths occur in Ohio with varying levels of government oversight, procedural requirements and liability limitation. Two cleanup tracks are described on the Ohio EPA website: 1) in a "classic" cleanup the certified professional submits a No Further Action letter to Ohio EPA to request a Covenant not to sue (CNS) under Ohio law, and 2) under the "MOA track" the volunteer follows Ohio VAP procedures and additional U.S. EPA requirements to obtain both state and federal liability limitations. In the "quiet remediation," a third type reported to occur in Ohio, a certified professional performs the cleanup to Ohio VAP standards without Ohio EPA review or official release from state law liability.

Under the classic cleanup, as long as a certified professional and laboratory verify the site cleanup as meeting Ohio VAP standards, the volunteer may request the certified professional submit a No Further Action Letter (NFA) for Ohio EPA to review and grant a CNS releasing the volunteer from civil liability for further cleanup and investigation under state law (Ohio EPA 2001: 1–2; Davis and Kwasniewski 2002: 813).[8] Ohio EPA has 30 to 90 days, depending on the complexity of the review, to grant or deny the CNS based on compliance with the VAP statute, adequacy in protecting public health and safety, or fraudulent

---

8   The NFA includes a Phase I property assessment of historical and current uses of the property, and if necessary, a Phase II assessment verifying that water, soil, and sediment samples are below the relevant standard, and that any remediation or controls that have been implemented are in compliance with the standards. *Ohio Administrative Code 3745 300. Ohio Rev. Code Ann. §3745.12.* The covenant does not impact the rights of the federal government or third parties to sue.

submission (ORC § 3746.13; Davis and Kwasniewski 2002: 815). In reality, however, six months is a more realistic timeline for Ohio EPA issuance of a CNS (Davis and Kwasniewski 2002: 815). Covenants and associated land use restrictions for the property are filed in the county recorder's office and run with the land, providing future purchasers, lenders, or others with notice of the terms (OAC § 3745-300-13(K); Davis and Kwasniewski 2002: 816). The form of the recorded CNS is subject to the requirements of the Ohio Uniform Environmental Covenant's Act (UECA) (ORC §§5301.80-5301.92) effective in 2004, which also provides that an environmental covenant is perpetual (unless terminated or limited by its terms), runs with the land, and is enforceable without regard to the common law covenant requirements such as privity of estate or contract. Although the UECA does not subject superior interest holders to the terms of the CNS, Ohio EPA's template UECA covenant requires any persons owning interests in conflict with its use and activity limitations to execute and record an agreement to subordinate their interests thereto (Ohio EPA 2005). Under ORC §3746.17 and OAC §3745-300-14, Ohio EPA shall audit at least 25% of No Further Action Letters to ensure the action complies with applicable standards and to review the qualifications and work of the certified professionals and certified laboratories.

For added assurance against CERCLA enforcement, Ohio EPA also offers an "MOA track" under an agreement with U.S. EPA. Volunteers fulfill the same VAP standards as the "classic" version of the VAP, but with greater Ohio EPA oversight and additional notice and comment (notice of entry, etc.) and public involvement requirements. The cleanup must have been granted a CNS, and must not be subject to certain other state and federal environmental actions, permits, etc. (U.S. EPA Region V and State of Ohio 2001; Ohio EPA 2001: 3).

"Quiet remediations" supervised by a certified professional to comply with Ohio VAP standards have provided enough assurance for some parties to go through with the finance or sale of property without pursuing the official CNS. Under the Ohio VAP, "state approval (i.e., CNS issuance) may be comforting but is not necessary" (National Brownfield Association Ohio Chapter 2005: 13). Davis estimates a "significant number" of these informal cleanups occur (Ohio EPA 2001: 2; Davis and Kwasniewski 2002: 815–16). Furthermore, volunteers and certified professionals are not obligated to notify Ohio EPA if the cleanup does not meet standards unless the volunteer pursues a CNS, so as not to discourage entry into the program. Nor is any information resulting from Ohio VAP participation admissible or discoverable in any legal action against the volunteer unless the certified professional finds a "sufficiently important" threat to human health or the environment (ORC §§3746.28, 3746.071).

Technical assistance: Ohio also offers financial and staff technical assistance to local governments, communities, Ohio VAP participants, and non-participants to promote brownfields development. Ohio EPA's Site Assessment and Brownfields Revitalization Program (SABR) serves as a first contact for questions or concerns about brownfields properties (Ohio EPA SABR 2005). SABR assists local governments with information, training, and advice to define community needs and identify available technical and financial assistance (Ohio EPA 2003: 1; Ohio EPA SABR 2005). SABR maintains the state's public record of scheduled cleanups as well as the voluntary Ohio Brownfield Inventory on its website to assist communities in attracting business and development to Ohio (Ohio EPA 2003: 5; Ohio EPA 2007: 23). For a fee, Ohio EPA staff also offers VAP technical assistance to private and public entities to review technical risk assessment groundwater reports and operation and maintenance, sampling and other plans, investigate sites and cleanup decisions, and provide legal guidance on deed restrictions (Ohio EPA 2001: 2). According to its website, Ohio EPA also has a mobile Site Investigation Field Unit (SIFU) that provides on-site sampling, assessment, and characterization assistance at the request of public entities.

Financial incentives: Multiple state agencies offer financial assistance and incentives for both public and private entities. Cleanup volunteers obtaining a CNS automatically receive a 10-year property tax abatement from the Ohio Department of Taxation, which may be further supplemented by local government tax abatements (ORC §5709.87). Ohio EPA's Water Pollution Control Loan Fund offers low interest loans for brownfield activities with water quality benefits (up to $3 million for VAP investigation or remediation projects; Ohio EPA 2001: 4; Reidy 2005; Ohio EPA 2007: 10). The Ohio Department of Development's (Ohio DOD) website reports that its $7 million Brownfield Cleanup Revolving Loan Fund, capitalized by a U.S. EPA grant, provides up to $2 million in below-market-rate funding to public and private entities to fund cleanup, removal, and public participation (odod.state.ohio.us). It is one of the largest brownfields loan programs in the country.

In 2000, voters approved the Clean Ohio Fund for $400 million of bond funding and extended it in 2008 for another $400 million. Half of the original fund was dedicated to brownfields, with $175 million dollars going to the Clean Ohio Revitalization Fund for grants and loans to public entities for evaluating, cleaning, and returning sites to productive use, emphasizing mixed-use activities, and $25 million to the Clean Ohio Assistance Fund for public entities for investigation or cleanup in specified economically distressed areas where the contamination threatens public health and where economic incentive to cleanup the property is inadequate.

Ohio's VAP program has been praised as "one of the most progressive and complete state voluntary cleanup programs" (Davis and Kwasniewski 2002: 819), and for "maximiz[ing] resources and expertise in the private sector" (Green 2004: 588–9). As of February 2009, 342 no further action letters had been submitted to Ohio EPA since the inception of the VAP program, with 239 receiving CNSs, 42 denied, 32 pending, and 17 withdrawn (Ohio EPA 2009). However, this number has been said to undercount cleanups as it excludes "quiet remediations" (Ohio EPA 2001: 2; Davis and Kwasniewski 2002: 815–16). As of 2005, Ohio EPA had provided technical assistance to over 300 more volunteers and owners. Additionally, 7,000 new jobs (part and full time) had been created at these sites. The Clean Ohio Fund had provided over $15 million in assistance funding and

$75 million in revitalization funding to 88 sites across the state, stimulating $930 million in investment and creation of 6,700 new jobs (Turner 2005: 21; Yersavich 2005: 20–21, 128). As of December 2008, 72 Clean Ohio Revitalization Fund and 117 Clean Ohio Assistance Fund projects had been awarded, totaling more than $215 million to over 76 communities in Ohio (Ohio DOD 2008). In 2008–2009, Ohio DOD has $34 million in loans and grants available for cleanup and demolition for projects in economic development priority areas.

Table 10.1   State comparison

	Pennsylvania	**Ohio**
State level cleanup standards	Yes	Yes
State law non-enforcement agreement for approved cleanups to standard	Yes	Yes
Lender liability limitation	Yes	Yes
Cleanup track for state and Federal non-enforcement agreements	Yes	Yes
State approach	Coordinate environmental and economic development (financing/permits)	Maximize limited government oversight resources; certify private sector professionals
Financing: loans, tax abatements and grants for assessment and cleanup	Yes	Yes
Brownfield development featured in state and local political campaigns	Yes	No
Reported results	Over 3,000 sites cleaned up to state standards under Land Recycling Program	239 Ohio EPA issued covenants not to sue since 1995; more uncounted quiet cleanups

Pennsylvania

Pennsylvania fosters brownfield redevelopment through its Voluntary Cleanup Program and its Brownfield Redevelopment Program, which includes the Brownfield Action Team's (BAT) remediation management and coordination support for projects sponsored by local governments (Penn DEP 2009). In 1995, the state legislature enacted Act 2 to create the Land Recycling Program, Pennsylvania's VCP (PA. Stat. Tit. 35, §6025.302-.305; "Penn LRP" or "Act 2"), to address the strict liability and cleanup standards that discouraged cleanup and redevelopment of industrial properties.[9] Like Ohio's VAP program, Penn LRP encourages private sector cleanup and redevelopment of industrial sites. Act 2 set broad goals of making contaminated sites safe, returning them to productive use and developing urban industrial sites to stimulate economic development and preserve farmland and greenspace by reducing pressure for development. The four basic elements of the Penn LRP are the uniform health and environment based cleanup standards, liability releases, standardized review procedures and time limits, and financial assistance (Penn. DEP 2005: 1).

Statewide standards: Prior to the implementation of the Penn LRP, Pennsylvania had required that brownfields be cleaned up to pristine pre-human conditions. Penn LRP provides for three categories of cleanup: 1) background standards, 2) statewide health standards, and 3) site-specific standards developed for the specific condition and uses of the site (Collings and Quimby 2006: 851–5). A person performing either a voluntary or state mandated cleanup has some flexibility to select a standard considering permitted controls, costs, allowed land uses, and liability limitations. *Background standards* require the cleanup of the increment of contamination from on-site sources to reach "background" levels determined through comparison with the concentration of substances offsite. Thus, the standards may be less stringent in areas of heavy industrial use. Engineering controls such as fencing to reduce exposure may not be used to meet background standards. *Statewide health standards* are set based on state and federal health based standards specific to the contaminant, the medium (ground water, soil), and the use (residential/nonresidential; drinking/agricultural water) (Collings and Quimby 2006). As required by statute (PA. Stat. Ann. Tit. 35 § 6026.303), statewide health standards cleanups must remove all air and surface water discharge sources, meet soil standards to a depth of fifteen feet, and may not apply engineering controls or use limits (fencing) to achieve standards. Different groundwater concentration standards may be requested for non-agricultural or non-drinking water aquifers (Collings and Quimby 2006: 858). Risk assessment tools are used to develop *site-specific standards* for cleanup and exposure restrictions that limit health risks for a particular site and use. For example, cleanup standards must ensure that health risk for cancer from carcinogens does not increase by more than 10^4 (Collings and Quimby 2006: 854).

Act 2 also allows DEP to develop limited cleanup standards for a "special industrial site"—an abandoned site in a state "enterprise zone" or one for which no financially viable responsible party can be identified to clean up the site (PA. Stat.

---

9   The three act package includes Act 2, the Land Recycling and Environmental Remediation Standards Act, which creates the standard setting process, sets a process and timeline for DEP approval of cleanup plans, grants full releases from state liability when standards are met, and provides assessment and cleanup funds. Act 3 is the Economic Development Agency Fiduciary and Lender Environmental Liability Protection Act. Act 4 is the Industrial Sites Environmental Assessment Act, which provides up to $2 million in incentives for assessment and cleanup.

Ann. Tit. 35 §6026.502). DEP assists a person, who has not caused or contributed to releases on the property, to acquire and remediate "only immediate, direct or imminent threats to public health or the environment, such as drummed waste, which would prevent the property from being occupied for its intended purpose." For example, groundwater contamination need not be remediated when it will not be used under the intended use, as it does not pose an immediate, direct, or imminent threat to health or environment (PA. Stat. Tit. 35 §6026.502, Collings and Quimby 2006: 855). In exchange, the agreement provides release from liability for all identified existing contamination (PA. Stat. Tit. 35 §6026.502).

Standardized procedures: The Land Recycling Program also standardized cleanup and review procedures and made state agency timing for approval predictable. State and local permits are not required for Penn LRP remediations unless required by the federal government (Penn DEP 2005: 2). For cleanups to background standards and state health standards that take less than 90 days, the party need only perform and document investigation, remediation, and attainment of cleanup to standard levels and submit the report to DEP. DEP must review and respond with approval or a report of deficiencies within 60 days or the cleanup is automatically approved (Collings and Quimby 2006: 865, PA §6026.303). Cleanups of more than 90 days require additional notice, including pre-cleanup publication of intent to cleanup, notification to DEP and the municipality, and notice following submission of final report. For site-specific standards, a thorough investigation report, exposure risk assessment, cleanup plans, notice requirements, municipality opportunity to comment, and DEP approval of cleanup plans and standards are required (PA. Stat. Tit. 35 §6026.302-.305). DEP has 90 days to review the final report. Deed notice is required for all cleanups leaving contamination above background or residential standards (Collings and Quimby 2006: 856). For "low risk sites"—smaller less contaminated sites—DEP accepts the reports sealed by a professional without further review and grants administrative approval within 10 days. While state enforcement is unlikely for voluntary cleanups conducted to state standards even without notice and approval procedures, the remediator would not receive documented liability relief (DePasquale 2006).

Liability releases: Persons who undertake voluntary or state-mandated cleanups are released from liability when DEP approves the final report documenting achievement of Act 2 cleanup standards (Penn DEP 2005; Collings and Quimby 2006: 856).[10] The release applies to the person completing the cleanup and to current and future owners, developers, occupants, successors and assignees, and public utilities (PA Stat. Ann. Tit. 35 §6026.501(a)). Performers of environmental

---

10   Liability release is subject to reopeners for fraud, contamination exceeding standards (failed remedy or newly-identified), and changes in risk due to land use changes, or new toxicity or exposure pathway information. The person performing the cleanup is responsible for additional remediation due to risk increases except when the increase is due to land use changes, in which case the person changing the use is responsible for additional cleanup.

site assessments are protected as long as the assessment is performed with "professional care." Act 3 provides for defense to liability for environmental cleanup for fiduciaries, lenders, economic development agencies, and financiers, if they did not cause the contamination (Penn DEP 2005; Collings and Quimby 2006: 851).

In April 2004, Pennsylvania and USEPA entered a MOA to create the One Cleanup Program, making the Land Recycling Program the first to provide a one-stop process for liability release under state standards and three key federal environmental laws—CERCLA, the Resource Conservation and Recovery Act (RCRA), and the Toxic Substances Control Act (TSCA) (U.S. EPA Region 3 and Penn DEP 2005). One Cleanup provides the advantage of incorporating RCRA federal notice and comment requirements, the federal law most likely to come into play in cleanups (Penn DEP 2005). The three paths of liability release are: 1) the Simple Approach, 2) Penn DEP lead, or 3) U.S. EPA lead. The degree of supervision and other requirements depend on the chosen state standard, path of liability release, and whether federal regulatory "triggers" such as contaminated groundwater exceeding maximum levels, waste in place requiring pathway elimination, subdivision of property prior to remediation of the whole, sites undergoing EPA corrective action, and remediations that do not include all known substances. If no federal triggers exist the cleanup is eligible for the *Simple Approach*, that is, routine Act 2 remediation to Pennsylvania background standards with updates, review, and final approval by DEP and U.S. EPA, and public notice of U.S. EPA decision. If any of the triggers exist, a Penn DEP or USEPA project manager becomes more involved in the process. Where Penn DEP oversees the remediation, cleanup is conducted to a state site-specific standard or health standard following state procedures and in compliance with RCRA, and following U.S. EPA requirements for public comment and statement of basis. If federal law triggers, a longstanding U.S. EPA relationship with the facility, or other concerns are identified, U.S. EPA takes the lead in coordination, oversight, and ensuring compliance with both Act 2 and RCRA standards. Both agencies review the final reports and U.S. EPA provides a RCRA statement of basis and public comment (U.S. EPA Region 3 and Penn DEP 2005).

Pennsylvania's Uniform Environmental Covenants Act (UECA), like Ohio's, standardizes the form of environmental covenants, their effect, and recording, and tracks future compliance with Act 2 controls and standards. According to the Penn DEP website, the agency is developing the "Pennsylvania Environmental Covenant Registry" under UECA §6512 to provide information about sites, copies of the covenants, and links to the county records office where the covenants are filed.

Financial and other assistance: Pennsylvania also provides an array of financial and technical assistance for brownfields development through several agencies, including the Department of Community and Economic Development (DCED) and Penn DEP. Penn DEP's Brownfield Action Team (BAT) provides a project manager to facilitate priority municipal brownfield redevelopment projects in Keystone Opportunity and other targeted zones by coordinating across state agencies to streamline regulatory and permit requirements and funding access and requests (Penn DEP 2004, 2009). DCED's Industrial Sites Reuse program (ISR) awards financial assistance to applicants who have not contributed to contamination for sites with pre-July 18, 1995 industrial activity (Penn DCED 2008). Award criteria include area economic distress, matching leverage, development potential, and local development priorities. Public entities are eligible for grants and low interest loans to cover up to 75% of assessment and remediation costs up to $200,000 and $1 million, respectively, while private parties may apply for the loan component (Penn DCED 2008: 5). Voters in May 2005 approved a $625 million bond issue for Growing Greener II to be awarded over six years to foster business investment, community revitalization, and environmental enhancement, with $50 million allocated to DCED for housing and mixed use community revitalization projects and $230 million to address environmental contamination including brownfields (Governor's Press Off. 2005; growinggreener2.com 2005). In addition, DCED's general economic development efforts, described on newPA.com, also target brownfield development are available: Tax Increment Financing (TIF) Guarantees, Keystone Opportunity Zones tax incentives to distressed socioeconomic areas with the infrastructure and energy to support additional development, Businesses in Our Sites funds to make a site shovel ready (acquisition, assessment, remediation, demolition, and infrastructure installation), and Elm Street Grants for planning, technical assistance and physical improvements to revitalize neighborhoods near central business districts.

A number of Pennsylvania initiatives focus on providing information on brownfields and remediated properties. Brownfield Inventory Grants (BIG) assist local public entities to inventory brownfields sites and gather information on infrastructure and environmental contamination, and are used to estimate the number of brownfields and populate the state's brownfields directory to attract investment to specific sites (Collings and Quimby 2006: 863). The state actively markets brownfields to potential buyers and developers. Previously, DEP operated PAsitefinder.com, including a database of brownfield and other sites targeted for redevelopment incentives, searchable by location, sale/lease price, property and building size, and information about technical and financial assistance for cleanup and development, including a directory of brownfields specialists—attorneys, appraisers, lenders, environmental consultants, and realtors. The brownfields information has now been integrated into the state's economic development efforts at pasitesearch.com, hosted by the Team Pennsylvania Foundation, a public private partnership including political, education, and business leaders focused on workforce development, job growth, and attraction of businesses.

As of June 2009, over 3,000 sites are listed on the Penn DEP webpage as having completed Penn LRP cleanups to background, statewide health, or site-specific standards, with most cleanups undertaken under the statewide health standards (Collings and Quimby 2002: 863). Pennsylvania claims that LRP cleanups resulted in creation or retention of over 76,000 jobs from 1995–2006.

The LRP was awarded the 1997 Ford Foundation/Harvard University "Innovations in American Government Award" and the 1997 Innovations Award from the Council of State Governments as listed on the DEP website (Kennedy School 1997). The American Legislative Exchange Council has endorsed the Land Recycling Act as a model for other states, and the American Planning Association (APA) named it one of the 25 most significant planning laws from 1972–2003 (APA 2003). In recognition of the state's leadership role in brownfields development, President Bush signed BRERA into law in Pennsylvania.

Comparing Local Brownfield Development

With similar size and industrial histories, Cincinnati and Pittsburgh face similar challenges to brownfield development. However, the state program approaches and local conditions have resulted in differences in the brownfields development in the two cities.

Pittsburgh:
"Information is Key to Redevelopment"—Urban Redevelopment Authority

The Penn DEP website lists 54 approved cleanups on 19 different sites in the city of Pittsburgh (multiple cleanups or contaminants are separately listed), including four high profile urban riverfront sites totaling over 400 acres redeveloped into successful mixed use properties. A numerical comparison with Cincinnati is difficult, because Ohio does not requiring reporting of cleanups (Fischer 2006; Page 2006). The amount of public dialogue and availability of information on brownfields development is in itself a difference between the approaches in the two cities. Pittsburgh appears to benefit from greater support from and collaboration between environmental and economic development institutions and public-private efforts to foster brownfields redevelopment. For example, a Boolean search of Westlaw's "allnews" database of general news sources for "Cincinnati" and "brownfield" in the same paragraph (Cincinnati /p brownfield!) from January 1, 2005 to May 4, 2006 recovered 17 references to Cincinnati brownfields development, while a similar search for "Pittsburgh" and "brownfield" (Pittsburgh /p brownfield!) found 48 relevant results. Searches for articles from 2001 to May 2006, and 1996 to May 2006 for Cincinnati revealed 83 and 162 articles, respectively, while a parallel search for Pittsburgh had 184 and 301 references in the corresponding periods. Although brownfields development was not always the main subject, even the more frequent mention likely indicates a greater degree of dialogue on the subject.

According to Deborah Lange, Executive Director of the Western Pennsylvania Brownfields Center (WPBC), "brownfields" has become a part of the political dialogue in Pittsburgh and Pennsylvania (Lange 2006). Politicians, including Pittsburgh's Republican mayor Bob O'Connor, Democratic Allegheny County Executive Dan Onorato, and US Senator Melissa Hart campaigned on the issue of brownfields development. Former Pittsburgh Mayor Tom Murphy, current Democratic Governor Ed Rendell, and former Republican Governor Tom Ridge, who signed Act 2, made brownfields development an administration priority. One of Murphy's first acts as mayor was to "snap up 1,000 acres of brownfields" (Sheehan 2002; Lange 2006). The visibility and energy from Pittsburgh's completed developments have bred further efforts and stimulated interest and political will (Lange 2006).

The degree of public information, discourse, and institutional support may also be seen in the number of websites discussing and promoting brownfields development. In this respect, Pittsburgh has the advantage of the state's open approach to dealing with contaminated properties. For example, the PAsitefinder website, now part of PAsitesearch.com, allowed brownfield owners to connect with interested purchasers via a searchable database. A May 2006 search of brownfield properties for lease or sale, ranging from less than one acre to 250 acres, resulted in 58 available sites in Allegheny County. The PAsitefinder website further offered a step-by-step guide to cleanup, public notice, and information on tax breaks and other incentives. Sellers could also use the site to create brochures and print out success stories to aid in marketing their sites. Note that a search of the newer PAsitesearch.com, a collaborative effort of environmental and economic development agencies, found only five Allegheny County brownfield properties. The Urban Redevelopment Authority of Pittsburgh (URAP; ura.org), the City of Pittsburgh's economic development agency, and the Western Pennsylvania Brownfields Center (cmu.edu/steinbrenner/brownfields/index.html) also have well-developed websites that document public and private brownfields efforts, success stories and awards and may stimulate dialogue and foster creativity and innovation.

In addition, the websites' content reflects an existing exchange of ideas and institutional support for local brownfields development. At the local level, the URAP fosters Pittsburgh's economic development by assembling sites for redevelopment, linking developers with available sites (with a website description of available URAP sites), partnering with community development corporations, and offering loans to businesses, developers, and homeowners. Brownfield redevelopment is featured prominently as part of its economic development mission, which includes job creation, business development, and neighborhood revitalization. URA has reclaimed thousands of acres of brownfields, playing a major role in public financing of several large, high-profile projects that were not economically feasible with private funding alone, including: Summerset at Frick Park, redevelopment of a 238-acre former waterfront slagheap into a residential development and 106-acre urban park extension and rehabilitation, and South Side Works, transforming a 123-acre former riverfront steel mill into a mixed use development with over 350 housing units, a 10-theater cinema, Steelers training facilities, and spaces for the Immigration Service offices, and other commercial, office, and retail tenants. These projects involved significant public funding including Tax Increment Financing bonds and State, County, and Water district funds, with $275 million in private investment leveraged by $83.5 million in public assistance in 2003. The $33 million invested in South Side leveraged $74 million in private financing and increased market value of the housing from $92,000 in 1994 to more than $300,000 (URAP 2006: 1, 25–7). Pittsburgh URAP projects have received national recognition from the American Institute of Architects, National Historic Planning Landmarks Award, and the American Society of Highway Engineers.

Pittsburgh brownfield development has also benefited from federal, state and local funding resources. For example, funding sources for the URAP's $103 million Southside Works project include $2.5 million in the U.S. Department of Housing and Urban Development Grants, $16 million in state grants, $25 million in TIF financing, $21 million from the City of Pittsburgh/URA, and $12 million from the city's water and sewer authority (URAP 2006). In 2005, Allegheny County brownfield projects received $1.7 million in state Growing Greener II grants (Governor's Press Office 2005). Pittsburgh and Allegheny County each have a development fund, which includes brownfields in their funding activities. Revenue from the 1% Regional Asset District sales tax is apportioned half each to Allegheny county and the county's local governments (Orfield and Luce 2001: 9). Pittsburgh dedicates part of its share toward service on the $60 million bond used to establish the Pittsburgh Development Fund (PDF). The county uses its share for regional facilities, such as the stadium, zoo, and libraries, including a similar county fund (URAP 2004: 6). These PDF and Allegheny County revolving loan funds have financed nearly 50 urban development projects, including industrial sites reuse, through 2003. While public funds have assisted in promoting voluntary cleanups, however, the majority of remediations are conducted with private funding (DePasquale 2006).

Pittsburgh's universities, such as Carnegie Mellon and the University of Pittsburgh, have attracted funding and fostered an interdisciplinary and collaborative approach for brownfields research, training, and technical assistance across economic development, public policy, engineering, architecture, and other departments. The Western Pennsylvania Brownfield Center (WPBC), affiliated with Carnegie Mellon, showcases brownfields case studies and develops brownfields information and decision-making tools. The WPBC received congressionally earmarked Small Business Administration funding to assist communities and small businesses to overcome redevelopment obstacles for small industrial sites with little appeal in terms of return on investment (Lange 2006). The center is reviewing and integrating land use and tax archives of former industrial sites into the Pittsburgh Neighborhood & Community Information System (PNCIS) mapping data. A five-year, $900,000 U.S. EPA Grant awarded in 2008 will enable WPBC to develop and provide user training for a site selection tool for communities, brownfields stakeholders, and public decision-makers to compare and assess the development impacts of various brownfields and greenfields sites on social, economic, and environmental health criteria (*Pittsburgh Post-Gazette* 2006). Other efforts include creative approaches to liability and funding for industrial sites redevelopment, such as a Charitable Trust Fund Model. Because investors with "less of an understanding about brownfields might be scared away by potential environmental contamination and the uncertainty of the cost to clean the property to a level that is consistent with the future use of the property," such research, knowledge, and experience may help to foster brownfields redevelopment (Carbasho 2005; Lange 2006).

Clearly, the synergy between federal, state, and local political leadership, financing incentives, technical assistance, academic research and outreach, marketing, and the general environment of openness have created a level of brownfields activity and dialogue that has furthered redevelopment of Pittsburgh's industrial properties. However, other factors have contributed to mobilizing interest. For example, the decline of Pittsburgh's steel industry in the 1970s and 1980s resulted in the near collapse of the local economy, with the loss of over 90,000 jobs and unemployment of over 14% (Rause 1989). These dire economic circumstances mobilized over 200 government, business, and community representatives to convene in the 1982 Allegheny conference, to strategize to diversify the city's economy, reduce its dependence on heavy manufacturing, and rebuild the local economy (Rause 1989).

According to Lange, these abandoned steel sites allowed redevelopment of large tracts of riverfront property to house biotech and other new industries (Carbasho 2005). For example, the 48-acre abandoned Jones & Laughlin Hot Strip Mill was transformed into the Pittsburgh Technology Center: a campus hosting several university offices, corporate research headquarters, and technology support industries (URAP website). The Urban Redevelopment Authority, as developer of last resort, has led many of these large scale public-private brownfields developments. By contrast, smaller sites present a greater challenge in terms of marketability and high assessment and cleanup cost as compared to the investment return (Lange 2006).

Although Penn DEP notes that most voluntary cleanups in the City of Pittsburgh since 1995 have been privately funded (DePasquale 2006), the success of large public-private projects has created visibility and energy behind Pittsburgh brownfields development (Lange 2006; URA website). Local redevelopment advocates hope to build on this momentum by providing information, marketing, and financial incentives to encourage further development of large and small sites. The Pittsburgh experience suggests "that the environmental issues related to brownfields are perhaps more tangible and, therefore, easier to solve than the social and economic issues" (Carbasho 2005).

Cincinnati

Cincinnati brownfield development faces some different challenges from other Rust Belt cities. Because of the city's hilly topography and history of small industries, many brownfields sites are on small plots. These sites are less attractive to developers as the assessment cost of a less than 10-acre site cuts far into expected returns (Fischer 2006). Port Authority and Hamilton County are examining landbanking as a means of assembling larger tracts for development, as the Ohio House (HB 313) considers expanding a 2009 authorization granted to Cuyahoga County to establish a County Land Reutilization Corporation as a separate legal entity from local government empowered to acquire through purchase or tax foreclosure, rehabilitate, and/or resell properties without liability under state environmental regulations (Fitzpatrick 2009: 5).

The Queen City Barrel site provides an example of the challenges of Cincinnati brownfields redevelopment. The City of Cincinnati purchased the 12-acre industrial site, one of the larger brownfields redevelopment projects in the city, for $1.6 million (Fischer 2006). Located in Lower Price Hill, one of the historic neighborhoods that developed along the canal and rail transport routes, the site housed the warehouse for Queen City Barrel, a company that cleaned industrial barrels for sale and reuse. In August 2004, the warehouse burned down in an explosive fire. The City received $100,000 from the Clean Ohio Fund for the phase I environmental assessment and spent $3.5 million to buy adjacent properties, hoping to assemble a 20- to 40-acre site more attractive for development (Eigelbach 2004; Fischer 2006; Monk 2006). In 2007, Cincinnati received $3 million for remediation and demolition from the Clean Ohio Fund to redevelop the 18.7 acre site, in partnership with two private sector developers, into the MetroWest Industrial Park. The plans include LEED certified buildings with 250,000 square feet of light industrial, service, and office space, and is projected to create more than 400 new jobs. This Ohio and Cincinnati government collaboration in funding and conducting assessment, site assembly, remediation and demolition leveraged $25 million in private sector resources for the project (Monk 2007). According to the developer, construction is scheduled to begin in February 2011 (resurgencegroupllc.com).

Unlike Pittsburgh, Cincinnati's economy never hit rock bottom like other Rust Belt cities, as its diverse economic base, with both large and small businesses, did not experience the boom and bust cycles of cities with concentrated technology or tourism sectors (Monk 2004; Fischer 2006). Because Cincinnati's industrial decline was gradual and cushioned by non-industrial sectors, it never faced a need for emergency economic restructuring that catalyzed Pittsburgh's brownfields movement (Fischer 2006).

Like the Urban Redevelopment Authority of Pittsburgh, Cincinnati's Port Authority is charged with brownfields development. While public attention has focused on Port Authority riverfront development efforts, it is also engaged in a number of larger brownfields redevelopment projects. According to the Port Authority's website, nine sites, totaling 155 acres, have been or are being redeveloped. Projects include remediation of the 155,000 square foot American Can Factory planned for market-rate housing and commercial space; assessment of the 60,000 square foot Harrison Terminal building which has housed warehouse, bakery supply, furniture manufacturing, transportation, music recording and other uses over its 115-year history; assessment and remediation of the 60-acre Center Hill landfill for redevelopment into office, manufacturing, and warehouse space, as well as remediation of smaller industrial properties for use as office and industrial space. Perhaps because of their smaller scale, less visible location, and end use of industrial or office space rather than residential or retail space, Cincinnati's projects have not attracted the same type of attention as Pittsburgh's large public urban brownfields developments.

In contrast to Pennsylvania, Ohio's culture as to brownfields information emphasizes protecting owners from the (mis)perception of contamination, which affects investment, lending, and value for the property (Fischer 2006). This cautious approach reflects the continued fear that followed strict CERCLA liability (Davis 2002). To some extent, Ohio's legal framework may have resulted from and contributed to this atmosphere. In particular, a 1997 Ohio Court of Appeals decision, *Dayton Power & Light Co. v. Schregardus*, 704 N.E.2d 589, criticized Ohio EPA's "Master Sites List" of properties suspected or known to be contaminated because the publication of the list affected the owner's legal right in its property value without an evidentiary hearing, thus exceeding the agency's authority (Ohio App. 10th Dist. 1997). According to Ohio EPA attorney Sue Kroeger, the decision led the agency to discontinue its list of Ohio brownfields sites (Greenlink 2001; Kroeger 2006). More recently, the agency has posted voluntarily provided information on publicly owned brownfield properties on its Ohio Brownfield Inventory webpage.

As discussed earlier, Ohio law does not require that volunteers notify the state or the public before or after a cleanup, since a volunteer who has used a certified professional to perform a cleanup that meets state standards is protected from liability even without obtaining a Covenant not to sue (Ohio VAP Sheet 2001). Since there is no required oversight by or reporting to the state, neither Ohio EPA nor the City of Cincinnati keep records on the private voluntary cleanups, other than those that request Covenants not to sue from the State of Ohio (Page 2006; Fischer 2006). As of May 2006, 19 Cincinnati sites were listed in the Ohio EPA database as enrolled in the Ohio Voluntary Action Program, while 157 sites had received Covenants not to sue as of April 2006.

There is disagreement regarding the extent to which the Ohio VAP has encouraged cleanups, as compared to other states' voluntary programs. A 2001 report by the Green Environment Coalition (GEC), an environmental advocacy organization, reviewed Ohio EPA VAP program files and criticized the program for failing to achieve its goals of encouraging voluntary cleanups and protecting health, environment, and property rights protection (Green Environmental Coalition 2001). The GEC report compared the 111 sites that had entered the Ohio program with the over 2,300 sites participating in New Jersey's Voluntary Cleanup program, noting that the Ohio program's incentives—technical assistance and Covenant not to sue—were attracting few participants. The GEC criticized the ability of persons conducting a cleanup to withdraw sites from the VAP program, or to apply for and be denied a Covenant not to sue without suffering any penalty or enforcement consequences, even after receiving state funding to participate in the program (Green Environmental Coalition 2001: 5–7). In addition, the group noted that VAP projects relied heavily on fencing and other exposure reduction measures rather than cleanup, provided minimal opportunity for community involvement or comment, and underserved low income and minority neighborhoods. The report recommended easing state restrictions on access to information on cleanup projects, increasing Ohio EPA staff and oversight resources, targeting communities for environmental justice, and requiring direct Ohio EPA oversight of cleanup projects (Green Environmental Coalition 2001: 6–8, 17–22).

Pennsylvania's higher numbers and acreage for state approved cleanups appears to significantly outpace that of Ohio, even though cleanups conducted to standards but outside Land Recycling Act procedures would also probably not be subject to enforcement. Despite the low number of official VAP cleanups, anecdotal evidence suggests that "quiet remediations" are occurring without notice to the state or requests for Covenants not to sue (Fischer 2006; Kroeger 2006; Wood 2006). The high estimated number of quiet cleanups in Ohio may be due to different prerequisites for private or public funding, or because the Ohio process may be more arduous or costly (DePasquale 2006). According to Bill Fischer of the Cincinnati Department of Community Development, cleanup volunteers often decide that requesting a Covenant not to sue is not worth the additional time and cost, because financial institutions are satisfied that they are protected from liability when an Ohio certified professional performs the cleanup and states that it meets state standards (Fischer 2006; Kroeger 2006). Some Lenders have developed lists of certified professionals whom they trust to certify that a site has been cleaned up to state standards. Under 2006 revisions to Ohio Administrative Code Rule 3745-300-03, the flat fee for Ohio EPA to review a phase I no-further-action (NFA) with no identified releases was $2,800.00, while a fee of $16,600.00 is assessed for review of an NFA letter which includes an O & M Plan and Agreement, an increase from the pre-2006 fee of $12,000.00. According to one Certified Professional, the review fee for NFA letters is likely to discourage clients further from seeking a Covenant not to sue. A list of NFA letters on file with Ohio EPA, withdrawn or otherwise without a request for a Covenant not to sue, may be found on the Ohio EPA website. This procedure of the filing of letters without requests might be compared with procedures for the "low risk sites program" in Pennsylvania, in which the sealed report is administratively approved without further Penn DEP review. According to Ann Wood of Ohio EPA, other reasons for quiet remediations include concern regarding negative publicity and the impact of the increased time for reporting and review on project timelines. However, according to Michael Weinstein, a Cincinnati-based Ohio EPA certified professional, his clients are often satisfied with a letter to financial institutions from a certified professional listing the assessment or cleanup activities performed and whether the cleanup is performed to the state standards (Weinstein 2006). As of 2006, Weinstein had successfully completed the CNS process for nine sites, but he had also written hundreds of letters to financial institutions for which his client did not want or need the notification of the state. From informal conversations with attorneys and environmental consultants, Kroeger believes that such quiet remediations may outnumber requests for Covenants not to sue by at least ten-to-one (Kroeger 2006). The activities covered by these letters range from a phase I assessment where no contamination was found to sites where actual cleanup activities were performed. While such a letter cannot be sealed as certified by the professional, Weinstein claims it often meets lender requirements. While Weinstein believes that it is the client's culture of risk tolerance and environmental stewardship that determines whether it will pursue a CNS, Fischer of the City of Cincinnati and Kroeger of Ohio EPA's Legal Office believe that large corporate owners or investors in sites with well-publicized or substantial contamination may still seek the official assurance of a Covenant not to sue, either to show accountability to the public for a higher-profile use or because it is a condition of public funding. However, in the case of smaller sites with minimal contamination (e.g. gas stations), many owners may not find the Covenant not to sue necessary (Greenlink 2001; Fischer 2006; Kroeger 2006).

Fischer and Weinstein agree that financial incentives are the key to promoting brownfields development. This is especially true for Cincinnati's small sites because the high cost of initial assessment and the risk of additional cleanup costs create an even greater barrier for profitability (Fischer 2006). In addition, funding, such as that provided by the Clean Ohio Fund and the Water Pollution Control Fund, has been helpful for financing assessment and assisting local government and private parties to dispel perceptions of contamination. For example, the Clean Ohio funded assessment of the Queen City Barrel site revealed that the contamination was less severe than expected. (Fischer 2006). State funding has also been important for larger public projects such as the City of Cincinnati's proposed Millworks project, which received 3 million dollars in Clean Ohio Revitalization funding for demolition and remediation of a former industrial facility to be redeveloped into retail and commercial space, including a grocery tenant (Ohio DOD, website, odod.state.oh.).

According to Weinstein, the legal framework for cleanup and reuse are a second key to promoting Brownfields development, noting that Ohio and Pennsylvania both employ risk-based cleanup standards, which tie the level of cleanup to the end use, in contrast to states which require cleanup to near pristine levels and see little remediation activity. Where unrealistic cleanup standards, state cleanup oversight, and inadequate financial incentives are combined, few cleanups will occur. For example, Weinstein suggests that even significantly fewer cleanups occur in Kentucky and Indiana, which have fewer financial incentives and higher standards (Weinstein 2006).

Ohio's risk-based standards and hands-off approach appear to encourage cleanups outside the VAP system. However, there is little opportunity for community involvement, or public comment or dialogue because there is no advance notice or final reporting to the public or the state in a quiet cleanup. According to Fischer, citizens have taken little interest in taking on a leadership role in brownfields cleanup and development. Although community members are often excited by nearby development, the long cleanup and development timeline is a challenge to sustained citizen involvement (Fischer 2006). Instead, most of Cincinnati's brownfield development occurs privately or fostered by the city in its targeted redevelopment activities.

Comparison and Contrast

Ohio's and Pennsylvania's programs have the two main components for successfully promoting brownfields development: state liability limitations for voluntary cleanups and significant financial incentives to make assessment, cleanup, and development of brownfields sites profitable. However, their approaches differ in that in addition to financial and technical assistance for assessment and cleanup, Pennsylvania has mobilized significant state staffing and financial resources to develop a marketplace for brownfields sites as a "green" economic development strategy for the state. Brownfields development has entered the public dialogue, and successive governors of both parties have taken significant leadership roles in moving brownfields legislation and funding forward. Pennsylvania's legal framework further encourages openness of information by requiring advance notice of large cleanups and a final report to the state for approval of release from liability. By contrast, Ohio's approach seems to have encouraged private assessments and quiet cleanups to state standards, supervised by certified professionals without any notice or final reporting requirements to the state. This approach minimizes state involvement and cost by making use of private expertise. Although public information may lead to greater accountability and possibly compliance and monitoring of remediated sites, it is uncertain whether Pennsylvania has actually fostered a greater number of sites or acreage redeveloped.

The local brownfields development climate for each city reflects the differences in state approach, as well as local history and politics. The crash of Pittsburgh's steel industry created an economic crisis, but also an opportunity for the city to diversify and restructure its economy while making use of large abandoned steel properties. As a review of news articles and websites on Pittsburgh brownfields development reveals, the success and visibility of large, high-profile brownfields projects initiated by the Urban Redevelopment Authority of Pittsburgh and supported by federal, state, and local funding have spurred private investment in the sites and further mobilized public, political, and private support behind additional regional financing for industrial sites reuse. Although smaller sites continue to pose a challenge because of their high costs and lower return, smaller, private brownfields efforts have also been catalyzed by combined state and local financial incentives and marketing opportunities. State and local agencies hope to build on proven and visible successes by documenting case studies and coordinating marketing of abandoned sites with targeted financing opportunities.

In Cincinnati, brownfields development is less prominent in the public dialogue. By the nature of Ohio's legal framework, documentation of brownfields sites and their cleanup is less public and more fragmented. Although the Greater Cincinnati Port Authority is engaged in important brownfields development, it has not attracted the same public attention as Pittsburgh's projects. While Cincinnati's riverfront "Banks" project is a topic of local discussion and news reporting, redevelopment of mostly smaller brownfields sites, located in less visible areas outside downtown and riverfront areas, are being redeveloped for industrial or office use, with job creation impact rather than creating retail or entertainment space for public use.

The actual difference in public participation between Pennsylvania, where notice is the norm, and Ohio, where quiet remediation is the norm, is also unclear. Although Penn LRP cleanups lasting more than 90 days or cleaning to site specific standards must comply with notice requirements and municipalities may request a public participation plan, there is generally little public interest in cleanups other than large scale redevelopments with direct impact on the community (DePasquale 2006).

The largely private nature of Ohio and Cincinnati's brownfields development contrasts with the open nature of Pennsylvania and Pittsburgh's industrial site development, making it difficult to precisely assess the results and success of state and local legal frameworks and programs. While Pittsburgh's brownfields remediations may be tracked as matters of public record, anecdotal information suggests that assessment, cleanup, and redevelopment is also occurring in Cincinnati, mostly outside the watch of the state and the public. If this anecdotal information is assumed to be true, both state legal structures seem to have had some success in overcoming the stigma and obstacles to financing, remediating, and developing brownfields. However, beyond a numerical cleanup comparison, there are important differences in the legal framework and norms for brownfield remediation. Ohio's freedom from reporting requirements enables the majority of cleanups and land transactions to remain private sector activities, supervised by state certified private actors. However, because private cleanups are not subject to the continued auditing and monitoring of the state, questions remain as to the adequacy of protection of public health and environment even where liability questions are settled. Further, while official state liability releases in both Pennsylvania and Ohio are recorded and run with the land, Ohio's quiet remediations and any use limitations or institutional or engineering controls required to meet state standards are not recorded. The long-term impact of information uncertainties from such privately documented cleanups on future land transactions and land markets remains to be seen. Although Ohio's approach may also maximize private assessment and remediation without the scale of government staff and other resources needed for marketing, notice, and review of applications that Pennsylvania's Land Recycling Program requires, Ohio and its cities may also be missing an opportunity to use public notice and government resources to mobilize public dialogue, attention, and resources around redevelopment of industrial sites.

References

American Planning Association. 2003. "Top 25 Most Significant Planning Laws (1978–2003)." Available at http://www.planning.org/25anniversary/laws.htm (accessed: January 3, 2010).

Bartsch, C. and E. Collaton. 1997. *Brownfields: Cleaning and Reusing Contaminated Properties*. Santa Barbara: Praeger Publishers.

Carbasho, T. 2005. "Learning from City's Experience in Developing Brownfields." *Pittsburgh Business Times*, July 11.

Child Policy Research Center and Cincinnati Children's Hospital Medical Center. 2007. Price Hill Profile. January.

Cohn, C. 2004. "The Brownfields Revitalization and Environmental Restoration Act: Landmark Reform or a 'Trap for the Unwary'"? *NYU Envir. Law Journal* 12: 672–709.

Collings, R.L. and J.V. Quimby. 2006. "Pennsylvania." In *Brownfields: A Comprehensive Guide to Redeveloping Contaminated Property*, 2d edition, ed. T.S. Davis. Chicago: American Bar Association, 850.

Davis, T.S. 2002. "Defining the Brownfields Problem." In *Brownfields: A Comprehensive Guide*, Chapter 1.

Davis, T.S. and J.A. Kwasniewski. 2002. "Ohio." In *Brownfields: A Comprehensive Guide*, 808.

DePasquale, E. 2006. Deputy Secretary, Office of Community Revitalization and Local Government Support, Penn DEP, to author (May 16, 2006) (on file with author).

Eigelbach, K. 2004. "Queen City Barrel Won't Reopen." *Cincinnati Post*. September 15.

Faberman, R.J. 2002. "Job Flows and Labor Dynamics in the US Rust Belt." *Monthly Labor Review* 3:5. September.

Fischer, B. 2006. Cincinnati Department of Community Development. May 4, 2006. Telephone Interview.

Fitzpatrick, T.J. VI. 2009. "Understanding Ohio's Land Bank Legislation." *Federal Reserve Bank of Cleveland, Policy Discussion Paper 23*. January.

Gibson, C. 1998. "Population of the 100 Largest Cities and Other Urban Places in the United States: 1790 TO 1990." *U.S. Bureau of the Census Population Division Working Paper*, No. 27 (June). Available at census.gov (accessed: January 15, 2010).

Green, E.A. 2004. "The Rustbelt and the Revitalization of Detroit: A Commentary and Criticism of Michigan Brownfield Legislation." *Journal of Law in Society* 5: 571.

Green Environmental Coalition. 2001. *Study Finds Ohio's 'Brownfield' Cleanup Program a Failure, Calls for Reforms*. Press Release, January 22, 2001.

——. 2001. *The State of Ohio's Voluntary Action Program: Findings and Recommendations*.

Hamilton County (Ohio) Regional Planning Commission (HCRPC). 2004. "State of the County Report: Economy and Labor Market." *Community Compass* 16:4 (November).

———. 2004. "State of the County Report: Land Use and Development Framework." *Community Compass Report* 16:11 (November).

Hird, D.B. 2002. The Brownfields Revitalization and Environmental Restoration Act. In *Brownfields: A Comprehensive Guide*, xxxvi.

Kennedy School of Government Harvard University. 1997. News release: *Land Recycling Program Re-Invigorates "Brown" Land*. October 8.

Kroeger, S. 2006. Attorney, Ohio EPA Legal Office. Email communication to author, May 17, 2006 (on file with author).

Lange, D. 2006. Director of the Steinbrenner Institute for Environmental Education and Research. May 11, 2006. Telephone interview.

LeBlanc, K. 2006. "Vacant Properties Forum Raises Key Issues for State's Future." *The Ohio Planner's News*, January/February, p. 4.

McGinty, K.A. 2003. Secretary, Department of Environmental Protection. 2003. Testimony before the Senate Environmental Resources & Energy Committee: Pennsylvania's Land Recycling Program. June 25.

Mintz, J.A. 2002. "New Loopholes or Minor Adjustments?: A Summary and Evaluation of the Small Business Liability Relief and Brownfields Revitalization Act." *Pace Environmental Law Review* 20: 405.

Monk, D. 2004. "A Look Back: During the Past 20 Years, Cincinnati's Economy Has Proved Truly Resilient." *Cincinnati Business Courier* December 31.

———. 2006. "Queen City Barrel Site Could Become Industrial Park." *Cincinnati Business Courier.* August 17.

———. 2007. "Neyer Wins MetroWest Bid." *Cincinnati Business Courier*, January 5.

National Brownfield Association Ohio Chapter—Technical Committee. 2005. White Paper: *Six Steps to Property Transaction Comfort in Ohio*. November 10.

Ohio Department of Development (Ohio DOD). 2008. Press Release: *Fisher Announces Clean Ohio Fund Revitalization Program Enhancements and Deadlines*. December 30.

Ohio Department of Natural Resources. *Overview of the Clean Ohio Fund*. Available at ohiodnr.com (accessed: January 15, 2010).

Ohio Environmental Protection Agency (Ohio EPA). 2001. Fact Sheet: *Ohio's Voluntary Action Program*. Available at epa.state.oh.us (accessed: January 15, 2010).

———. 2003. Fact Sheet: *Community Revitalization Support from Ohio EPA*. May.

———. 2005. *VAP Environmental Covenants Guidance and Environmental Covenant Template*. Available at epa.state.oh.us (accessed: January 15, 2010).

———. 2005. *Site Assessment and Brownfield Revitalization Program (SABR) Brochure*, October. Available at epa.state.oh.us (accessed: January 15, 2010).

———. 2007. *Ohio Brownfield Redevelopment Toolbox*, November. Available at epa.state.oh.us (accessed: January 15, 2010).

——. 2009. *Summary of NFAs Received and Covenants Issued* (updated February 11).

Orfield, M. and T. Luce. 2001. *Cincinnati Metropatterns: Update of Key Findings.* Metropolitan area Research Corporation. Minneapolis, MN.

Page, J. 2006. Ohio EPA, Division of Emergency and Remedial Response. May 3, 2006. Telephone Interview.

Pennsylvania Department of Environmental Protection (Penn DEP). 1998. *Ridge Administration Announce Key Sites Initiative*, April 17.

——. April 2004. *Brownfield Action Team Program Guidelines and Application for Assistance.*

——. 2004. Press release: *DEP Unveils Brownfield Action Team Application as Next Step in Brownfields Enhancement*, April 21.

——. 2005. Fact Sheet: *Overview of the Land Recycling Program.*

——. 2005. Daily update: *Governor Rendell Announces Streamlined Process for Joint State-Federal Cleanup of Industrial Sites*, Sept 30. Available at depweb. state.pa.us (accessed: February 1, 2010).

——. 2008. Uniform Environmental Covenants Act. Available at depweb.state. pa.us (accessed: February 13, 2010).

Pennsylvania Governor's Press Office. 2005. Governor Rendell says PA Investing in the Future with Environmental Grants; Safeguarding Communities, Attracting Business Investment, November 2. Growing Greener Website. Available at growinggreener2.com (accessed: February 1, 2010).

Pennsylvania. 2006. *Revitalizing Pennsylvania: A Report on Brownfield Investment.* 2003–2006.

Pennsylvania Department of Community and Economic Development. 2008. Industrial Sites Reuse Program Guidelines, May. Available at newpa.com (accessed: February 1, 2010).

Pittsburgh Post-Gazette. 2006. "Brownfield Center at CMU." Business Briefs Column. March 3.

Port of Greater Cincinnati Development Authority, *Milestone: Brownfields Redevelopment.* Available at cincinnatiport.org (accessed: May 18, 2006).

Rause, V. 1989. "Pittsburgh Cleans up its Act." *New York Times*, November 26, 1989, Section 6, New York Edition.

Reidy, J. 2005. "Energizing Ohio Brownfield Development." *Heartland Real Estate Business*, October.

Sheehan, C. 2002. "Residential Development Revitalizes Brownfields Areas." *Nation's Cities Weekly*. June 3.

Turner, M.R. 2005. "The Ohio Experience: What Can Be Done to Spur Brownfield Redevelopment in America's Heartland." Opening statement of Chairman Rep. Michael Turner, Hearing before the House Subcommittee on Federalism and the Census of the Comm. on Gov't Reform. 109th Cong. 3.

U.S. Bureau of the Census. 1998. "'Rust Belt' Rebounds." *Census Briefs.* CENBR/98-7 (December).

U.S. Conference of Mayors. 2003. *Recycling America's Land: A National Report on Brownfields Redevelopment IV.*

U.S. Economic Development Administration. 2005. Remediation Activities Account for a Small Percentage of Total Brownfield Grant Funding. GAO-06-7, October 27.

U.S. EPA. 2002. *The New Brownfields Law.* October 2002, EPA Brochure 500-F-02-134.

———. 2004. *The Brownfields Program Setting Change in Motion*, EPA Brochure 560-F-04-258.

———. 2004. *State Brownfields and Voluntary Response Programs: An Update from the States.*

———. 2008. *Brownfields Training, Research and Technical Assistance Grant Fact Sheet: Carnegie Mellon University.* Available at cmu.edu/steinbrenner/brownfields/index.html (accessed: January 27, 2010).

———. 2009. *About Smart Growth.* Available at epa.gov (accessed: January 27, 2010).

U.S. EPA Region V and State of Ohio. 2001. *Superfund Memorandum of Agreement Brownfields and Voluntary Action Program MOA Track.* Available at epa.gov (accessed: January 27, 2010).

U.S. EPA Region 3 and Penn DEP. 2005. *Streamlining the Process for the One Cleanup Program under RCRA*, September 2005.

U.S. General Accounting Office (GAO). 1996. *Barriers to Brownfield Redevelopment.* GAO report RCED-96-125.

Urban Redevelopment Authority of Pittsburgh (URAP). 2004. *Pittsburgh Redevelopment News*, May 2004, 1.

———. 2004. URA ANNUAL REPORT 2003–2004, 25–7.

———. 2006. *About the URA: Showcase Projects: Southside Works.* Available at ura.org (accessed: 4 January 2010).

Weinstein, M.D. 2006. Certified Professional and Senior Vice President, SRW Environmental Services, Inc. May 12, 2006. Telephone Interview.

Yersavich, A. 2005. The Ohio Experience: Testimony of Amy Yersavich, Ohio EPA VAP Program Manager. Hearing before the House Subcommittee on Federalism and the Census of the Comm. on Gov't Reform. 109th Cong. 3.

Chapter 11

Institutional Network Management in Brownfield Cleanup and Redevelopment

Rob Alexander

Introduction

Much has been made in the public administration literature about how the past thirty years of government downsizing and privatization in combination with "wicked" policy problems reaching the policy agenda have led to increasing reliance on nongovernmental actors for government problem-solving and service delivery (Rittel and Webber 1973; Lowndes and Skelcher 1998; Agranoff and McGuire 2001; Weber and Khademian 2008). Accompanying this rise of networked government has been an equal expansion of public policies, or policy tools, designed to aid in the leveraging of these networks to attain public sector goals (Peters 2002; Salamon 2002; Howlett 2005). As a result, researchers have increasingly focused on the development, selection, and performance of policy instruments at the state and regional levels, particularly in the areas of economic development, energy policy, and environmental regulation (Bressers and O'Toole Jr. 1998; Agranoff and McGuire 2003; Feiock, Jeong and Kim 2003; Howlett, Kim and Weaver 2006; Feiock, Tavares and Lubell 2008). At the same time, research examining how public management strategies at the individual and interpersonal levels have adapted to this new governance environment has grown, creating a theoretical foundation for network management practice (Agranoff and McGuire 2003; McGuire 2006; Ansell and Gash 2008).

However, a gap exists regarding the interface between networked government and the policy tools presumed to aid its management (Peters 2000). Peters (2000) explains this disconnect as deriving from different emphases made by policy tool researchers on the role of structure and by public management researchers on the role of agency in explaining network outcomes. As a result, much of the existing research only explains part of the management story.

This chapter builds upon the work of scholars bridging the gap between policy tools and public management, particularly those writing about the strategic use of policy tools to influence the behaviors of network actors (Bressers and O'Toole Jr. 1998; Agranoff and McGuire 2003; Klijn and Koppenjan 2004). These authors predict that program outcomes stem from neither the proper design of policy tools nor the appropriate application of management strategies but rather a combination of both. But what does the combination look like? Is policy tool design more influential than public management behavior, or vice versa? How might the interaction between these two components of network governance explain network outcomes?

A useful venue in which to explore these questions is that of brownfield— or contaminated property—remediation and redevelopment. The integrated environmental and economic nature of this policy problem requires the involvement of multiple public and private actors whose relationships are governed by a range of policy instruments (McCarthy 2002; DeSousa 2005; Bartsch 2006). Brownfield projects also exist at the local level where actions are focused on real, empirically available properties. Yet, despite the availability of multiple policy instruments designed specifically for brownfield situations, individual projects experience varying ranges of success (ICMA 2005) even, as this research describes, between similar projects within the same municipality.

This chapter asks the question "why do municipally-led brownfield cleanup and redevelopment projects experience different ranges of success despite common policy tool availabilities and public management contexts?" and applies an institutional network management framework in order to explain outcomes for two brownfield projects in Rochester, NY, selected for their similar characteristics.

Policy Tools and Network Management

Policy Tool Theory

Policy tool theorists examine why policy actors select a particular mix of multiple policies, or "tools" when developing strategies for influencing groups of actors towards a particular set of policy goals. Through this lens, public policies are not maps to be followed but goods and activities shaping relationships within the service delivery system (Salamon 2002). This configuration is a distinct departure from the traditional policy implementation literature which considers public policy as a set of top-down rules or directives instructing policy targets as to the behaviors desired of them (Pressman and Wildavsky 1973; Goggin 1986).

Policy tools themselves vary by activity, degree of coerciveness, ease of use, and visibility (de Bruijn and ten Heuvelhof 1997; Salamon 2002; Howlett 2005). Conceptually, they consist of a range of attributes including an activity type, such as a cash payment or a prohibition, a delivery vehicle, such as a grant, a delivery system, like a government agency or private firm, and a set of rules defining the relationships facilitated by the tool (Salamon 2002). They also vary along dimensions of visibility, readiness for use, directness, and coerciveness (Salamon 2002). Examples of policy tools include those that promote or restrict relationships—such as subsidies, taxes, loans, grants, and public enterprises—and those that impact relationship processes—such as rule reforms and information control programs (Howlett 2004, 2005).

Much of the early policy tool research focused primarily on how governments select policy tools for specific problem-solving scenarios. Economists weighing in largely predicted the selection of tools that minimize transaction costs, preserve market integrities, and avoid direct regulation of private sector counterparts while political scientists predicted the use of less visible, more politically acceptable tools (Salamon 2002; Howlett 2004). As case studies accumulated, researchers began to present more and more complex ideas about the development and selection of policy tools that drew attention to variation in the policy context as a potential explanation of tool selection and, presumably, policy outcomes. It is this piece of the policy tool puzzle that is best explained by the growing research on public network management.

Public Network Management Theory

Public management research has increasingly utilized institutional theories of organizational behavior to describe strategic management in situations of organizational networks (Klijn and Teisman 2003; Skelcher, Mathur and Smith 2005; Edelenbos and Klijn 2007; Kort and Klijn 2009). These theories draw attention to the formal and informal rules that govern interactions between individuals and, ultimately, their organizations, as they seek out shared meanings, manage conflicts, and make resource allocation decisions. Through this lens, effective network management involves the strategic manipulation of rules and norms while simultaneously reacting and responding to the same rules and norms.

A primary challenge for effective institutional management of networks stems from the risks and uncertainties associated with multiple organizations coming together within a problem-solving arena. The conflicting priorities of self-interest and network interest quickly complicate abilities both to predict and to strategically influence actor behaviors (Kickert, Klijn and Koppenjan 1997; Rethemeyer and Hatmaker 2008). One way to address these uncertainties is to engage in "framing," or changing the rules of the network (Koppenjan and Klijn 2004; Skelcher, Mathur and Smith 2005; Fung 2006). Framing includes selecting or barring actors to and from the network, adjusting the rules shaping how members engage and make decisions, invoking process transparency, applying deadlines, and developing and assigning roles (Agranoff and McGuire 2001; Ansell and Gash 2008).

Policy Tools and Network Management

A clear link exists between institutional management of networks and the application of policy tools. In policy networks where government agencies serve as a lead network organization, public managers may apply certain policy tools when addressing specific uncertainties and seeking specific actor behaviors. For example, in many public-private partnerships, public managers activate contracts as policy tools to address transaction-cost concerns while simultaneously minimizing potential for private actors acting outside the public interest. However, management has been shown to matter when evaluating the success of this particular policy tool. While well-written contracts have been shown to transfer financial risks effectively to the private sector, managers must still manage the relationship in order for contracts to reach their full potential[1] (Fischbacher and Beaumont 2003; Klijn and Teisman 2003; Hodge 2004).

In practice, network management procedures become more explicit the more frequently public managers apply the policy tool. As public managers encounter uncertainties both for their own home organizations as well as for citizen and private partners, they are likely to practice adjustment-seeking behaviors with state and federal actors to modify policy tools to address these uncertainties better (Agranoff and McGuire 2004; Amirkhanyan 2009). With citizen and private partners, public managers are also likely to focus on information sharing as a basis for relationship building, generating trust through which local government can gain legitimacy of action and can increase the potential for these horizontal actors to contribute the additional resources necessary for project completion (Provan and Milward 2001).

What, then, does public management research have to say about how these strategies impact network outcomes? Research implies that centralized network integration, network stability, and resource-rich environments all factor in to positive performance outcomes (Provan and Milward 1995; Milward and Provan 1998). Public managers finding themselves in the lead organization have been shown to have their influence enhanced when their network position overlaps with the fiscal management system for the network (Rethemeyer and Hatmaker 2008), and when this system is not fragmented amongst network members (Milward and Provan 1998). If public managers have the capacity to manage their networks to states of high levels of integration, stability, and resources, they should then be successful at addressing the policy problem for which the network convened.

An Integrated Framework—Contextual Interaction Theory

The data collection and analysis in this paper is guided by an integrated framework bridging policy tool and network management theories, termed "contextual

---

1   Given the prominent role of contracts in brownfield projects, it is relevant to explore the literature on contract management. Factors emerging from this research shown to positively impact contract implementation and management include adequate levels of competition for the contract, sufficient government and private contractor resources and capacities, thorough performance management and well-trained contract management staff. Negative impacts on effectiveness include political influence of interest groups undermining contract manager authority, overly complex subcontractor relationships, and too much risk transferred to the contractor (Romzek and Johnston 2002). In addition, contract relationships may evolve to a more relational, trust-based one of principal-steward, but that is contingent on low levels of market competitiveness and sufficient management capacity in both parties (Van Slyke 2007; Amirkhanyan 2009; Fernandez 2009).

interaction theory," or CIT (Bressers and Klok 1988; Bressers and O'Toole Jr. 1998; Bressers, Klok and O'Toole Jr. 2000; Bressers 2004; Bressers and O'Toole 2005). The CIT framework is built on the premise that stakeholders enacting and responding to policy tools have their own agendas and issues. The study of policy tool application must therefore capture not only the cause and effect of the policy itself, but the large number of incentives and constraints caused by social interactions in the management environment (Bressers and O'Toole 2005).

The CIT framework captures both the policy tool mix and its management context by discerning between "core circumstances" and "external circumstances" leading to tool implementation. Core circumstances are those that directly affect project outcomes and include the *motivations* of network actors, their relative degrees of *power*, and the extent to which they have *information* about tools and tasks. The arrangement of these core circumstances relative to one another at any given point in time during implementation determines the quality of policy tool implementation. External circumstances impact the likelihood and adequacy of policy implementation to the extent to which they impact the core circumstances and include all other variables (Bressers, Klok and O'Toole Jr. 2000).

Narrowing the analytical focus to these core circumstances enables the researcher to form testable propositions incorporating three fundamental qualities of policy tool use. First, tool use is not only about success, but also about attempts by stakeholders to thwart and alter processes. Second, key actors have influence upon one another, oftentimes due to interactions that occurred far prior to the implementation at hand. Finally, new policies and policy instruments add new layers of possibility to the social interactions between key actors in the form of new information and cognitions. Therefore, network outcomes based on policy tool use are not so much due to the design and structure of policies as they are about the social relationships between the key actors, thereby tying together policy tool and network management theories.

Policy Tools and Brownfield Cleanup and Redevelopment

What are Brownfields?

The term "brownfield" gained policy meaning in the United States after early federal legislation designed to address orphaned hazardous waste sites was found to have erected significant liability barriers for those who would voluntarily redevelop properties containing far less severe contamination (Page and Rabinowitz 1994; Yount 2003; Heberle and Wernstedt 2006). Recognizing this negative side-effect, the United States Environmental Protection Agency (USEPA) enacted its Brownfields Action Agenda in 1995 to clarify liability issues and to support efforts underway in various states to provide incentives for voluntary cleanup of brownfield properties (Yount 2003; Heberle and Wernstedt 2006). The resulting revisions of federal laws freed states to innovate policy instruments intended to lessen market constraints on brownfield redevelopment. These laws also empowered local governments to pool existing resources and proactively address local brownfield problems.

Brownfield projects integrate environmental remediation with property planning and redevelopment, reflecting elements of both professional practices throughout each project phase. Table 11.1 describes the general model of project evolution.

Table 11.1   Phases of brownfield cleanup and redevelopment

Phase	Stage	Description
Cleanup I	Site Identification	Potential developers (public and private) identify contaminated sites of interest with assistance from public brownfield directories or through marketing by current property owners.
	Initial Site Assessment—Phase I Investigation	Assessing to determine whether contamination is present through historical records and examination of neighboring sites.
	Detailed Site Assessment—Phase II Investigation Remedial Assessment	Environmental engineers sample and analyze chemical parameters of site if Phase I Investigation suggests potential for contamination.
Redevelopment I	Property Assembly Economic Assessment and Planning	Assembling property parcels (if necessary.) Assessing for potential economic return vs. cost of restoring site to productive use. Sites categorized into viable, threshold, and nonviable groups according to this potential/cost ratio. End use plans generated.
Redevelopment II (Overlap) Cleanup II	Project Development and Financing	Assuming financial feasibility studies are complete, developers arrange financing for redevelopment. This is a likely stage for meetings between multiple stakeholders.
	Cleanup Planning and Execution	Selecting and implementing a clean up plan in compliance with regulations.
Redevelopment III	Redevelopment of Site	Altering the site for suitability to its new use.

Source: Derived from Dennison (1998: 142–7).

The challenges facing successful implementation of an individual brownfield project largely depend upon the marketability of the property, the assignment of liability for cleanup, and the political support for public investment (McCarthy 2002). Private sector actors seek positive returns on investment and list cleanup uncertainty, liability concerns, the time required for regulatory compliance, and funding availability for remediation as primary reasons for avoiding brownfield properties (Meyer and Lyons 2000). On the other hand, public actors pursue a variety of public health, economic development, and sustainability goals that require high levels of financial resources (Greenberg and Schneider 1995). These costs provide strong incentives for local governments to focus their efforts on diminishing barriers to potential private investors in order to distribute the financial risks (Page and Rabinowitz 1994). As a result, federal, state, and local governments innovated public policies that create artificial brownfield markets in which private investors may participate at low cost and with low risk (Meyer and Lyons 2000; Wernstedt et al. 2006). Table 11.2 summarizes these policy tools.

Table 11.2   Common brownfields policy instruments

Category of Instrument	Barrier Addressed	Target Actor	Mechanism
Environmental insurance	Risk of future liability	Public, private, nonprofit developers	Protect developers from third party liability and provide cost cap protection
Tax relief Tax increment financing	Cost of assessment, cleanup and redevelopment	Private developers	Financial incentive to commit to specific property
Low-interest loans	Cost of assessment, cleanup and redevelopment	Private, nonprofit developers	Revolving loan fund enabling more affordable clean up
Technical assistance	Cost of assessment and cleanup Citizen outreach	Public, private, nonprofit developers	Provide information enabling more efficient processes
Liability waiver	Risk of future liability	Public, private, nonprofit developers	Statutory protection
Assessment and cleanup grants	Cost of assessment and cleanup	Public, nonprofit developers	Project-specific block grant
Redevelopment authority	Timeliness of government service delivery	Public, private, nonprofit developers	Administrative entity with greater flexibility

Brownfield research indicates a range of environmental, economic, and social factors shaping project outcomes. Environmental factors include the type and extent of contamination, the technology available for proposed remedies, and the flexibility of regulatory agencies in allowing project managers to reach remediation goals (Wernstedt and Hersh 2003). Economic factors range from prevailing market strength to the ability of policy instruments to impact marketability and available financial resources (DeSousa 2005; Wernstedt et al. 2006). Social factors include the effective use of partnerships by public managers (Wernstedt 2001; Silverstein 2003; Bartsch 2006; Dair and Williams 2006), the entrepreneurial abilities of project champions to perceive and obtain opportunities, the capacity of municipal governments to compete for remediation grants (Greenberg and Issa 2005), and the abilities of managers to communicate relevant data effectively to appropriate stakeholders (Nijkamp, Rodenburg and Wagtendonk 2002). Each of these factors implies a central role for policy tools in moving projects forward, if they are implemented correctly.

A Policy Tools Public Management Model of Brownfield Projects

Tying together these brownfield-specific findings with theories of policy tools and network management creates the opportunity to model the interactions between public management and policy tools in networked governance of complex problems (Figure 11.1).

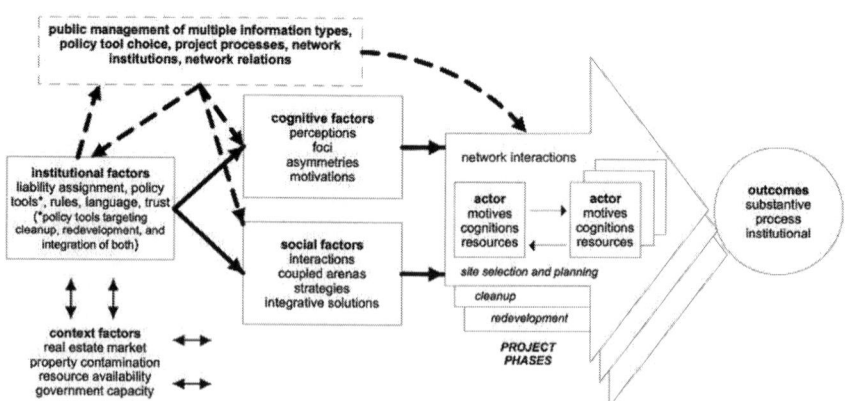

Figure 11.1   Network management model

As indicated in this model, public management shapes and is shaped by the rule-setting qualities of policy tools (top left boxes) at the same time that public managers manage the information and social relations between the other project actors throughout project processes. This then leads to the following propositions:

Proposition 1: The greater the match between policy tool design and the social and market constraints they are to address, the greater the influence of policy tools on network outcomes.

Proposition 2: The more information public managers have about network actors, their motivations, and the factors constraining their actions, the greater the abilities of public managers to influence network outcomes through policy tool application.

Methodology

The larger study behind this chapter investigated the relationship between public management and performance in brownfield cleanup and redevelopment projects in the cities of Buffalo and Rochester, NY. These two cities were selected based upon differences in management capacities, while projects were selected within each city that matched on a range of project characteristics[2] but varied according to measures of successful project outcomes (Table 11.3). Success measures were derived from perceptual data, from surveys and interviews indicating project actor opinions about project economic impacts, timeliness, costs, and remediation adequacies.

---

2   Projects were narrowed to those initiated by municipal governments with market-rate residential end uses. It was important to match end uses as the designated end use of a property impacts who will be involved with a project and the types of policy tools for which the project will be eligible. Housing plays a key role in sustainable community development (Bromley, Tallon and Thomas 2005) and market-rate housing more readily attracts private partners, broadening the scope of potential project networks. While market-rate housing projects represent a minority of total brownfield properties addressed by New York state policy programs, they are more likely to be municipality-led (Page and Berger 2005), considered integral to drawing employers to downtown areas (Bromley, Tallon and Thomas 2005), and play an important role in increasing tax bases for government services (Strom 2008).

Table 11.3   Case selection framework

	High Management Capacity City	Low Management Capacity City
High Project Success	Case #1	Case #3
Low Project Success	Case #2	Case #4

For this paper, data and analysis derive from Case #1 and Case #2 in the city of Rochester, the city exhibiting higher levels of management capacities. Examining two city-initiated projects within the same jurisdiction ensured a close match in both public management capacities and policy tool options against which differences in outcomes may be compared. Table 11.4 includes a summary of general property characteristics.

Table 11.4   Case descriptions

Case	Success Level	Acreage/ Size	Cleanup Costs	Timeframe	End Use
#1	High	6.85 acres	$4.05 million	1996–2004	27 single family homes
#2	Low	1 acre	$600,000	2003–present	Proposed—8 townhouses, 32 condos

After case selection, project-specific data were collected between 2007 and 2009 from federal, state, and local sources regarding policy implementation and management processes. Primary data were derived from a series of semi-structured interviews with key actors involved in the planning and/or implementation of each project. Initial interview subjects were selected through data provided by the New York State Department of Environmental Conservation (NYSDEC) brownfield database, with subsequent interview subjects identified through a snowball sampling technique and verified by their appearance in project documents and media reports, until all primary network actors were identified. Ultimately, Case #1 and Case #2 consisted of 15 and 14 interviews respectively, with stakeholders including environmental and planning officials at the federal, state, and local levels, elected municipal officials, private developers, community activists, media representatives, and private consultants. Secondary data came from extensive documentation about the projects including contracts, interdepartmental memos, public meeting minutes and announcements, technical environmental reports, regulatory communications, city council meetings, grant applications, RFPs, submitted proposals, marketing materials, and other communications. Additional document data external to the partnership came from media reports and citizen blogs.

Data were coded utilizing the qualitative software TAMS Analyzer for the actors involved, policy tools applied during project phases, network management behaviors on behalf of public managers, and evidence of actor motivations, situation comprehension, and political and financial resources bases. These coded data were then used to build stories of what project decisions were made when and why and how these decisions led to project outcomes.

Findings

With a history of a downsized and abandoned industrial core leaving behind a legacy of polluted properties and a city government committed to addressing them, Rochester, NY, is a rich site for the study of public management of brownfield cleanup and redevelopment projects. Prior to 1995, several key environmental cases in the region, including the infamous Love Canal in Niagara, NY, raised general civic awareness of brownfield problems, but inadequate public policy and declining financial resources made it difficult for the city to address them.

Rochester's city government became involved in contaminated property redevelopment in 1993 with a $7 million remediation of a municipal emergency-training property. With no brownfield-specific policy programs available, a team of city managers successfully coordinated the appropriate remediation and obtained $27 million of local, state, and federal funding to clean up the property and build a new emergency training facility. The success of this project laid the foundation for steady evolution of the Division of Environmental Quality (DEQ) in the city's Department of Environmental Services (DES) into a group of seven full-time environmental scientists, engineers, and geologists working to obtain state and federal money for cleaning up city-owned properties and to manage subsequent redevelopment projects in conjunction with individuals in the Community Development, Economic Development, and Law Departments whose work related to property development. While no precise inventory of contaminated properties within the city exists, by 2010, 36 projects had been accepted into New York State brownfield policy programs and eight different EPA grants had been secured for site assessments and remediation, indicating a large brownfield population as well as a strong ability to secure brownfield-specific resources.

Two Divergent Projects

Case #1, or the project with high success, lies in the southeast quadrant of the city of Rochester at the boundary between an established, stable neighborhood and a neighborhood experiencing the challenges associated with older housing stock and a lower-income, transient population. The property itself originally hosted two active construction companies grandfathered into a residential zone. One company provided general contracting services and the other performed asphalt services for road and infrastructure construction with both utilizing the property for waste disposal and equipment storage. At the time that citizens living adjacent to these properties began filing complaints, the city government had an ongoing policy of not foreclosing or otherwise acquiring environmental properties that might pose undue liability and financial risk. Therefore, as long as political pressure remained low, there was little incentive for the city real estate department to invest time and money into property acquisition.

As the number of nuisances from the property accumulated, citizen pressure on their city council representative and also the city council president grew. Ultimately, the timely convening of state-level remediation funding, a willing private development partner, and increasingly organized community activism pushed city council to direct the city's Bureaus of Real Estate (BRE) to proceed with property acquisition and the Division of Environmental Services (DES) to commence with site investigation, remediation, and redevelopment. Today, the property consists of 27 market-rate single-family homes assessed between $160,000 and $270,000, in a neighborhood of older houses assessed between $50,000 and $90,000.

Case #2, or the project with low success, lies within the boundaries of downtown Rochester on the city's east side and is bordered to the west by a mix of new condominium-rental residential and light commercial use, to the north and east by 19th-century brick commercial buildings and a few single family homes, and to the south by a downtown freeway. Prior to the city's property acquisition, the land had a variety of industrial, commercial, and residential uses, including auto body shops and a dry cleaning business. Seeking to capitalize on the momentum of earlier success remediating a neighboring brownfield and working with private developers to create rental condominiums and a popular restaurant/coffee shop, city economic development and real estate managers obtained city council permission to purchase and assemble the parcels for Case #2. Utilizing a multi-stakeholder design charette regarding this particular section of downtown, these public managers determined that the best use of this property would be for market-rate townhomes and condominiums.

Pressured by the need to maintain development momentum, the city therefore pursued a more time-efficient set of state policy tools for assessing and remediating the property, but found the situation complicated by an off-site groundwater contamination plume that originated from a property not owned by the city. As a result, in order to comply with county health requirements, the request for proposals sent to developers included a set of institutional controls on any resulting condominium design. These controls required that the developer create a vapor barrier between the ground and the units, ideally in the form of a first floor garage above which the town homes would sit.

It was at this point in time that financial institutions across the country began absorbing the impacts of the mortgage problems plaguing real estate markets and creating a highly risk-averse lending environment. It was not until the third round of submitting an RFP that the city identified what they felt was a financially feasible proposal for developing the property. However, subsequent negotiations between the developer, the city, and lender broke down as the city experienced turnover of key project personnel, the lender could not work out terms, and the developer shifted priorities to a different downtown development project. As of 2010, the property remains on hold.

Policy Tool Use

The following table summarizes the policy tools brought into play by the city managers leading each project management team and describes the implementer/ target roles surrounding each tool. Tools are organized by project phase to differentiate between those utilized during property assessment, redevelopment planning, cleanup, and construction.

Table 11.5   Policy tools by phase and tool target

Case #1: High Success			
Project Phase	*Policy Tools*	*Tool Implementer*	*Tool Target*
Cleanup I Environmental Assessment	Contract Public information	DES DES	Environmental consultant Neighborhood group
Redevelopment I Site Assembly and Economic Assessment	Eminent domain/ Direct action Public information	BRE DES, BRE	Nuisance property owner Neighborhood group
Cleanup II Planning and Execution	Regulation Grant Contract Liability release Public information	DEC, NYDOH, MCDOH DEC city council, DES DEC DES, BRE	City of Rochester City of Rochester* Subcontractors City of Rochester*, homebuilders Neighborhood group
Redevelopment II Planning and Financing	Contract	City council, DCD	Homebuilders Association
Redevelopment III Construction	Contract (streets) Regulation Public information	City of Rochester MCDOH DES, BRE	Subcontractors Homebuilders Association Neighborhood group
Case #2: Low Success			
Project Phase	*Policy Tools*	*Tool Implementer*	*Tool Target*
Cleanup I Environmental Assessment	Contract Public information	City council, DES DES	Environmental consultant Public
Redevelopment I Site Assembly and Economic Assessment	Direct action	BRE	property owners
Cleanup II Planning and Execution	Grant Contract Public information Liability release Regulation Permit	EPA city council, DES DES City of Rochester DEC, MCDOH DEC	DES* Subcontractors Public Private developer City of Rochester City of Rochester
Redevelopment II Planning and Financing	Request for Proposals	BRE	Private developer community
Redevelopment III Construction	N/A		

Notes: *City of Rochester initiated policy tool. BRE: City of Rochester Bureau of Real Estate. DCD: City of Rochester Department of Community Development. DEC: New York Department of Environmental Conservation. DES: City of Rochester Division of Environmental Services. EPA: U.S. Environmental Protection Agency. MCDOH: Monroe County Department of Health. NYDOH: New York Department of Health.

Comparing policy tool use between Case #1 and Case #2 reveals three significant differences between the projects. First, Case #1 featured prominent use of public information by the city to shape neighborhood group behaviors. Second, liability release in Case #1 applied to both the City of Rochester as well as the homebuilders, whereas in Case #2, liability release only applied to potential developers. Third, Case #2 utilized a competitive request for proposal (RFP) process to identify prospective developers, whereas Case #1 relied upon developers found through pre-existing social networks.

Public Management Strategies and Policy Tool Use

The heavy use of public information in Case #1 generated strong positive relationships between city officials and community activists by establishing strong precedent for reciprocity, an important antecedent for trust (Oliver 1990). Prior to this project, a primary complaint by neighborhood residents was that the city rarely provided adequate information about actions taken to address the nuisance. During the project, public managers, particularly those associated with the Division of Environmental Services, attended every neighborhood group meeting and copied the group leader on every report generated about the property, its contamination, and progress towards its remediation and redevelopment. Building these relationships appeased what had been building up as contentious political backlash and enabled city officials to gain entrance to neighboring properties for testing and procuring small parcels toward fulfilling design parameters.

These same officials implemented public information as a strategy in Case #2 only when required by the EPA and the DEC to fulfill grant and permitting obligations, largely because residents and business owners that were adjacent to the site were not advocating strongly for site remediation. While there was much less political demand for public information tools, the flip side was that there ended up with less political attention focused on the project and fewer political champions motivated to engage with slow-moving, but essential, network partners.

The different applications of liability releases between Case #1 and Case #2 reflect the different priorities of city officials in each project. In Case #1, public managers were initially motivated by pressure from city council and the mayor's office, which were, in turn, acting on pressure from neighborhood residents. As a result, there was a greater impetus to seek complete cleanup of the site, once it was determined that all contamination was local, and to maximize external funding for that cleanup. The appearance of the state Environmental Restoration Program, a set of policy tools developed to help municipalities address brownfield properties directly, provided a strong incentive for the city to move forward. By being the first property accepted into that program, Case #1 received significant political attention at the state level that probably helped to move required paperwork forward. In addition, it created a sense of adventure and innovation for the actors involved in the project network. As a member of the homebuilder group involved with the redevelopment described: "It was a common goal for everybody, nobody was in it to make any money, in some ways we actually lost money on paper but we felt it was worth it. It was an investment." Operating the remediation under a policy program that provided both the city and the homebuilders protection from future liability enabled this alignment of motivation and level of participation.

In Case #2, on the other hand, city officials were motivated more by the economic development momentum established by adjacent residential development and felt time pressure in terms of preparing the site for developer action. Therefore, the decision was made to not pursue the Environmental Restoration Program for cleanup funding and liability release or the more investor-oriented Brownfield Cleanup Program that would provide tax credits the city could pass on to private developers. Instead, the city pushed to clean up the property utilizing an existing EPA grant for cleanup and to register the site with the state Petroleum Spills Program. This program provided a quick six-week approval process for the remediation plan and a "no further action" letter regarding cleanup requirements, but without any financial incentives or the opportunity for liability release at the state level. The trade-offs resulting from this management decision were subsequent difficulties negotiating an agreement with potential developers because of the liability issue.

The difference between Case #1 and Case #2 regarding the relationships between city officials and private development partners was a primary factor impacting eventual project outcomes. In Case #2, the project was largely approached as an economic development opportunity first, with city officials moving forward on the cleanup process before having a committed private development partner. By committing to a traditional competitive RFP process before cleanup was complete, city officials did not fully allow the uncertainties associated with the remediation process and their resulting impact on relationship uncertainties to unfold. The decision to submit to state policy tools that did not provide comprehensive liability protections impacted developer trust that the project was within acceptable risk limits. In Case #1, the private development partners were recruited outside the purview of any specific policy tools through preexisting trust-based relationships. This not only provided more time for public managers to cultivate this relationship by providing home builders with the same public information provided to community activists, it allowed the home builders to gain a sense of ownership in the process—an alignment of motivations.

Sufficiency of Explanations

The sufficiency of institutional management of a brownfield project network through the application of policy tools to explain disparities in project outcomes is tenuous. In Case #2, the low success project, many indications point to factors beyond the control of managers or policy tools. The changing market and lending environment as well as the offsite groundwater contamination presented two uncertainties that were not realized until after remediation policy tools were selected. This is no surprise as the behaviors of private partners in public management networks are often more contingent on changing resource availabilities and market forces than they are on project relationships (Weschler and Mushkatel 1987; Hodge 2004; Warner and Hefetz 2007). However, the potential for public managers to influence private actor motivations within these contextual constraints increases with the application of policy tools designed to address them.

Continuing Thoughts

The data collected in this paired-case analysis support the propositions that policy tools and public management interact to create positive outcomes to networked government operations. However, the ability to explain differences in project outcomes based on strategic application of policy tools becomes quickly complicated by fundamental differences in problem characteristics. If Case #2 had the potential for complete remediation, would there have been greater response to the RFP process? If the real estate market and lending environment had not suddenly constricted at mid-project, would the city public managers have been more successful at matching policy tools to developer concerns?

In addition, network management and policy tool theory suggests that the strategic management focus of policy tools is that of tool implementer on tool target. However, the most important decision in both of these cases that impacted subsequent network actor motivations was the decision of which state-remediation program city officials pursued in order to be the target of the tools they offered. This suggests that, in networked government, management emphasis is not only about influencing behaviors of other actors by targeting them with policy tools. It is also about strategically submitting to additional policy tools as their target in anticipation of subsequent opportunities to transform their benefits into additional leverages of the motivations of other project actors.

References

Agranoff, R. and M. McGuire. 2001. "Big Questions in Public Network Management Research." *Journal of Public Administration Research and Theory* 11(3): 295–326.

——. 2003. *Collaborative Public Management: New Strategies for Local Governments*. Washington, DC: Georgetown University Press.

——. 2003. "Inside the Matrix: Integrating the Paradigms of Intergovernmental and Network Management." *International Journal of Public Administration* 26(12): 1401.

——. 2004. "Another Look at Bargaining and Negotiating in Intergovernmental Management." *J Public Adm Res Theory* 14(4): 495–512.

Amirkhanyan, A. 2009. "Collaborative Performance Measurement: Examining and Explaining the Prevalence of Collaboration in State and Local Government Contracts." *Journal of Public Administration Research and Theory* 19(3): 523–54.

Ansell, C. and A. Gash. 2008. "Collaborative Governance in Theory and Practice." *J Public Adm Res Theory* 18(4): 543–71.

Bartsch, C. 2006. *Promoting Brownfield Redevelopment: The Role of Public-Private Partnerships*. Northeast-Midwest Institute.

Bressers, H. 2004. "Implementing Sustainable Development: How to Know What Works, Where, When and How." In *Governance for Sustainable Development*, edited by W.M. Lafferty. Cheltenham: Edward Elgar.

Bressers, H., P.-J. Klok and L.J. O'Toole Jr. 2000. "Explaining Policy Action: A Deductive but Realistic Theory." In *International Political Science Association*. Quebec City, Quebec.

Bressers, H. and L.J. O'Toole. 2005. "Instrument Selection and Implementation in a Networked Context." In *Designing Government: From Instruments to Governance*, edited by P. Eliadis, M.M. Hill and M. Howlett. Montreal: McGill-Queen's Press.

Bressers, H. and P.-J. Klok. 1988. "Fundamentals for a Theory of Policy Instruments." *International Journal of Social Economics* 15(3/4): 22–41.

Bressers, H.Th.A. and L. O'Toole Jr. 1998. "The Selection of Policy Instruments: A Network-based Perspective." *Journal of Public Policy* 18(3): 213–339.

Bromley, R.D.F., A.R. Tallon and C.J. Thomas. 2005. "City Centre Regeneration through Residential Development: Contributing to Sustainability." *Urban Studies* 42(13): 2407–29.

Dair, C.M. and K. Williams. 2006. "Sustainable Land Reuse: The Influence of Different Stakeholders in Achieving Sustainable Brownfield Developments in England." *Environment and Planning* 38: 1345–66.

de Bruijn, J.A. and E.F. ten Heuvelhof. 1997. "Instruments for Network Management." In *Managing Complex Networks: Strategies for the Public Sector*, edited by W.J.M. Kickert, E.-H. Klijn and J.F.M. Koppenjan. Thousand Oaks, CA: Sage Publications.

DeSousa, C. 2005. "Policy Performance and Brownfield Redevelopment in Milwaukee, Wisconsin." *The Professional Geographer* 57(2): 312–27.

Edelenbos, J. and E.-H. Klijn. 2007. "Trust in Complex Decision-Making Networks: A Theoretical and Empirical Exploration." *Administration & Society* 39(1): 25–50.

Feiock, R., A. Tavares and M. Lubell. 2008. "Policy Instrument Choices for Growth Management and Land Use Regulation." *Policy Studies Journal* 36(3): 461.

Feiock, R.C., M.-G. Jeong and J. Kim. 2003. "Credible Commitment and Council-Manager Government: Implications for Policy Instrument Choices." *Public Administration Review* 63(5): 616.

Fernandez, S. 2009. "Understanding Contracting Performance: An Empirical Analysis." *Administration & Society* 41(1): 67–100.

Fischbacher, M. and P.B. Beaumont. 2003. "PFI, Public-Private Partnerships and the Neglected Importance of Process: Stakeholders and the Employment Dimension." *Public Money and Management* 23(3): 171–6.

Fung, A. 2006. "Varieties of Participation in Complex Governance." *Public Administration Review* 66: 66.

Goggin, M.L. 1986. "The 'Too Few Cases/Too Many Variables' Problem in Implementation Research." *The Western Political Quarterly* 39(2): 328–47.

Greenberg, M. and L. Issa. 2005. "Measuring the Success of the Federal Government's Brownfields Program." *Remediation* Summer: 83–94.

Greenberg, M. and D. Schneider. 1995. "Hazardous Waste Site Cleanup and Neighborhood Redevelopment: An Opportunity to Address Multiple Socially Desirable Goals." *Policy Studies Journal* 23(1): 105–12.

Heberle, L. and K. Wernstedt. 2006. "Understanding Brownfields Regeneration in the US." *Local Environment* 11(5): 479–97.

Hodge, G.A. 2004. "The Risky Business of Public/Private Partnerships." *Australian Journal of Public Administration* 63(4): 37–49.

Howlett, M. 2004. "Beyond Good and Evil in Policy Implementation: Instrument Mixes, Implementation Styles, and Second Generation Theories of Policy Instrument Choice." *Policy and Society* 23(2): 1–17.

——. 2005. "What Is a Policy Instrument? Policy Tools, Policy Mixes, and Policy-Implementation Styles." In *Designing Government: From Instruments to Governance*, edited by P. Eliadis, M.M. Hill and M. Howlett. Montreal, Quebec: McGill-Queen's University Press.

Howlett, M., J. Kim and P. Weaver. 2006. "Assessing Instrument Mixes through Program- and Agency-level Data: Methodological Issues in Contemporary Implementation Research." *Review of Policy Research* 23(1): 129.

ICMA. 2005. "Successful Brownfield Redevelopment Relies on Government and Private Commitments." In *Outstanding Partnerships: Public-Private Partnerships*. Kalamazoo, MI: ICMA.

Kickert, W.J.M., E.-H. Klijn and J.F.M. Koppenjan. 1997. *Managing Complex Networks: Strategies for the Public Sector*. Thousand Oaks, CA: Sage Publications.

Klijn, E.H. and J.F.M. Koppenjan. 2004. "Institutional Design in Networks: Elaborating and Analysing Strategies for Institutional Design." In *Eighth International Research Symposium on Public Management*. Budapest, Hungary.

Klijn, E.-H. and G.R. Teisman. 2003. "Institutional and Strategic Barriers to Public-Private Partnership: An Analysis of Dutch Cases." *Public Money and Management* July: 137–46.

Koppenjan, J.F.M. and E.-H. Klijn. 2004. "*Managing Uncertainties in Networks*." New York City, NY: Routledge.

Kort, M.B. and E.H. Klijn. 2009. "Public Private Partnerships in Urban Regeneration Projects: Arms Length or Managerial Capacity?" In *13th IRSPM Conference*. Copenhagen, Denmark.

Lowndes, V. and C. Skelcher. 1998. "The Dynamics of Multi-Organizational Partnerships: An Analysis of Changing Modes of Governance." *Public Administration* 76 (Summer 1998): 313–33.

McCarthy, L. 2002. "The Brownfield Dual Land-use Policy Challenge: Reducing Barriers to Private Redevelopment While Connecting Reuse to Broader Community Goals." *Land Use Policy* 19(4): 287–96.

McGuire, M. 2006. "Collaborative Public Management: Assessing What We Know and How We Know It." *Public Administration Review* 66: 33.

Meyer, P.B. and T.S. Lyons. 2000. "Lessons from Private Sector Brownfield Redevelopers." *Journal of the American Planning Association* 66(1): 46.

Milward, H.B. and K.G. Provan. 1998. "Measuring Network Structure." *Public Administration* 76(2): 387–407.

——. 1998. "Principles for Controlling Agents: The Political Economy of Network Structure." *Journal of Public Administration Research and Theory* 8(2): 203.

Nijkamp, P., C.A. Rodenburg and A.J. Wagtendonk. 2002. "Success Factors for Sustainable Urban Brownfield Development: A Comparative Case Study Approach to Polluted Sites." *Ecological Economics* 40 (Special Section: Economics of Urban Sustainability): 235–52.

Oliver, C. 1990. "Determinants of Interorganizational Relationships: Integration and Future Directions." *The Academy of Management Review* 15(2): 241–65.

Page, G.W. and R.S. Berger. 2005. "Property Characteristics of Contaminated Land in Environmental Cleanup Programs in New York State." *Public Works Management and Policy* 10(2): 157–69.

Page, G.W., and H.Z. Rabinowitz. 1994. "Potential for Redevelopment of Contaminated Brownfield Sites." *Economic Development Quarterly* 8(4): 353–63.

Peters, B.G. 2000. "Policy Instruments and Public Management: Bridging the Gaps." *Journal of Public Administration Research and Theory* 10(1): 35–47.

——. 2002. "The Politics of Tool Choice." In *The Tools of Government: A Guide to the New Governance*, edited by L.M. Salamon. Oxford: Oxford University Press.

Pressman, J.L. and A. Wildavsky. 1973. *Implementation*. Berkeley, CA: University of California Press.

Provan, K.G. and H. Brinton Milward. 1995. "A Preliminary Theory of Interorganizational Effectiveness: A Comparative Study of Four Community Mental Health Systems." *Administrative Science Quarterly* 40(1): 1–33.

——. 2001. "Do Networks Really Work? A Framework for Evaluating Public-Sector Organizational Networks." *Public Administration Review* 61(4): 414–23.

Rethemeyer, R.K. and D.M. Hatmaker. 2008. "Network Management Reconsidered: An Inquiry into Management of Network Structures in Public Sector Service Provision." *J Public Adm Res Theory* 18(4): 617–46.

Rittel, Horst W.J. and M.M. Webber. 1973. "Dilemmas in a General Theory of Planning." *Policy Sciences* 4: 155–69.

Romzek, B.S. and J.M. Johnston. 2002. "Effective Contract Implementation and Management: A Preliminary Model." *Journal of Public Administration Research and Theory* 12(3): 423–53.

Salamon, L.M. 2002. "The New Governance and the Tools of Public Action: An Introduction." In *The Tools of Government: A Guide to the New Governance*, edited by L.M. Salamon. Oxford: Oxford University.

Silverstein, J.D. 2003. "Mechanics of the Deal: Assembling the Brownfields Team." *Environmental Practice* 5(1): 53–7.

Skelcher, C., N. Mathur and M. Smith. 2005. "The Public Governance of Collaborative Spaces: Discourse, Design and Democracy." *Public Administration* 83(3): 573–96.

Strom, E. 2008. "Rethinking the Politics of Downtown Development." *Journal of Urban Affairs* 30(1): 37–61.

Van Slyke, D.M. 2007. "Agents or Stewards: Using Theory to Understand the Government-Nonprofit Social Service Contracting Relationship."" *Journal of Public Administration Research and Theory* 17(2): 157–87.

Warner, M. and A. Hefetz. 2008. "Managing Markets for Public Service: The Role of Mixed Public/Private Delivery of City Services." *Public Administration Review* 68(1): January/February 155–66.

Weber, E.P. and A.M. Khademian. 2008. Wicked Problems, "Knowledge Challenges, and Collaborative Capacity Builders in Network Setting." *Public Administration Review* 68(2): 334–49.

Wernstedt, K. 2001. "Devolving Superfund to Main Street: Avenues for Local Community Involvement." *American Planning Association. Journal of the American Planning Association* 67(3): 293.

Wernstedt, K. and R. Hersh. 2003. *Brownfields Redevelopment in Wisconsin: Program, Citywide, and Site-Level Studies*. Washington, DC: Resources for the Future.

Wernstedt, K., P.B. Meyer, A. Alberini and L. Heberle. 2006. "Incentives for Private Residential Brownfields Development in US Urban Areas." *Journal of Environmental Planning and Management* 49(1): 101–19.

Weschler, L.F. and A.H. Mushkatel. 1987. "The Developer's Role in Coprovision, Cofinancing, and Coproduction of Urban Infrastructure and Services." *Nonprofit and Voluntary Sector Quarterly* 16(3): 62–9.

Yount, K.R. 2003. "What Are Brownfields? Finding a Conceptual Definition." *Environmental Practice* 5: 25–33.

Chapter 12

From Blighted Brownfields to Healthy and Sustainable Communities: Tracking Performance and Measuring Outcomes

Laurel Berman, Christopher A. De Sousa, Terri Linder and David Misky

Efforts to manage the cleanup and redevelopment of potentially contaminated sites in the US are approaching a 30th anniversary and continue to evolve in scope and character. Initial actions in the late 1970s were spurred by pollution disasters such as Love Canal and the Valley of the Drums, which forced government to understand the human and environmental risks posed by contaminants better, develop suitable methods for efficient site remediation, and begin to tackle high risk sites. In the mid-1990s, the focus shifted to developing, testing, and implementing tools to promote the economic redevelopment of all types of "brownfields" in an effort to bring jobs, investment, and taxes back to ailing cities. Over the last few years, however, there has been an emerging shift in attention to redeveloping brownfields in a manner that continues to address contamination and economic development issues, while also seeking broader goals of improving community health and sustainability. But what does this mean? Unlike earlier periods where "success" could be easily measured on the basis of a few variables (e.g. acres of land remediated, number of jobs created, and property values), defining and measuring well-being and sustainability outcomes, both on-site and neighborhood-wide, is a significantly more complicated proposition.

A review of the literature identifies very few efforts that address public health and sustainability impacts associated with brownfields redevelopment specifically. There are, however, several methods for establishing indicators and accessing impacts related to community or environmental health in general, which typically incorporate health and environmental impact assessment methodologies. Community and sustainability indicator frameworks have also been utilized for decades to assess the state of an issue or location using benchmarks that relate to quality of life, development, and the environment, among other things (Maclaren 1996; Besleme and Muffin 1997; Tyler Norris Associates 1997; Leitmann 1999; Maclaren 2001). This indicator movement is motivated primarily by a need to educate the public, to gather background information for policy development, and to monitor and evaluate the performance of government plans and programs in achieving a broader goal; in this case brownfields redevelopment. This, in turn, reflects sensitivity to preserving the most cherished characteristics and priorities of stakeholders and communities at large.

This chapter examines issues and efforts aimed at linking brownfields redevelopment to public health and sustainability via benchmarking and indicator reporting. It begins with a brief introduction to brownfields policy in the United States. Indicator efforts from relevant sectors (public health and planning) and different scales (international to project level) that deal directly with, or have potential for, brownfields redevelopment are then briefly described and assessed. Two community-wide efforts in Milwaukee, Wisconsin that are administered by two of the authors (i.e. The Menomonee Valley Benchmarking Initiative and the 30th Street Corridor baseline characterization framework) are then examined in greater detail. The description of these initiatives considers the procedures employed for addressing sustainability, community involvement, and public health issues, and the methods for establishing benchmarks to examine them. The final part of the chapter addresses the challenges associated with moving from performance assessment in public health and sustainable development to managing brownfields redevelopment activity "on the ground." The overall objective is to reveal where we have been successful and where we have not, and to recommend ways to link goals, data, and development outcomes in a more comprehensive, cost effective, and applicable manner.

The Evolving Nature of Brownfields Redevelopment

The fact that toxic materials were discovered under homes in Love Canal brought to people's attention the grave risks to human health and environmental integrity that contaminated sites can pose. Indeed, the Love Canal incident marked the first time in US history that federal emergency financial aid was approved for something other than a natural disaster. Soon after, Congress passed the *Comprehensive Environmental Response and Liabilities Act* (CERCLA 1980). Commonly referred to as *Superfund*, the law gave the US Environmental Protection Agency (EPA) and other federal agencies the regulatory authority to respond to a release, or threat of a release, of a hazardous substance or "any pollutant or contaminant which may present an immediate and substantial danger to public health or welfare." The legislation enabled the federal government to recover the costs of cleanup actions from responsible parties or to clean up sites at their own expense and also established a remediation fund financed primarily by a tax on crude oil and certain chemicals for the EPA to use to clean up property. In 1986, the *Superfund Amendments and Reauthorization Act* (SARA) was passed, broadening the EPA's mandate to include research and remediation activities and to increase state involvement in negotiating with responsible parties. Within the CERCLA framework, individual state administrations were assigned responsibility for enacting and implementing their own contaminated site legislation, which most states had implemented by the mid-1990s (Rogoff 1997).

During this initial period, the focus of policy-making was clearly on safeguarding health and reducing environmental risk by identifying high risk sites and managing contamination. The term *contaminated site* was generally used to refer to property that had soil, groundwater, or surface water containing contaminants at levels that exceeded those considered safe by regulators. The distinction was often made between *known* contaminated sites, which had undergone testing, and *potentially* contaminated sites, which were suspected of being contaminated because of previous land-use practices (i.e. waste disposal, manufacturing, military, petroleum-based activities, etc.).

Beginning in the early 1990s, reaction against the *Superfund* apparatus intensified on the part of policy-makers and the private sector (Stroup 1997). Indeed, an international study conducted by the Business Roundtable (1998) comparing the approaches of various countries to contamination problems concluded that "no other country has adopted a program that is as cumbersome, inefficient, slow and costly as the US Superfund program." Consequently, the EPA and many state and local governments started addressing these inefficiencies in specific ways, which led to a substantial refocusing on brownfields redevelopment throughout the US. At this point both policy and discourse shifted from public health under the banner of "contaminated land management" to economic development and urban renewal under "brownfields redevelopment."

In 1995, the EPA introduced the *Brownfields Action Agenda* to ignite interest in the redevelopment of brownfields. This *Agenda* contained four main components (i) it provided funds for pilot programs to test redevelopment models and to facilitate stakeholder cooperation, (ii) it clarified the liability of prospective purchasers, lenders, property owners, and others regarding their association with a site, (iii) it fostered partnerships among the different levels of government and community representatives aimed at developing strategies for promoting public participation and community involvement in brownfields decision making, and (iv) it incorporated job development and training opportunities into brownfields efforts. In addition, the agenda put into place the administrative structures for linking brownfields redevelopment with other relevant socioeconomic issues, while allowing the EPA to concentrate on the management of high-risk contaminated properties via CERCLA.

By June 2000, all 50 states had participated in the federal government's brownfield program. Over 45 individual states had also implemented so-called *Voluntary Cleanup Programs* (VCPs), up from 30 in 1997, to loosen the prescriptive structures that both federal and state Superfund-style legislation had imposed. (Jenner and Block 1997; Meyer and Lyons 2000; and Simons 1998). These states have implemented policies offering more flexible cleanup options and affording the private sector more leeway to work on its own terms on marketable projects, while at the same time providing technical assistance, financial support, and protection from liability. Nationwide, the Brownfields Action Agenda and the Voluntary Cleanup Program approach culminated in the 2002 passage of the federal *Small Business Liability Relief and Brownfields Revitalization Act* (2002). It should be noted that the merit of the brownfield projects carried out under the guise of these programs has been judged largely on the basis of a few economic benchmarks such as real estate value and job creation/retention. Simons and El Jaouhari (2001) surveyed brownfield program managers in 63 cities in 1997–98, finding that local governmental policies also aimed primarily to encourage private sector driven investment and redevelopment, making cleanup standards more flexible, and providing public funds and tax incentives to support redevelopment activity.

Interest in linking sustainability with brownfields redevelopment started to gain momentum in the late 1990s when the EPA issued a series of reports aimed at guiding communities and governments in the planning, implementation, and evaluation of sustainable brownfields redevelopment (U.S. EPA 1999, 1998). Drawing on the Bruntland Commission's definition of sustainability, the EPA (U.S. EPA 1998: 1) defined a sustainable brownfields project as one in which "redevelopment and growth are maintained over the long-term and occur within the limits of the environment so that the current needs of the citizens are met without compromising the ability of future generations to meet their needs." Despite this effort, a review by the EPA's Office of Program Evaluation in 2002 found, ironically, that its own measures to track brownfields performance were designed only to take into account development and economic outcomes, failing to contribute to the EPA's role in protecting human health and safeguarding the environment (U.S. EPA 2002). That said, interest in linking brownfields, sustainability and community health have continued to pick up steam as evidenced by the focus of recent national conferences and EPA funding initiatives. Sustainability is also gaining Response Programs momentum at the local and state level. In a recent report of state Brownfields and Voluntary Cleanup Programs (SRA International 2008) for instance, it was concluded that "in addition to focusing on a project's end use and economic development, more states are focusing on sustainability" and that "many states are incorporating sustainability into cleanup and end use decisions and developing sustainability initiatives."

Tracking Performance and Measuring Outcomes: Emerging Efforts and Best Practices

A comprehensive review of the literature reveals that there are few documented efforts of incorporating public health and sustainability into measuring success in brownfields redevelopment, despite growing calls from public, academic, and professional organizations (Whiteman and Groeneveld 2002) and the growing use of indicators in public health and sustainable development sectors. Examples of indicator efforts related to brownfields redevelopment and public health are presented here, from larger- to smaller-scale initiatives, and additional resources are provided in the bibliography at the end of this chapter.

RESCUE Initiative

The Regeneration of European Sites in Cities and Urban Environments (RESCUE) initiative is a large-scale effort to integrate sustainable practices into brownfields redevelopment that was launched through a 36-month research project. The major goal of RESCUE was to develop and test a systematic holistic approach for sustainable regeneration of European industrial brownfields by diverse brownfield stakeholders (Edwards et al. [RESCUE] 2005: 5, 9).

RESCUE analyzed eight brownfields regeneration case studies in industrial regions in France, Germany, Poland, and the United Kingdom to develop best practice guidance for sustainable brownfield regeneration (RESCUE 2005). RESCUE developed administrative tools and incentives, end-user tools, and web-based training resources all of which focused on sustainable brownfields regeneration; these are available through the RESCUE Web site at http://www.rescue-europe.com/index_mf.html.

The RESCUE project developed sustainability indicators that are typically associated with brownfields redevelopment objectives, such as risk management from reuse of contaminated soil and debris; management and reuse of existing buildings and infrastructure; land use and urban design on brownfield sites; promotion of land use functions that suit the natural and man-made environment of the site and its neighborhood; planning processes and methods for citizen participation; and management of brownfield projects. RESCUE also identified indicators to measure, in both quantitative and qualitative terms, the degree of compliance with these objectives (Franz et al. 2006).

The site-specific RESCUE Sustainability Assessment Tool (SAT) for brownfield regeneration projects was developed because the research coordinators felt that sustainability objectives are not equally relevant for all brownfields sites, locations, and stakeholder perspectives. The SAT can be used to ensure that publicly-funded regeneration projects are likely to constitute sustainable development. The SAT process involves three steps (RESCUE 2005: 130):

- Actor collaboration to set priorities;
- Funding applications by the project developer; and
- Quantified assessment model for decision making.

These steps are implemented through a combination of questionnaires, workshops, and funding applications. Participants weigh sustainability objectives, where possible, and using checklists, allocate points to projects that have the most benefit for the community (RESCUE 2004: 207–308). During the quantified assessment model stage of SAT, decision makers can use the weightings of the actors along with weightings of evaluators to calculate a total project score, which allows the sustainability of the project to be benchmarked as a whole, without benchmarking individual indicators (RESCUE 2005: 30).

Strengths/Weaknesses of RESCUE

The primary strength of the RESCUE initiative is that it addresses the concept of sustainable brownfields redevelopment from a multi-stakeholder perspective, clearly outlining the steps that should be taken and issues considered by key participants in the brownfields redevelopment process in order to achieve sustainability. The guidance materials and best practice descriptions provide a plethora of information for those interested in applying this model and help the reader obtain an awareness and understanding of the broad spectrum of topics considered in RESCUE. The SAT tool also empowers stakeholders and participants by allowing them to prioritize sustainability objectives via a comprehensive weighing scheme.

One weakness of the RESCUE tool is that it is primarily directed to projects that seek government funding, although the authors do note that the procedures could also be used to assess planning applications. Another drawback is that the scale of community involvement, the time commitment for the assessment model (one to two years), and the ensuing costs seem onerous for publicly-funded projects, let alone privately-driven ones. Given that sustainability requirements do not exist for greenfield projects and are not mandated nationally in the US (as they are for many EU programs), it is likely that the time and costs involved with this model will scare away both developers and economic development officials. The high level of community involvement in making project-specific decisions may also deter developers in the US who tend to shy away from public consultation generally.

Health Impact Assessment

Health Impact Assessment (HIA) is another indicator framework that has been widely used in Europe, Canada, and Asia to evaluate the health impacts of a project before, during, or after it is implemented. HIA is defined by the World Health Organization as "A combination of procedures, methods, and tools by which a policy, programme or project may be judged as to its potential effects on the health of a population, and the distribution of those effects within the population (WHO 1999)." The HIA process can provide an evidenced-based assessment of potential health impacts associated with a project and may be useful in assessing the impacts of brownfields redevelopment on community health.

The HIA approach is participatory, with the involvement of multiple stakeholders. The major steps in conducting an HIA include (CDC 2009):

- Screening—identifying projects or policies for which an HIA would be useful;
- Scoping—identifying which health effects to consider;
- Assessing risks and benefits—identifying who might be affected and how;
- Reporting—presenting results to decision makers; and
- Evaluating—determining the affect of the HIA on the decision process.

While a relatively new methodology, HIA is currently being viewed as a tool that could be valuable for use by decision makers in the US (Dannenberg et al. 2006). For example, HIA has been applied to develop a measurement tool for land use development plans in San Francisco (SFDPH 2008). The San Francisco Department of Public Health (SFDPH) Program on Health, Equity, and Sustainability conducted the Eastern Neighborhoods Community Health Impact Assessment (ENCHIA) to assess the health benefits and burdens of development in several San Francisco neighborhoods. The HIA resulted in a final report, "Impacts on Community Health of Area Plans for the Mission, East SoMa, and Potrero Hill/Showplace Square: An Application of the Healthy Development Measurement Tool" (SFDPH 2008: 17), which in general describes the HIA process and introduces the Healthy Development Measurement Tool (HDMT)—a methodology that was validated through a community process to evaluate land use development, policies, and projects in the Eastern Neighborhoods area. The HDMT provides land use planners, public agencies, and community stakeholders with a set of metrics to assess the impacts of urban development on community health (HDMT SFDPH 2006; SFDPH 2008).

The ENCHIA process extended over a three-year period, beginning with the HIA and culminating in the HDMT. Multiple stakeholders formed a community council and met monthly over 18 months to participate in the HIA. As of spring 2009, the HDMT was available through the ENCHIA Web site at http://www. thehdmt.org/. The HDMT provides information on six elements (overarching goals), 28 objectives, and 125 indicators related to each element and each objective. The six elements are: Environmental Stewardship, Sustainable and Safe Transportation, Social Cohesion, Public Infrastructure/Access to Goods and Services, Adequate and Healthy Housing, and Environmental Stewardship (HDMT SFDPH 2006). The HDMT may be used by selecting an element or by viewing the entire list of elements, objectives, and indicators. When an objective is selected, the user is linked to a table that lists the relevant indicator(s), development targets with benchmarks, and policies/design strategies. An abbreviated example is provided in Table 12.1 (HDMT SFDPH 2006).

Table 12.1   Example: the ENCHIA healthy development measurement tool

Objective PI.4 Assure spaces for libraries, performing arts, theatres, museums, concerts, and festivals for personal and educational fulfillment

Indicators	Development Targets	Policies/Design Strategies
Primary		Require 1–2% of total construction costs in high density public areas (i.e. C-3 zone downtown) be allocated for public art. Design parks and open spaces to be accessible and usable for arts and cultural activities.
a. City-serving art/ cultural facilities within 1/2 mile of a regional transit stop	***Benchmark #1:*** Does the project protect and maintain existing art work on site in accordance with applicable state and federal laws AND/OR create space for murals, public art, or public performances? **AND** ***Benchmark #2:*** If project is a new art/cultural facility, is it sited within 1/2 mile of existing or proposed regional transit stop?	

Source: HDMT SFDPH, 2006.

Strengths/Weaknesses of HIA

In their multidisciplinary and multinational assessment of HIA, Krieger et al. (2006) noted key HIA strengths, such as the recognition of social determinants of health; engaging multiple stakeholders in strategic planning; and encouraging interdisciplinary work and responsibility to promote health and reduce health disparities. Weaknesses of HIA included the lack of theoretical frameworks to guide implementation; a lack of interdisciplinary expertise; an erroneous impression that impacts can be precisely measured or predicted; an implication that health is the key arbiter of all policies; high costs of implementation; and an impediment to action if an emphasis on "evidence based policy" precluded informed analysis of policies to improve population health and reduce health disparities (Krieger et al. 2006). Cole et al. (2005) recognized HIA as a promising approach for introducing information about potential health impacts into policymaking, but questioned the feasibility of HIA, including uncertainties about the evidence base upon which to estimate health impacts. Wright et al. (2005) cited difficulties of HIA that included public participation and constraints of time and resources.

We recognize HIA as an effective means to assess community health associated with land redevelopment. However, the HIA structure appears to limit the assessment process to one change within a community, such as a transit, housing, or health care system, as opposed to the assessment of numerous changes and the effects on public health within a community. The HIA process also requires training, and consultants are often hired to conduct the HIA, which removes the process from community implementation. However, community feedback and involvement can still occur.

PACE EH

In the US, the National Association of County and City Health Officials (NACCHO) Protocol for Assessing Community Excellence in Environmental Health (PACE EH; NACCHO 2000, 2008: 1) is a step-by-step methodology to assist local public health officials in facilitating a community-based environmental health assessment. The process focuses on: 1) characterizing and evaluating local environmental health conditions and concerns; 2) identifying populations at risk of exposure to environmental hazards; 3) identifying and collecting meaningful environmental health data; and 4) setting priorities for local action to address environmental health problems.

The PACE EH methodology has 13 tasks and calls for the development of an assessment team to identify local environmental health priorities, establish locally relevant indicators, and coordinate significant short- and long-term interventions (NACCHO 2000, 2008: 13). The 13 tasks range from determining community capacity to evaluating progress and planning for the future, creating and ranking issues, and devising an action plan, among other activities.

The PACE EH requirement to develop locally-appropriate indicators focuses on environmental health measures that reflect local conditions and concerns (NACCHO 2000, 2008: 13). In the PACE guidance, NACCHO stresses that development of an indicator list is one of the most important aspects of the assessment process for ensuring long-term successes and results. Reasons include having meaningful measurements to define areas of concern and mark change; establishing indicators that will highlight pre-existing issues related to data availability and collection; and promoting effective communication by including indicators with reliable data (NACCHO 2000, 2008: 38).

The PACE EH methodology was piloted by local health departments in 10 demonstration communities across the US that were selected through a competitive process (NACCHO 2002: II-1). In general, the demonstration sites adapted the methodology to meet specific local needs, which was an expected outcome of

PACE EH. Pilot-site coordinators reported that conducting community-based environmental health assessment was worth the time and effort, and in most communities the assessment process was going to become an ongoing community activity (NACCHO 2002: IV-1).

Strengths/Weaknesses of PACE EH

Pilot-site communities described several benefits from the PACE EH process, such as improved leadership roles in the community regarding environmental health issues, new professional partnerships, new work skills, confidence to take on large initiatives, broader and more flexible working definitions of "environmental health practice," local environmental health database development, and new relationships between local public health agencies and communities (NACCHO 2002: V-1–V-2). Lessons learned across pilot sites included both strengths and weaknesses of the process, namely that (NACCHO 2002: V-1–V-2):

- Community collaboration is time consuming;
- A PACE EH process requires effective facilitation and meeting-management skills;
- Communities respond favorably to inclusion in a PACE EH process;
- A PACE EH process requires commitment of time and skills;
- A PACE EH process is most effective when combined with additional support and guidance; and
- The effectiveness of PACE EH is a direct result of its adaptability.

The pilot communities implemented the PACE EH process in 1998. Since then, PACE EH has been advocated for use by the EPA's Community Action for a Renewed Environment (CARE) program as one tool that may be used for community environmental health assessment.

We recognize PACE EH as an effective environmental health assessment tool to build capacity within local health agencies for community involvement. However, smaller local health agencies might not have the budget or staff to invest in learning and implementing a PACE EH process. Based on the pilot-site results, the PACE EH process, while adaptable, took 12 to 24 months to move through the 13 tasks, which can be a burden to community volunteers. In addition, brownfields and land reuse issues often require moving beyond environmental health assessment to a more holistic focus on overall community health status, which is a limitation of the PACE EH process.

LEED

One of the more popular tools for defining and tracking "green" development in the United States is the Leadership in Energy and Environmental Design (LEED) rating system released in 2000 by the US Green Building Council (USGBC 2005a). The assessment tool is divided into five categories related to location, water conservation, energy, materials, and indoor environmental quality, plus an innovation and design category. Each category contains a specific number of credits and each credit carries one or more possible points. Brownfield redevelopment is allocated points within the Sustainable Sites category. A project that earns enough points can become "LEED Certified," or acquire more points to move on up the ladder to Silver, Gold, and Platinum LEED levels.

According to the USGBC, the overarching goals of the green building tools are to achieve sustainability-oriented benefits including (USGBC 2005a):

- Environmental benefits—reduce the impacts of natural resource consumption;
- Economic benefits—improve the bottom line;
- Health and safety benefits—enhance occupant comfort and health; and
- Community benefits—minimize strain on local infrastructures and improve quality of life.

Although the initial rating system focused solely on the new construction of commercial buildings, numerous LEED rating products have since been developed or are under development, including: New Commercial Construction and Major Renovation projects; Existing Building Operations and Maintenance; Commercial Interiors projects; Core and Shell Development projects; Homes, Neighborhood Development, and Guidelines for Multiple Buildings and On-Campus Building Projects; LEED for Schools; LEED Retail for New Construction, LEED Retail for Commercial Interiors; and LEED for Healthcare.

While most rating systems are building specific, the USGBC, the Congress for the New Urbanism (CNU), and the Natural Resources Defense Council (NRDC) have been involved in developing the Leadership in Energy and Environmental Design for Neighborhood Development (LEED-ND) rating system. The LEED-ND system focuses on the neighborhood scale and rates location and design based on the combined principles of smart growth, new urbanism, and green building (Ewing et al. 2006). A preliminary draft of the LEED-ND rating system was produced in September 2005, and identifies four target categories (USGBC 2005b):

- Location Efficiency;
- Environmental Preservation;
- Compact, Complete, and Connected Neighborhoods; and
- Resource Efficiency.

A LEED ND pilot program was initiated in the summer of 2007 and has been tested on nearly 240 projects. The pilot system is wrapping up and the post-pilot version is expected to launch in late summer of 2009.

As of early 2009, 2,476 projects have been certified and 19,524 projects registered under the various LEED programs, and it is predicted that by 2010 approximately 10% of all commercial construction starts will be "green" (USGBC 2009). While an early study of 38 first-generation LEED projects in the US found that only three of the 38 buildings were allocated points for brownfields redevelopment (Cassidy 2003: 11), over half (31) of the certified LEED-ND projects as of January 2010 have received brownfields credits (Bogaerts 2010).

Strengths/Weaknesses of LEED

One of the strengths of LEED is that unlike RESCUE, many feel that green objectives are equally relevant for all brownfield projects, locations, and stakeholders. This system thus provides a broadly acceptable standard and measurement system that is quantifiable and easy to track in many different locations and for many types of projects. The rating system thus provides very specific direction for those seeking to build in a more sustainable manner, which makes green building accessible to a much broader audience. The third party certification process also provides substance to the developer's green claims, in addition to a "badge of honor." Also important is that certification is voluntary and often carried out without the need for financial incentives or planning approvals from government.

One of the historic criticisms of LEED is that both the rating system and the greener projects are more detailed, time consuming, and expensive to carry out. Another common criticism, often referred to as the bike rack paradox, is that equal points are given to building features or mechanisms that are not equal in terms of environmental impact or cost. That is, you get the same amount of points for building on a brownfield as you do for installing a bike rack. The fear is that builders will concern themselves with achieving easy points as opposed to building the most sustainable projects in the most sustainable locations. The system is also criticized for focusing too heavily on environmental issues, while giving social issues limited attention. It should be noted, however, that the US Green Building Council has been extremely proactive in acknowledging and addressing key concerns. For example, it has increased the speed and reduced the cost of certification, created rating systems for alternative building types, established regionally-specific points, and funded research to assess the array of benefits associated with green building. As such, it could be argued that it is better to make improvements to what has quickly become a well-known and publicly-embraced program, as opposed to starting from scratch.

Community-wide Brownfields Tracking Efforts in Milwaukee, Wisconsin: The Menomonee Valley Benchmarking Initiative

The Menomonee Valley Benchmarking Initiative (MVBI) was initiated in 2001 by the University of Wisconsin-Milwaukee's Center for Urban Initiatives and Research and the Sixteenth Street Community Health Center (SSCHC), a local nonprofit, to link brownfields redevelopment and sustainability via indicator reporting. Milwaukee's Menomonee Valley is Wisconsin's largest brownfields district, covering roughly 1,400 acres in the heart of the city. In the late 1800s, large industrial complexes dotted the entire Valley, while dense residential communities spread extensively along its bluffs. As in many mid-western cities, urban development patterns and economic disinvestment during the second half of the 20th century hit the Valley's industry hard. What was once a jobsite for tens of thousands became an isolated and environmentally degraded site perceived by many Milwaukeean's as a blight rotting away at the urban core.

Starting in the mid-1990s, government officials, community activists and business interests devised various proposals to remedy the situation, embarking upon a slate of initiatives aimed at improving the economic, social, and environmental sustainability of the Valley and its surrounding neighborhoods. The lack of public knowledge on the status of the Valley and the paucity of information available for public use, along with the absence of a framework for measuring progress against clearly identified sustainability objectives, constituted the primary motivation behind the MVBI.

Although some preliminary work in establishing benchmarks had been carried out by the SSCHC in 1999, a partnership was created with the University of Wisconsin-Milwaukee in order to draw on the scholarly expertise, impartial perspective, and information dissemination capabilities of the university. The core objectives of the MVBI defined by the partners at the outset of the project were as follows:

- To raise awareness in the community regarding the current state of the Menomonee Valley and the progress made towards its revitalization;
- To create an information clearinghouse on data related to environmental, economic, and social indicators;
- To promote the principles of sustainability in an urban context by exploring issues and assembling data in a more holistic manner that considers economic, environmental and social concerns;
- To generate a practical synthesis of the raw data for the benefit of a wide variety of users; and
- To stimulate research interest in the Valley as a complex laboratory for studying urban environments.

The first step in devising the study involved identifying the most suitable benchmarks. Three small (three-hour) "Indicator Work Group" meetings (one social, one environmental and one economic) were organized in the fall of 2001 in which over 40 people representing key stakeholder and community groups participated. The meetings involved ascertaining key "issues of concern" for the Valley first, and then selecting specific "indicators" for investigating those issues based on their overall relevance; that is, whether they were: (1) understandable to a larger public, (2) perceivably responsive to change, (3) scientifically valid, and (4) able to support or trigger action. The coordinators of the study and the stakeholders agreed that the MVBI should not focus on historical trends and legacies, but evaluate the Valley's future progress based on its conditions at the start of the new millennium.

Prior to the meetings, work group participants were sent a description of the project and a long list of hundreds of potential indicators to consider. While the selection of issues of concern was rather straightforward, the selection of specific indicators to study those issues was more difficult and required a voting scheme to narrow the list to around 50. The Economic Work Group identified four key issues and 21 benchmarks, the Social/Community Work Group identified four key issues and 18 benchmarks, and the Environmental Work Group identified four key issues and 12 benchmarks (see Table 12.2). The number of indicators was expanded to 57 for the 2005 State of the Valley report because they complemented and refined existing indicators.

Preparing the first MVBI report involved identifying stakeholders willing to supply existing data or gather and supply new data, and then to report the results. While some of the data could be gathered from existing sources (e.g. the 2000 US Census, city records), a significant amount had to be collected from scratch. For this reason, it was felt that establishing a protocol and making arrangements for future data collection was an important aspect of the MVBI process. To organize the data collection effort, the coordinators split the benchmarks between themselves and were aided by a graduate assistant. Overall, the data gathering process was very extensive for both studies and those contemplating a similar initiative should note that each indicator is an individual research project with specific literature contexts, experts, and data collection and reporting protocols.

Measuring and tracking the state of economic activity in the Valley was a central focus of the MVBI. Given that much of the information on business activity and employment for the Valley was not available, a survey designed by stakeholders from the Economic Work Group was administered to Valley businesses by mail and then followed up with telephone calls. As for environmental benchmarks, the partners worked with a number of key scientists from the university to establish a water-quality monitoring network to conduct research on biotic integrity and physical water quality in the Menomonee River. They also worked with the Wisconsin Department of Natural Resources (DNR) to analyze data from local air-monitoring stations, while information on land coverage and bird activity was collected by graduate students and an array of stakeholders and volunteers from local organizations and nonprofits. For the community indicators, data on recreation and art were gathered by university students as part of independent study and fieldwork classes. Housing and crime data were obtained from relevant city departments, while health and pollution data (e.g. fertility rates, lead poisoning rates, Ozone Action Days) were gathered via local and state health agencies.

The results of the first State of the Valley study were disseminated through a short summary pamphlet and a project website (www.mvbi.org), while a more formal hard-copy and web report were produced for the 2005 study. The report commences with an overall introduction to the Valley and the MVBI, and includes maps of the study areas. Indicator analyses are then sorted into three sections (Economy, Environment, and Community), and each section commences with an introductory page that highlights the most important results from the section and presents an index of the issues and indicators examined. The analysis of each indicator addresses three fundamental questions: (1) what has been measured? (i.e., benchmark, sources of data, and methodological approach); (2) why is it important? (i.e., explains the indicator's role in achieving sustainability); and (3) how are we doing? (i.e., describes the performance of each indicator). The analysis of each indicator is summarized on a single page, and tables, figures, and/ or maps are employed to help clarify the results by providing a snapshot view of performance. Following the indicator analyses, a section entitled Vital Signs presents raw data by census tract intended for those stakeholders, particularly community groups that might benefit from more detailed data for their planning and programming activities.

Overall, MVBI has been attempting to educate the public, inform policy-making efforts, and monitor the performance of renewal activities by gathering analytical information reflective of overall redevelopment in the area. It has generated a useful synthesis of data for interested stakeholders, helped promote principles of sustainability in an urban brownfields context, and brought together stakeholders in a collaborative effort. Nevertheless, it is difficult to gauge the impact of the project on achieving its goals "on the ground." The challenges involved in incorporating the results of this type of information are discussed later in the chapter.

Table 12.2   2005 MVBI issues and indicators

Community			

Health
- Birth Rates
- Child Lead
  Poisoning Rates
- Ozone Action Days
	Crime
	- Number of
	  Selected Offenses
		Housing
		- Owner-Occupancy
		- Number of
		  Residential
		  Housing Units
		- Housing Values
		  near the Valley
		- Household Income
		- Household
		  Ethnicity
			Arts and Events
			- Public Art
			  Installations
			- Community
			  Recreation

Environment			

Water Quality
- Index of Biotic
  Integrity
- Physical Water
  Quality Parameters
	Air Quality
	- Particulate Matter
	  (PM 2.5)
	- Air Toxics
	- 1-Hour Ozone
	- 8-Hour Ozone
		Land Cover and
		Habitat
		- Percentage of
		  Impervious
		  Surfaces
		- Percentage of
		  Canopy Cover
			Flora and Fauna
			- Breeding Bird
			  Population
			- Native and Non-
			  Native Tree
			  Species

Economy			

Commercial
Property
- Amount of
  Developed
  Property
- Land Utilization
- Average Value per
  Square Foot
- Total Assessed
  Value
- Average Net Rent
	Employment
	- Employment by
	  Business Activity
	- Total Number of
	  Employees
	- Number of Jobs
	  per Acre
	- Average Salary
	- Residential
	  Location of
	  Employees
	- Provision of Health
	  Insurance
		Business
		- Type of Business
		  Activity
		- Total Annual Sales
		- Local Sales and
		  Expenditures
		- Adv. / Disadv.
		  of the Valley for
		  Business
		- Percentage of
		  Local Ownership
			Infrastructure and
			Access
			- Road Access
			- Rail Access
			- Linear Feet of
			  Sidewalks
			- Bus Routes, Bus
			  Stops and
			  Ridership

Milwaukee's 30th Street Corridor:
Establishing Baseline Community Health Measures

The Menomonee Valley Benchmarking Initiative was one of the first efforts ATSDR encountered that tracked changes in community health across a large brownfields area. Because of the momentum and success of this project, ATSDR wanted to continue a benchmarking indicators project in Milwaukee that could be used as a model for brownfields communities across the US. ATSDR recognized that while HIA and PACE EH may have rigorous evidence-based approaches and are effective measurement frameworks, both methods require training and/ or financial expenditure to implement, which may be a deterrent to community mobilization and involvement. One principle of the ATSDR benchmarking project was to develop a grass-roots community-level model that could be flexible enough to be used in various communities and at different scales, whether community-wide or site-specific in approach, without incurring large expenditures of volunteers' time or financial resources.

Preliminary discussions with the City of Milwaukee directed the focus of the project to the 30th Street Industrial Corridor. During the summer of 2006, ATSDR began a project with the City of Milwaukee, in partnership with the US Environmental Protection Agency (EPA), the Wisconsin Department of Natural Resources, the Milwaukee Department of Community Development, the Milwaukee Health Department, the Wisconsin Division of Public Health, University of Wisconsin-Milwaukee, non-profit agencies, and community groups. These partners comprised the "Development Community"—all those with a strong interest in the integration of public health in the redevelopment plans for the 30th Street Corridor. The goal was to determine whether community health outcomes can be measured as redevelopment occurs, from baseline or pre-development status through completion of redevelopment.

The 30th Street Industrial Corridor lies two miles northwest of downtown Milwaukee, in what is one of the city's most segregated areas. In the first half of the 20th century, this area was home to many businesses, including foundries, tanneries, motor manufacturers, and breweries, which employed nearly 40% of the local residents. Like many other industrial cities in the Midwest and Northeast, changes in the structure of industry resulted in an exodus of about 60% of businesses from the Corridor, and from 1990–2006, the number of neighborhood residents employed by manufacturing decreased dramatically from 40% to less than 15% (WDNR and City of Milwaukee 2006). The Corridor is ripe for redevelopment, and city leaders have taken a keen interest in the revitalization of this area.

Through the implementation of this project, ATSDR created the Brownfields/ Land Revitalization Action Model, a four-step community health assessment tool. The Action Model is used to facilitate community involvement and to assess the impacts of redevelopment on public health, with a goal of achieving positive, sustainable improvements in overall community health. The Action Model was designed to be simple and easy to use by any member of the Development Community, with minimal expenditure of time and resources by the community.

In the 30th Street Corridor project, members of the Development Community worked together to assess community issues, redevelopment approaches, community health benefits, and data collection needed to characterize health outcomes. The formation of a diverse Development Community was critical to the successful implementation of the Action Model. While the process can be implemented by community residents, data are often needed for benchmarking indicators, which can be supplied by partnering with local health agencies, city agencies, and other governmental bodies. For example, in the 30th Street Corridor, Development Community members from state and local health agencies provided data on health statistics.

The Action Model framework encouraged the Development Community to focus on broad public health topics connected to community health, such as physical and mental health; environment; education and economy; planning; safety and security; and communication and risk communication. Changes in the community related to redevelopment can then be tracked over time.

The Action Model is built around four steps or questions:

1. What are the issues in the community?
2. How can redevelopment address these issues?
3. What are the corresponding community health benefits?
4. What data are needed to measure change?

Through a series of two or more workshops, 30th Street Corridor Development Community members brainstormed around the four questions, ultimately creating a framework for tracking and assessing overall community health status associated with community redevelopment.

In the Milwaukee 30th Street Corridor Action Model process, seven community issues (Step 1) and 19 corresponding baseline measures as indicators of community health status (Step 4) were selected in public health categories of Health, Community, Land and Environment, and Buildings and Infrastructure. In addition, community partnerships and data sharing occurred. ATSDR published the results of this project in a report entitled *Building Healthy Communities: A Baseline Characterization of Milwaukee's 30th Street Corridor* (ATSDR 2008). A summary of the 30th Street Corridor Action Model exercise is provided in Table 12.3.

Building Healthy Communities:
A Baseline Characterization of Milwaukee's 30th Street Corridor

Prepared by
Agency for Toxic Substances and Disease Registry

June 12, 2008

Figure 12.1   30th street corridor baselines assessment process

Table 12.3   Results of ATSDR's brownfields/land revitalization action model

Category	STEP 1	STEP 2	STEP 3	STEP 4
	What are the community issues?	How can redevelopment address the issues?	What are the community health benefits?	What data are needed to measure change?
Health	Exposure to harmful substances in the environment, such as those at brownfields sites or in old housing stock, is one of many risk factors for diseases and adverse health effects (e.g., asthma, high blood lead levels).	Environmental cleanups at brownfields sites may reduce risk of exposure to harmful substances. In addition, renovation of old housing stock and construction of newer homes may help to further reduce exposures to harmful substances.	Reduced blood lead levels, reduction of learning disabilities in children, fewer hospitalizations for asthma, fewer infant deaths, and fewer low birth weight infants. May also reduce exposures to carcinogens.	Hospitalizations for asthma Infant mortality rate Lead and copper in tap water Lead poisoning in children Low birth weight
Community	Elevated crime rates are detrimental to the overall health and well-being of the community.	Development of abandoned sites, vacant lots, and vacant buildings may reduce areas where certain crimes occur and create a better sense of community among local residents.	Reduced crime-related injury and death. Reduced fear of crime, likely resulting in increased mobility of local residents.	Acreage of vacant lots Violent crimes
	Because of lower educational attainment levels, local residents may not be competitive in the labor force and thus not receive the benefits from full-time employment.	Improvements at existing educational facilities and development of new educational centers (e.g., vocational schools, community centers) may promote the educational development of youth in the community.	Increased educational attainment, employability, health insurance coverage, and understanding of health topics and information.	Education of adults Third grade reading comprehension
	A lack of jobs is contributing to a high poverty rate, leaving residents with limited resources to access medical care and improve the residential infrastructure.	Whether through renovating abandoned or deserted buildings or constructing new ones, redevelopment activities designed to attract business can bring jobs into the community.	Lower unemployment rates and poverty may increase health insurance coverage. People may be able to afford better housing and crime rates may decrease.	Percentage of adults with health benefits Percentage of people employed Percentage of people living in poverty

Table 12.3 Continued

	STEP 1	STEP 2	STEP 3	STEP 4
Land and Environment	Opportunities for physical activity are limited, in part, by a lack of usable parks and "green space."	Providing recreational facilities (e.g., basketball courts) at parks and converting vacant lots into "green space" may increase physical activity and strengthen the sense of community.	Increased physical activity, decreased likelihood of disease and health problems related to a sedentary lifestyle.	Acreage of parks People using parks
	Community members may be exposed to physical and environmental hazards when brownfields sites are not cleaned up.	Cleanup activities at brownfields sites and other sites with contaminated land will reduce harmful exposures in the community.	Reduced disease and injury as a result of harmful exposures. Increased opportunity for redevelopment.	Contaminated land
Buildings and Infrastructure	Vacant and poorly maintained buildings can expose residents to health hazards and increase the perception of blight in the community.	Redeveloping commercial buildings can create new jobs, and new or renovated housing units can reduce exposures to harmful environmental contaminants (e.g., lead) and improve residents' quality of life.	Decreased exposure to environmental contaminants, physical hazards, and decreased childhood blood lead levels. Reduction in crime and increase in employment.	Commercial properties Number of lead abatements Number of new construction permits Residential properties

Source: ATSDR 2008: 7.

Moving from Assessment to Action

There are many challenges associated with moving from performance assessment in public health and sustainable development to managing brownfields redevelopment activity "on the ground." Here we present responses to questions regarding such challenges from two of the authors of this chapter who work in the municipal economic development sector at the regional (economic development) and city (local health department) levels. These authors address the various challenges associated with implementing a tracking program, including whether or not brownfields stakeholders are considering the broader public health and sustainability indicators in the redevelopment process.

What were the strengths and weaknesses of benchmarking in terms of both the process of tracking (long-term) and the utility of information tracked in affecting redevelopment activity?

An obvious strength of health indicator benchmarking is that it brings the community into the discussion. One of the challenges with brownfield development is the inclusion into the process of surrounding community members. Frequently, the redevelopment is strictly a financial deal between the municipality and the private developer. It is seldom the case where the community has a part in the process. Both the Menomonee Valley and 30th Street initiatives were started by the municipality and began as brownfield redevelopment projects. One could argue that the Menomonee Valley began as a planning initiative with the City drafting three different plans for it. However, none of the plans gained any momentum until brownfield redevelopment became a part of economic development in the late 1990s. The success of the Menomonee Valley over the last 10 years and numerous other brownfield projects throughout the City helped formulate the recent 30th Street initiative. In both initiatives, the public has been given several opportunities to contribute to the process through various public meetings; however, the benchmarking initiative provides the community with another forum. The benchmarking information collected can contribute to the planning process and could be included in the City's overall redevelopment initiative.

A second strength of the health indicator benchmarking is that over time, the benchmarking process could provide the City with another benefit of brownfield redevelopment besides economic improvement. In order to justify any brownfield program at city, state, or federal levels, qualitative results can be extremely helpful. One of the challenges our community has had with our brownfield program is quantifying the economic benefits of a brownfield development. This additional benchmarking of health indicators could show positive outcomes possibly resulting in additional financial resources for brownfield programs.

The weaknesses of a benchmarking initiative are simply the time and effort required. As was acknowledged with the Menomonee Valley initiative, the level of effort required to obtain the baseline information was a significant undertaking.

In that initiative, it was necessary for the health indicators to be split between the various stakeholders. Obviously, this can lead to some differences in how the information was collected and ultimately presented. The City of Milwaukee does not have the capacity to track the process of data collection nor the utilization of the specific data.

Are stakeholders truly that interested in having their activities tracked with such detail via a framework that links environmental and social factors with traditional economic ones?

The reality is that brownfield development is viewed at state and federal levels as primarily an economic driver for a community. It is assumed that a net benefit of a redevelopment project would be improved public health. A catalytic redevelopment project in an area will provide the benefit of an increased tax base, and it will also drive improvements to public health and the environment. The inverse is seldom the case (i.e. public health issues drive brownfield redevelopment). Consequently, the focus of brownfield development in the City of Milwaukee has generally centered on economic factors.

While the public health indicators could assist a community in making decisions on brownfield redevelopment, and some of the data used as indicators are routinely collected by City and State agencies, many of the indicators require time and personnel to collect and analyze. Even prior to the severe economic strain of 2008, city budgets were tight and maintaining data on indicators is not likely to be included as a priority. The alternative for the City of Milwaukee is to use the city's brownfield experience and identify catalytic projects that will be economic drivers *and* incidentally result in a benefit to public health.

Why aren't stakeholders (e.g. the city and Menomonee Valley Partners) more involved in the indicator projects? How is the information from the indicator projects used?

To date, the City of Milwaukee has not used the information from the indicator projects. Due to financial and staffing constraints, the City of Milwaukee is not in a position to develop new procedures that would tie together the health parameters with economic development. Several years ago the City of Milwaukee had an employee who split his time between the Milwaukee Health Department working on public health issues and the Milwaukee Department of City Development working on brownfield redevelopment. This employee's background and strengths were primarily in public health, but due to the brownfield movement in the State of Wisconsin in the early 1990s, he was asked to support this movement which was housed in the DCD and dealt with the economics of brownfields. Once this employee retired, he was replaced by an individual whose strengths were in economic development rather than health parameters. For the past seven years, the City of Milwaukee has been extremely successful in brownfield redevelopment.

Nearly 100 brownfield sites have been redeveloped in the community, which has eliminated threats to public health and the environment. These efforts have positively impacted the surrounding community, but it is difficult to tie together the benefits of both since the City of Milwaukee does not have an individual or team that delves in both the economics and public health of brownfield redevelopment.

How could the studies be designed to make them more useful? What are barriers to their use? Are these useful for tracking community changes associated with redevelopment?

Public health is a very data-driven practice, and epidemiological data such as morbidity and mortality are the basis for how we develop programs and deploy resources. Some of the indicators such as land use, business and employment are not typically considered and although they are certainly valuable in assessing the overall health and desirability of a community they rarely inform public health decisions. One barrier to employing the indicators from a Public Health perspective is simply the cost of collecting and analyzing the data. Another, perhaps more significant, barrier could be the lack of political will in an organization or a community to prioritize the use of indicators in guiding redevelopment.

The Milwaukee Health Department has worked with the Department of City Development in assessing individual properties being considered for foreclosure for potential environmental threats and liabilities. The Health Department also works with State partners such as the Department of Health Services and the Department of Natural Resources to assess the risk posed by contaminated properties and to inform citizens of risk related to existing brownfield properties. The health department does not, however, weigh in on city development planning and priorities. Creating a change in a political culture that exists between multiple city departments and between the departments and elected officials can be a very long and challenging process.

The indicators could be very useful in describing a neighborhood both before and after redevelopment. The problem is that redevelopment happens over an extended period of time, often decades. In a city health department, we tend to be more focused on year-to-year trends in health outcomes than on street grid layout, green space, or business development and how these change from year to year.

How can the process be more useful to municipal governments?

The City of Milwaukee has made it clear in the past five years that brownfield development is an economic driver for the community and the redevelopment efforts will result in a benefit for the surrounding area and will include public health. One challenge is that the background study generally focuses on a large area (e.g. miles), whereas most brownfield development projects involve individual parcels or possibly a few assembled parcels. A connection would have to be made between the overall study results and site-specific issues.

Conclusion

As brownfield redevelopment continues to expand beyond simply focusing on economic development to also improving community health and sustainability, indicator frameworks offer a system to measure changes in communities over the course of redevelopment. To incorporate overall sustainability, we can borrow from previous large-scale efforts, such as the RESCUE project Sustainability Assessment Tool or the LEED rating systems. We can also use existing frameworks to characterize community health status with indicators, for example through Health Impact Assessment, PACE EH, or the ATSDR Brownfields/Land Revitalization Action Model. All efforts and methodologies hinge on substantial stakeholder or community involvement. Nevertheless, both the Menomonee Valley and 30th Street Corridor projects reveal that there are challenges to supporting indicator tracking efforts and incorporating their results into the day-to-day operations of local government. Some of these challenges result from a lack of capacity in terms of staff and resources at the municipal level. Other barriers include the cost of collecting and analyzing the data or the lack of political will in an organization or a community to prioritize the use of indicators in guiding redevelopment. If these challenges can be overcome, the use of indicator frameworks may prove to be a benefit to shaping the sustainable redevelopment of brownfields and other properties with the goal of broader public health improvement.

Bibliography and References

Bibliography

Many efforts have addressed the characterization of communities or sustainability initiatives by indicators. Much literature exists, of which only a small sample is described here.

Carlin, S. and D. Sprintzen. 2000. *Indicators of Community Sustainability*. Long Island, New York: The Institute for Sustainable Development at Long Island University.

Commission for Environmental Cooperation (CEC). 2006. *Children's Health and the Environment in North America. A First Report on Available Indicators and Measures.* Montréal (Québec) Canada: CEC.

Fraser, E.D.G., A.J. Dougill, W.E. Mabee, M. Reed and P. McAlpine. 2006. "Bottom Up and Top Down: Analysis of Participatory Processes for Sustainability Indicator Identification as a Pathway to Community Empowerment and Sustainable Environmental Management." *Journal of Environmental Management* 78: 114–27.

Galster, G., C. Hayes and J. Johnson. 2005. "Identifying Robust, Parsimonious Neighborhood Indicators." *Journal of Planning Education and Research* 24(3): 265–80.

McCool, S.F. and G.H. Stankey. 2004. "Indicators of Sustainability: Challenges and Opportunities at the Interface of Science and Policy." *Environmental Management* 33(3): 294–305.

Meter, K. 2004. "Fifty-year Vision and Indicators for a Sustainable Minneapolis. Minneapolis Sustainability Roundtable." Minneapolis, Minnesota: Crossroads Resource Center, September 16. Available at http://www.crcworks.org/msi/indicators.pdf (accessed: January 13, 2010).

Mindell J., C. Onyekere, J.A. Roberts, C.L. Ross, C.D. Rutt, A. Scott-Samuel and H.H. Tilson. 2006. "Growing the Field of Health Impact Assessment in the United States: An Agenda for Research and Practice." *American Journal of Public Health* 96(2): 19–27.

Ryan-Nicholls, K.D. and F.E. Racher. 2004. "Investigating the Health of Rural Communities: Toward Framework Development." *Rural and Remote Health* 24 (Electronic Journal). Available at http://rrh.deakin.edu.au (accessed: January 15, 2010).

Scott, F., P. Fleiszer, N. Day and E. Kennedy. 2005. *Public Health in Toronto, 2004: Program Profiles and Indicators*. Toronto: Toronto Public Health, Health Education, January. Available at http://www.city.toronto.on.ca/health (accessed: January 13, 2010).

Yuan, W., P. James, K. Hodgson, S.M. Hutchinson and C. Shi. 2003. "Development of Sustainability Indicators by Communities in China: A Case Study of Chongming County, Shanghai." *Journal of Environmental Management* 68: 253–61.

References

Agency for Toxic Substances and Disease Registry (ATSDR). 2008. *Building Healthy Communities: A Baseline Characterization of Milwaukee's 30th Street Corridor*. Atlanta: ATSDR, June 12.

Besleme, K. and M. Muffin. 1997. "Community Indicators and Healthy Communities." *National Civic Review* 86(1): 43–52.

Bogaerts, M. 2010. Associate, Neighborhood Development, U.S. Green Building Council, personal communication, January 8, 2010.

Business Roundtable. 1998. *The Business Roundtable Comparison of Superfund with Programs in Other Countries*. Washington, DC: The Business Roundtable, 1.

Cassidy, R. 2003. *Building Design and Construction White Paper of Sustainability*. Oak Brook, IL: Building Design and Construction, Reed Business Information.

Centers for Disease Control and Prevention (CDC). 2009. *Health Impact Assessment*. Available at http://www.cdc.gov/healthyplaces/hia.htm (May 13) (accessed: January 13, 2010).

Comprehensive Environmental Response, Compensation, and Liability Act of 1980 (CERCLA). 42 U.S.C. 9601-9675, 96th Congress 1980.

Cole, B.L., R. Shimkhada, J.E. Fielding, G. Kominski and H. Morgenstern. 2005. "Methodologies for Realizing the Potential of Health Impact Assessment." *American Journal of Preventive Medicine* 28(4): 382–9.

Dannenberg, A.L., R. Bhatia, B.L. Cole, C. Dora et al. 2006. "Growing the Field of Health Impact Assessment in the United States: An Agenda for Research and Practice." *American Journal of Public Health* 96(2): 262–70.

Edwards, D., G. Pahlen, C. Bertram and P. Nathanail. 2005. *The RESCUE Manual. Best Practice Guidance for Sustainable Brownfield Regeneration.* (Regeneration of European Sites in Cities and Urban Environments [RESCUE], 2005), 5, 9. (See also footnotes cited as "RESCUE, 2005).

Ewing, R., L. Frank and R. Kreutzer. 2006. *Understanding the Relationship between Public Health and the Built Environment: A Report Prepared for the LEED-ND Core Committee,* May.

Franz, M., G. Pahlen, P. Nathanail, N. Okuniek and A. Koj. 2006. "Sustainable Development and Brownfield Regeneration. What Defines the Quality of Derelict Land Recycling?" *Environmental Sciences* 3(2): 135–51.

Healthy Development Measurement Tool (HDMT). 2006. (HDMT SFDPH). Available at http://www.thehdmt.org/tool.php (accessed: May 13, 2009).

Jenner & Block, Roy F. Weston Inc. 1997. *The Brownfields Book.* Chicago: Jenner & Block, Roy F. Weston Inc.

Krieger, N., M. Northridge, S. Gruskin, M. Quinn, D. Kriebel et al. 2004. "Assessing Health Impact Assessment: Multidisciplinary and International Perspectives." *Journal of Epidemiology and Community Health* 58: 169–74.

Leitmann, J. 1999. "Can City QOL Indicators be Objective and Relevant? Towards a Participatory Tool for Sustainable Urban Development." *Local Environment* 4(2): 169–80.

———. 1996. "Urban Sustainability Reporting." *Journal of the American Planning Association* 62(2): 184–202.

Maclaren, V.W. 2001. "Blighted or Booming? An Evaluation of Community Indicators and their Creation." *Canadian Journal of Urban Research* 10(2): 275–91.

Meyer, P. and T. Lyons. 2000. "Lessons from Private Sector Brownfield Redevelopers: Planning Public Support for Urban Regeneration." *Journal of the American Planning Association* 66(1): 46–57.

NACCHO. 2002. *PACE EH In Practice. A Compendium of Ten Pilot Communities.* NACCHO, II-1.

National Association of County and City Health Officials (NACCHO). 2000. *Protocol for Assessing Community Excellence in Environmental Health.* May. Reprint, Centers for Disease Control and Prevention (CDC), NACCHO 2008, 1.

RESCUE. 2004. "Administrative Tools and Incentives for Sustainable Brownfield Regeneration D 2-5.2" (RESCUE), 207–308.

Rogoff, M.J. 1997. "Status of State Brownfield Programs—A Comparison of Enabling Legislation." (paper presented at the 90th Annual Meeting and Exhibition Air and Waste Management Association Toronto, Ontario, Canada).

San Francisco Department of Public Health (SFDPH). 2008. "Impacts on Community Health of Area Plans for the Mission, East SoMa, and Potrero Hill/ Showplace Square: An Application of the Healthy Development Measurement Tool. Final Report and Addendum." San Francisco Department of Public Health Program on Health, Equity, and Sustainability, October.

Simons, R. 1998. *Turning Brownfields into Greenbacks*. Washington, DC: Urban Land Institute, 1998.

Simons, R. and A. El Jaouhari. 2001. "Local Government Intervention in the Brownfields Arena." *Commentary* 25: 12–18.

Small Business Liability Relief and Brownfields Revitalization Act of 2001. 2002. H. R. 2869, 107th Congress 2002.

SRA International. 2008. State *Brownfields and Voluntary Response Programs: An Update from the States*. Washington, DC: Prepared by SRA International, Inc. Contract No. EP-W-07-023, for the U.S. Environmental Protection Agency, Office of Solid Waste and Emergency Response, Office of Brownfields and Land Revitalization. Available at http://www.epa.gov/brownfields/pubs/st_ res_prog_report.htm (accessed: March 12, 2009).

Stroup, R.L. 1997. "Superfund: The Shortcut that Failed." In *Breaking the Environmental Policy Gridlock*, edited by T.L. Anderson. Stanford: Hoover Institution Press, 115–39.

Tyler Norris Associates. 1997. "Redefining Progress, and Sustainable Seattle. The Community Indicators Handbook: Measuring Progress toward Healthy and Sustainable Communities." San Francisco: Redefining Progress.

U.S. EPA. 1998. *Characteristics of Sustainable Brownfields Projects*. Washington, DC: U.S. EPA, Office of Solid Waste and Emergency Response, EPA-R-98-001.

——. 1999. *A Sustainable Brownfields Model Framework*. Washington, DC: U.S. EPA, Office of Solid Waste and Emergency Response; Report # EPA500-R-99-001.

——. 2002. *Observations on EPA's Plans for Implementing Brownfields Performance Measures*. Washington, DC: U.S. EPA, Office of Inspector General, Final Memorandum Report; Report Number 2002-M-00016.

U.S. Green Building Council (USGBC). 2005a. An Introduction to the U.S. Green Building Council and the LEED Green Building Rating System. Presentation, Washington, DC: U.S. Green Building Council.

——. Congress for the New Urbanism (CNU), and Natural Resources Defense Council (NRDC). 2005b. *LEED for Neighborhood Developments Rating System—Preliminary Draft*. Partnership of USGBC, CNU, NRDC, September 6.

——. 2009. *Green Building by the Numbers*. Washington, DC: Fact Sheet U.S. GBC, April.

Whiteman, K. and T. Groeneveld. 2002. *Measuring Success in Brownfield Redevelopment Programs.* International City/County Management Association (ICMA).

Wisconsin Department of Natural Resources and City of Milwaukee Department of City Development. 2006. *Wisconsin's Urban Reinvestment Initiative: Targeting Milwaukee's 30th Street Industrial Corridor Area. Brownfields Petroleum Assessment Grant.* WDNR and City of Milwaukee.

World Health Organization (WHO). 1999. "Health Impact Assessment: Main Concepts and Suggested Approach." In *Gothenburg Consensus Paper.* Edited by WHO Regional Office for Europe, European Centre for Health Policy. Brussels, 4.

Wright, J., J. Parry and J. Mathers. 2005. "Participation in Health Impact Assessment: Objectives, Methods and Core Values." *Bulletin of the World Health Organization* 83(1): 58–63.

Chapter 13

Michigan Brownfield Redevelopment Innovation: Two Decades of Success

Robert A. Jones and William Welsh

Michigan's Brownfield Redevelopment Legacy

As a result of its unique geographic location in the Great Lakes region, its 19th- and 20th-century industrial heritage, and recent state, national and global economic transformations, Michigan has been left with a significant number of contaminated brownfield sites throughout its Great Lakes coastal areas. These brownfield sites are defined by the federal Environmental Protection Agency as "real property, the expansion, redevelopment, or reuse of which may be complicated by the presence or potential presence of a hazardous substance, pollutant, or contaminant" (U.S. EPA 2006). The term "brownfield" came into use during the 1970s among planners and others involved in economic development work in the US. However, the term originally referred to any previously developed property, irrespective of contamination issues (Yount 2003: 26–7). The current official use of "brownfield" as a contaminated site came into use in 1992 at a US congressional field hearing hosted by the Northeast Midwest congressional Coalition. Since then the federal government has promoted the re-use of brownfields, largely because of the existing infrastructure and buildings already in place for such previously utilized properties.

Meanwhile the State of Michigan was also very active in supporting brownfield redevelopment, recognizing that cleaning up and reinvesting in these properties takes development pressures off undeveloped or open land, and both improves and protects the environment. Michigan's growing awareness of issues related to urban sprawl and the development of valuable open space and agricultural resources has led to increasing demands for the redevelopment of industrial brownfields (Michigan Land Use Leadership Council 2003). Whereas the Comprehensive Environmental Response, Compensation, and Liability Act (CERCLA) of 1980 emphasized the notion that property owners should pay the cost of remediating brownfield properties, irrespective of who may have done the polluting, Michigan realized that such an approach would lead to very little actual site remediation. Rather, landowners of contaminated properties were abandoning them and/or letting them become tax delinquent.

A growing awareness of the issues and potentials associated with brownfield remediation and redevelopment led to the passage of a remediation and redevelopment bond measure by Michigan voters in 1988. Known as the Environmental Protection Bond fund, it included $45 million specifically targeted for site redevelopment purposes. Now more than two decades later, two programs still remain active from this early bond measure: the Site Assessment Grant Program, originally funded with $10 million; and the Site Reclamation Grant Program, established with $35 million in bond funds.

By the mid-1990s the concern for brownfield redevelopment led Michigan to become a leader in crafting innovative brownfield policies. Through both administrative and legislative action, Michigan cast aside the singular federal focus on cleanup of toxic sites and the imposition of strict liabilities placed on property owners. The new Michigan approach was specifically targeted to encourage redevelopment, relying on a combination of private initiative and public support (Hula 1999; Hula and Bromley 2008).

Michigan propelled itself to the forefront of brownfield redevelopment through the implementation of policies and programs that:

- limit the liability of those who purchase contaminated property;
- allow flexibility in cleanup standards based on the redeveloped use of the site;
- rely heavily on voluntary cleanup and redevelopment action;
- recognize economic redevelopment as a primary brownfield policy goal;
- enhance public funding for site assessment and redevelopment activities; and
- expand the definition of brownfield to include an array of blighted properties.

To aid with this last point, state voters approved a second bond measure, the Clean Michigan Initiative (CMI) in 1998, authorizing $675 million in general obligation bond funds for environmental cleanup efforts, with a significant portion of the funding dedicated to programs supporting local redevelopment efforts. Among the CMI brownfield redevelopment programs are the Brownfield Redevelopment Grant Program, established with $37.5 million; the Brownfield Redevelopment Loan Program, established with another $37.5 million in CMI funds; and a $50 million allocation for a Waterfront Redevelopment Grant Program. In addition to these bond-funded programs, the state established a Revitalization Revolving Loan Fund in 1996, with an initial legislative allocation of $5 million. Together the six brownfield redevelopment programs funded by the two bond initiatives represent over $155 million of the approximately $1.4 billion that the state has expended for brownfield remediation and redevelopment work as of September 30, 2008 (Michigan Department of Environmental Quality 2008).

Now, more than two decades past the original Environmental Protection Bond and over a full decade after implementation of CMI programs, bond monies available to local communities for brownfield redevelopment through the six Michigan Department of Environmental Quality (MDEQ) programs have become scarce, and the state's long-term financial prospect to supplement CMI funds does not look bright. Thus, it is critical to assess the success of Michigan's brownfield redevelopment efforts in order to understand the causes, consequences, and potential correctives of brownfield redevelopment,with an emphasis on common elements of "successful" redevelopment projects.

A significant problem that MDEQ has had in undertaking an assessment of the brownfield redevelopment projects it has supported over the past 20 years is that, with the exception of the Waterfront Redevelopment Grant Program, funding from the brownfield redevelopment programs supports remediation activities but cannot be applied to the actual redevelopment effort. Further, although the six programs require that redevelopment take place according to plans articulated in the grant or loan application, the MDEQ has very little authority and even less capacity to evaluate and monitor the redevelopment effort of projects it helps to fund.

Table 13.1   Site inventory

DNRE Program	Start Date	Funding Source	Allocation	Remaining Funds (end of FY 2008)
Site Assessment Grant	1989	1988 Environmental Protection Bond	$10 million	$240,000
Site Reclamation Grant	1989	1988 Environmental Protection Bond	$35 million	$1.15 million
Revitalization Revolving Loan	1996	Michigan State Legislature	$5 million (initial allocation)	$2.1 million
Brownfield Redevelopment Grant	1999	1998 Clean Michigan Initiative	$37.5 million	$7.1 million
Brownfield Redevelopment Loan	1999	1998 Clean Michigan Initiative	$37.5 million	$9.2 million
Waterfront Redevelopment Grant	1999	1998 Clean Michigan Initiative	$50 million	$0

Note: Michigan Department of Environmental Quality brownfield redevelopment programs included in the study.

Source: MDEQ Consolidated Report, FY-08.

Research Questions

This study of Michigan's brownfield redevelopment efforts began with a question formulated in meetings between representatives on the Michigan Sea Grant Program, MDEQ, and interested stakeholders. As refined by these parties, the primary research questions are: what are the causes, consequences and potential correctives of brownfields located on Michigan's Great Lakes coasts with an emphasis on common elements of "successful" brownfield redevelopment projects, and how can those elements be incorporated into prospective future projects? How should brownfield redevelopment projects be properly assessed? What makes a brownfield redevelopment project successful, or what common elements are found in successful projects?

This research also looks at possible metrics that the state might use in assessing the relative level of success for individual brownfield redevelopment projects. Through a series of case studies of state-assisted brownfield redevelopment projects, it looks at some of the possible ingredients that help lead to a successful redevelopment effort. These concerns are particularly important given that Michigan's brownfield funds are limited and given the current austerity of the state's budget outlook.

Because the sponsorship of this research project has come from the Michigan Sea Grant Program, the majority of the brownfield sites examined are in coastal communities of the state. However, several of the case studies have been from non-coastal communities such as Lansing, Mt. Pleasant, Grand Rapids, and Ypsilanti. Inclusion of projects in these and other non-coastal areas allow conclusions to be drawn about brownfield redevelopment success apart from the coastal areas of the state and their associated scenic and commercial amenities.

Federal Policy Context

The 1980 Comprehensive Environmental Response, Compensation, and Liability Act (CERCLA) and its 1986 reauthorization, the Superfund Amendments and Reauthorization Act (SARA), established federal priority in the area of environmental cleanup, setting the standards for containment and remediation of contaminated sites in the US. Some of the more salient and sometimes controversial aspects of federal policy in this arena are:

- The imposition of cleanup costs on those responsible for contamination and subsequent property owners, including retroactive liability for contamination caused before such contamination became illegal.
- A demand that contaminated sites effectively be restored to "greenfield" status.
- A highly centralized decision-making structure located in the federal bureaucracy.

Neither of these acts showed significant concern for site redevelopment or post cleanup use, the idea being that remediation standards would prepare the site for any future use that might come along. Program goals were defined primarily in terms of protecting public health interests. Site cleanup was the desired end and final policy goal. The 1972 Coastal Zone Management Act (CZMA) provides a notable exception to the remediation emphasis of CERCLA. Section 303 of the act specifically calls for cleanup, restoration, and redevelopment efforts.

Despite the coastal brownfield redevelopment policy established in CZMA, it was not until the mid-1990s that the federal Environmental Protection Agency took administrative action to establish a brownfield redevelopment grant program, followed two years later by support for states to set up revolving loan funds to aid in local brownfield cleanup and redevelopment efforts. Because EPA's brownfields program was administratively under CERCLA, which is intended to address the nation's worst hazardous waste sites, many of its requirements are not appropriate in the context of funding for state and local brownfields assessment, remediation, and redevelopment. To address some of the limitations and problems with the EPA's administrative brownfield cleanup and redevelopment program, Congress passed the Brownfield Revitalization and Environmental Restoration Act of 2001 (BRERA) to provide support for state and local efforts to revitalize communities through the assessment, remediation, and redevelopment of brownfield sites.

By the early- to mid-1990s the federal CERCLA paradigm was facing significant challenge from a variety of fronts, including from several states, from congress, and from within the EPA itself. Arguments were repeatedly made to transfer policy responsibility and authority to the states, not only because of the mounting opposition and growing unpopularity of the federal approach, but also because state environmental remediation capacity had increased significantly in the years following passage of the federal CERCLA legislation.

States began to experiment with a variety of alternatives to the CERCLA model. Voluntary programs for site remediation that relied more on incentives than the coercion of federal law were developed, as more efficient cleanup strategies. Such voluntary programs have caught on relatively quickly. From 1993 to 1998, the number of state-level brownfield cleanup programs grew from 14 to 44, with a much greater emphasis being placed on redevelopment following remediation. To be sure, though, there is a great deal of variation between states in terms of policy and administration, as well as in actual program successfulness (United States General Accounting Office 1997; Hula and Hemond 2003).

Michigan Policy Context

In 1996, Michigan became one of the first states to break from the federal policy lead in the area of environmental cleanup and brownfield redevelopment. Through both administrative and legislative action, Michigan cast aside the singular federal focus on cleanup of toxic sites and the imposition of strict liabilities placed on property owners. Prior to 1996, Michigan statutes followed the federal lead in imposing a strict liability framework for site contamination. This essentially meant that property ownership carried with it the liability for cleaning that site, irrespective of those who may have actually been responsible for contamination. Michigan's changed approach allows subsequent purchasers to limit their liability for contamination for which they are not directly responsible. New owners of potentially contaminated property can secure exemption from contamination and cleanup liability by filing a Baseline Environmental Assessment (BEA) with the MDEQ within 45 days of their purchase of a brownfield site. The site-specific information developed for the BEA serves as the basis from which to evaluate liability claims against previous and current landowners. The completion of the BEA is largely a private action, with only limited state oversight, although the state may provide grant or loan funds to the property owner to conduct the assessment. The baseline assessment process provides a much more efficient and streamlined approach to limiting liability than the lengthy and cumbersome process of developing a "covenant not to sue," which was previously the only way for new property owners to limit liability for pre-existing contamination.

The BEA process does not provide complete relief from liability for new owners, however, as it clearly establishes full liability for contamination beyond that reported in the baseline assessment, even if contamination preceded new ownership. Owners have the option of filing a petition with MDEQ requesting written documentation that they qualify for the liability exemption. While new owners no longer bear full liability for site cleanup, they are required to meet "due-care" requirements that the public be protected from any existing contamination. These "due-care" requirements have been extended to all owners of contaminated sites, which represents an extension of past liability in that potentially responsible parties now have an affirmative responsibility to show "due care," regardless of any identification of the site as a potential public health threat by a public agency.

Site cleanup standards for Michigan are strikingly different from those of the CERCLA-type framework that requires remediation to a single, "greenfield" standard; cleanup standards are tied to proposed redevelopment use. MDEQ has created a three-tiered system with different standards for industrial, commercial, and residential redevelopment projects. Not surprisingly, industrial standards are less stringent than commercial, and both of these are less demanding than those for residential projects. Another part of this reconfiguration has reduced overall risk standards for various types of contamination. The state has also recognized local institutional controls on land use and restrictive deed covenants as acceptable alternatives to cleaning a site to the highest possible standard. Thus, local zoning of a site restricting it to industrial uses is considered adequate for application of the least stringent remediation standard.

Another important element of the Michigan initiative is the explicit linking of redevelopment to cleanup goals. This linkage has been reinforced through numerous state funding mechanisms. Thus, projects proposed for funding using state environmental bond funds such as the Clean Michigan Initiative are required to demonstrate a viable redevelopment plan as well as a cleanup strategy. In fact, redevelopment concerns overshadowed environmental priorities in 2001, when the state legislature modified the legal definition of brownfields to include "blighted" or "functionally obsolete" properties with specific reference to the state's core cities such as Detroit and Flint. This expanded definition of a brownfield allows environmental bond funds to be expended for the redevelopment of properties that may in fact have no contamination issues at all, real or perceived (Lang 2001).

Michigan has also developed a number of sources of financing for local brownfield projects that typically come not from state general funds, but rather from specific revenue streams that are targeted to local brownfield redevelopment efforts. Four revenue sources have been of particular importance in the state: the now defunct brownfield Single Business Tax credit, replaced with a Michigan Business Tax Credit program; local Brownfield Redevelopment Authorities with tax increment financing authority; the 1988 Environmental Protection Bond; and the 1998 Clean Michigan Initiative (both of the bond measures have been substantially depleted, however).

Michigan law permits municipalities to create a brownfield redevelopment authority (BRA), effectively creating a specialized local institutional structure that promotes planning for and implementation of brownfield redevelopment. The Brownfield Redevelopment Financing Act of 1996 provides BRAs with a variety of fiscal powers, including paying or reimbursing private or public parties for brownfield remediation activities; the leasing, purchasing, or conveying of brownfield properties; accepting grants and donations of property, labor or "other things of value" from public or private sources; investing the authority's money; borrowing money; and engaging in lending and mortgage activities associated with the brownfield property it acquires (Davis and Margolis 1997). These authorities may also create revolving loan funds to finance projects. Each authority must develop a plan for redeveloping eligible properties within its jurisdiction.

Brownfield redevelopment authorities have the legal authority to raise revenue through a number of tools provided in the state legislation. These include tax increment financing authority to capture increases in state and local taxes, including school taxes, resulting from the redevelopment of a brownfield sites within the BRA district. These TIF funds can be used for a range of purposes by the BRA, including evaluation and feasibility studies of specific sites, onsite demolition of buildings, necessary onsite construction, and the combining of contaminated property with adjacent parcels to create larger redevelopment properties.

Apart from creating additional funding sources for brownfield redevelopment, BRA-controlled moneys such as TIF funds provide a level of flexibility that is not available for support derived from MDEQ sources. Although a number of MDEQ brownfield redevelopment programs require a redevelopment proposal, funds can only be applied to assessment and remediation efforts, and for some types of public infrastructure improvements. MDEQ funds cannot be applied towards actual project construction costs. As a consequence of such funding limitations, MDEQ has very little authority or capacity when it comes to monitoring and enforcing actual redevelopment efforts. Once a site has been remediated to the appropriate state standard and state funds are fully accounted for, MDEQ is essentially done with the project.

Funding for remediation and redevelopment projects also comes from direct revenue streams. In 1998, Michigan voters approved a $675 million environmental bond issue, the Clean Michigan Initiative (CMI), thereby providing a second important funding source for brownfield redevelopment. The CMI included $335 million targeted directly to local brownfield redevelopment efforts. CMI programs included direct funding for brownfield redevelopment projects by the state, with projects selected from among those nominated by local authorities to be directly funded by the state. Other funds are allocated through several assessment, remediation, and redevelopment programs administered by the Michigan Department of Environmental Quality. In addition, grant funds from a number of other state agencies have often been used to support specific projects. Two things are of importance here. First, although a number of CMI-funded programs required a redevelopment proposal, funds can only be applied to assessment and remediation efforts, and for some types of public infrastructure improvements. Funds cannot be applied towards actual project construction costs. Second, after nearly a decade of funding cleanup efforts, CMI funds are nearly depleted. A revolving loan program is ongoing, but DEQ officials estimate that other programs will run out of funds within the next year or two, with only slim prospects for legislative replacement.

Michigan Land Use Leadership and Brownfield Redevelopment

Recent political leadership in the State of Michigan has expressed a level of dissatisfaction with local land use decisions from a variety of perspectives, particularly in terms of the increasing levels of sprawl in urban areas throughout the state and the unchecked consumption of open land for new development. There is a strong desire to concentrate more new growth within existing urbanized areas and to preserve valuable open space and agricultural resources vital to Michigan's economic well-being.

In 2003, Governor Jennifer Granholm signed an executive order creating the bipartisan Michigan Land Use Leadership Council (MLULC), charged with the task of identifying "the trends, causes, and consequences of unmanaged growth and development in Michigan and [to] provide specific recommendations that address those issues" (Michigan Land Use Leadership Council 2003: 11). Among its observations, the Council noted that new development is consuming land at a rate as much as 13 times greater than the rate of population change. Moreover, virtually none of this land has previously been developed with urban uses.

To begin to address these and other issues the MLULC adopted the 10 basic tenets of the "smart growth" movement, to:

1. Create a range of housing opportunities and choices,
2. Create walkable neighborhoods,
3. Encourage community and stakeholder collaboration,
4. Foster distinctive, attractive communities with a strong sense of place,
5. Make development decisions predictable, fair, and cost-effective,
6. Mix land uses,
7. Preserve open space, farmland, natural beauty, and critical environmental areas,
8. Provide a variety of transportation choices,
9. Strengthen and direct development towards existing communities and,
10. Take advantage of compact development design (MLULC 2003: 27; ICMA nd).

This represents an attractive set of principles for which brownfield redevelopment projects can play significant roles (particularly in relation to number nine), but it also requires a strong political will for implementation in a state with overwhelming support for individual property rights and home rule, and with an almost single-minded devotion to the private automobile. In their report to the state, the MLULC showed a marked reluctance to get involved with regulatory issues related to the development rights of individual property owners, such as zoning and subdivision regulation that encourages greenfield development over brownfield redevelopment projects. The reluctance on the part of state leadership to address regulatory issues related to land development and redevelopment is unfortunate, for it means that local jurisdictions, and particularly those located at the suburban periphery, are free to continue to expand in the same land-exploitative fashion they have shown for the past several decades. Little is done to encourage the adoption of local regulations that would encourage redevelopment of brownfield sites.

Under Michigan law, local governments have nearly exclusive control over development, and are generally not required to follow any county or regional planning effort, so long as they have their own planning and zoning in place. Thus, Michigan is left playing something of a double hand when it comes to brownfield redevelopment. On one side are state brownfield environmental and redevelopment policies that are touted as significant contributions in the effort to promote cleanup efforts and provide economic stimulation for local communities in need of help in dealing with brownfield sites. Such activities were provided an administrative boost with the informal adoption of canons of "smart growth" by the Michigan Land Use Leadership Council. At the same time, the state has been very reluctant to provide the leadership necessary to limit policies and local attitudes that support new "greenfield" development and suburban competition to attract growth away from existing urban areas.

The Michigan Land Use Leadership Council noted several areas where public policy and legal concerns contribute to a sprawling pattern of growth discouraging redevelopment efforts; land division and zoning requirements favor large-lot single-family residential development at the urbanized periphery. Government spending patterns throughout the state also encourage so-called greenfield development over the redevelopment of previously built areas. Furthermore, the state-mandated process for clearing land titles in urban areas is so time-consuming and cumbersome that it discourages land assembly and redevelopment efforts in urban areas. Additionally, intergovernmental competition is a serious issue in Michigan, where more than 1,800 units of local government have legal authority to engage in land-use planning and zoning. There is little coordination of planning and development efforts between these units. The state has enabling legislation in place to support regional land-use planning efforts in the event that two or more local governments might desire to engage in mutual planning efforts, but it is seldom called into play in any but a very limited manner.

Like many other states, Michigan is a strong home-rule state that has long relied upon local governments to make decisions that are of primarily local concern, including those related to development and land-use planning. State statutes granting this authority to Michigan cities have existed since the early 1900s. Over time, much of this same authority has been extended to townships, making Michigan one of only a few states to grant this level of decision-making power and authority to townships. In Michigan, townships, much like cities, have the authority to make a wide range of decisions related to development, taxation, public service provision, and a host of other issues, even in unincorporated and rural areas. State planning enabling statutes reflect this home-rule concept, granting to townships the same authority for autonomous land use and zoning decision making as that typically enjoyed by cities. In rural and other unincorporated areas of Michigan, counties play little more than an advisory role in shaping development and in creating long-range comprehensive growth plans that can encourage brownfield redevelopment as a critical component of land use strategy.

Public Benefits of Brownfield Redevelopment

The issue of how to define and evaluate the benefits of brownfield redevelopment projects is key in developing successful policies to support actual redevelopment efforts. However, the issue of proper evaluation is difficult, not because of a lack of studies in this area, but because a standardized methodology to assess public benefits has been little tested. Nevertheless, there are a growing number of studies being done in this arena, both case studies of individual sites or sets of actual projects and analyses of specific assessment mechanisms.

Assessing the benefits of brownfield redevelopment is complicated. Projects vary greatly in their redevelopment objectives, extent of public sector involvement, and character of environmental contamination. Furthermore, state and local initiatives to promote brownfields differ widely across the US Given such variation, is there a standard set of metrics than can be used to measure the public benefits of brownfield redevelopment? This question is further complicated because the evaluative metrics chosen imply particular definitions of the goals of a project and, therefore, of the character of "success."

In the case of brownfield redevelopment, different stakeholders can have very distinct goals. Private, for-profit real estate developers involved in brownfield redevelopment define successful brownfield projects in terms of acceptable profit given the level of risk involved. Cities have a different perspective. For example, the President of the U.S. Conference of Mayors speaks about successful brownfield redevelopment in terms of economic vitality, the utilization of existing infrastructure such as roads and public utilities, and the easing of pressure to develop open spaces and farmland (U.S. Conference of Mayors 2000, 2008). Community groups and environmental activists have a different focus. For example, the Center for Public Environmental Oversight (1998) has insisted that a number of social justice concerns be addressed with brownfield redevelopment, and that those most directly impacted by the redevelopment effort define its success, rather than leaving it up to project proponents. Despite the difficulties in measurement due to varied project types and goals, there have been efforts to evaluate the benefits of brownfield redevelopment. However, most of these studies have generally taken either a purely qualitative approach or a narrowly defined quantitative approach to measuring benefits. Numerous case studies have been and continue to be written about brownfield redevelopment (Bartsch and Collaton 1997; Pepper 1997; Dennison 1998; Simons 1998a; Meyer 2007). Generally these include a qualitative description of the benefits of the project. These descriptions vary by case and are not organized into any standard or consistent format.

Several quantitative studies have been conducted to measure the benefits of brownfield redevelopment, but indicators of benefit are relatively narrow, compared to the full range of possible benefits identified in the qualitative studies discussed above. For example, the Federal EPA estimates that its brownfield program has helped create more than 61,000 jobs across the country and leveraged $18.68 of additional investment for every federal dollar spent on brownfield redevelopment (U.S. EPA 2010).

The U.S. Conference of Mayors regularly conducts a survey of US cities with regard to their brownfield properties and found that cities reported that redeveloping their brownfields would collectively result in between $1.3 and $3.8 billion in additional annual tax revenues, 550,000 new jobs, and capacity for 5.8 million new people in cities without adding new infrastructure (U.S. Conference of Mayors 2006, 2008). The Conference of Mayors also found that tax-base growth, followed by job creation and neighborhood revitalization, were among the most commonly expected benefits of brownfield redevelopment. Neither the EPA study nor the U.S. Conference of Mayors survey show how their assessment measures could be normalized to enable project-by-project comparisons of project benefits. Rather, they aggregate these benefits across a plethora of projects, suggesting the potential benefits of brownfield redevelopment on a broad, national scale. Thus, these assessments do little to explore project benefits at the local level.

A more detailed project-level study on the benefits of brownfield redevelopment was published by the Council for Urban Economic Development (CUED, now the International Economic Development Council), in 1999. CUED's explicit goal was to focus on the economic development impacts of brownfield redevelopment. The authors developed two benchmarks (i.e. discrete, measurable elements) to evaluate a broad variety of projects in terms of their economic benefit. These benchmarks are powerful in that they can be used to measure the impact of a wide variety of projects, and are relatively simple to compute and understand. CUED measured public cost per job created in each project and private sector funds leveraged per dollar of public investment for each project. The authors concluded the median public cost per job created was $14,003, and that the median leverage for a typical project was $2.48. However, the CUED study measures only a very limited aspect of public benefit that can be realized from brownfield redevelopment.

A report published jointly by the Center for Public Environmental Oversight (CPEO) and the Urban Habitat Program (UHP) in 1998 stressed that the most common measures of public benefit from brownfield projects can neglect the impact of a project on the local neighborhood. The understanding of successful brownfield redevelopment is most typically defined by government agencies as the number of jobs created, the amount of private investment leveraged, and the new tax revenue created. However, redevelopment assessments using these simple metrics are not able to show that a project provided benefits to those who were negatively affected by the brownfield property. Environmental justice and community advocates argue that evaluations of brownfield projects should measure the benefits provided to the local community, not just project proponents. "Success cannot be merely defined in terms of dollars and cents. Rather it should be judged by the effectiveness of a community's ability to drive and benefit from the redevelopment process" (CPEO and UHP cited in Dyke 2000: 68).

CPEO executive Lenny Siegel has suggested that brownfield projects requiring public subsidies be evaluated across a number of metrics (Siegel 2001), specifically the extent to which each project would:

- involve the local community in planning,
- protect public health,
- generate local jobs and business,
- provide needed services or housing for the community,
- expand open space or otherwise improve the local quality of life,

- generate additional tax revenues for local agencies,
- retain the existing community and its cultural base,
- provide any of the above in a particularly blighted area.

To date there is no generally accepted method for evaluating this broader class of public benefits of brownfield redevelopment. CUED's rationale for focusing solely on economic development impacts may be indicative of the general reluctance to quantify the spectrum of public benefits created by brownfield redevelopment. CUED considered environmental and social dimensions of brownfield redevelopment to be more difficult to measure than economic impacts. Further, CUED notes that economic development is often cited as a primary goal of brownfield redevelopment. Data on economic benefits are often used by legislators and policy makers as a basis for allocating funding between projects and for measuring project success. Finally, as CUED states, "economic statistics are often seen as more rigorous than qualitative measures, which are often discounted as mere subjective pronouncements." Although CUED did briefly describe environmental and social benefits of the projects, there was no standardized method of evaluating these benefits (CUED 1999).

While this rationale for limiting project scope is understandable, there is a need for additional quantifiable benchmarks that address dimensions of brownfield redevelopment along the lines suggested by Siegel (2001, 2008). Meaningful assessment of brownfield redevelopment projects must reflect the primary goals and desired outcomes of a brownfield program. While job creation and leveraging private investment are primary goals in many projects, other projects have primarily social or environmental goals, or focus on other dimensions of economic benefit, such as increasing utilization of existing infrastructure or providing jobs to local residents.

Brownfield policies are new, experimental, and constantly evolving. Even the most established federal and state brownfield programs have been in existence for relatively few years. The wide variety of approaches to establishing incentives for redevelopment reflects the experimental nature of these policies: each jurisdiction is, in effect, a test case. The most successful approaches will only become evident as more projects are completed. Therefore, it is important to monitor the success of brownfield projects and policies in order to provide a feedback mechanism for policy evaluation and improvement. One way to monitor the success of brownfield policies is to develop standardized metrics that reflect the broad array of public benefits that are used as the rationale for promoting brownfield redevelopment. Metrics that can be applied across state boundaries could enable a comprehensive evaluation of the success of brownfield policies, both individually and as a collective, in making progress toward explicit local, state, and federal policy objectives.

Case Study Methodology

Researchers from both Eastern Michigan University and MDEQ conducted detailed case studies of 57 brownfield redevelopment sites representing 63 MDEQ projects throughout the State of Michigan (a listing of the 57 case study sites is found in Appendix 13.1). These projects were selected from among the 365 projects for which the state has provided funding support for brownfield assessment, remediation, and due care activities from the five 1988 and 1998 bond supported programs and the revolving loan program. In accordance with requirements of the six grant and loan programs, state support requires that fund recipients redevelop remediated sites. Of these, eight are currently undergoing redevelopment effort, but have completed much of their remediation and undergone sufficient redevelopment to indicate a strong likelihood of success. Another eight brownfield projects are considered delayed in their remediation and/or redevelopment effort such that little, if any, redevelopment effort is evident at the sites. The remaining 41 sites have successfully completed or substantially completed all remediation and redevelopment work.

The project research team used an iterative, non-linear case study approach to examine the successful redevelopment of brownfield sites in coastal areas of Michigan. "Coastal Michigan," defined as being within approximately one mile of the coast, was selected due to the mission of the funding authority, Michigan Sea Grant. However, with additional staff and intern support from MDEQ, the research team was able include 14 projects located in communities away from the coastal areas of the state. Projects were selected based on the knowledge of MDEQ staff, with no attempt to randomize selection from the approximately 365 projects that received funding from the six MDEQ programs of interest. Further, no attempt was made to isolate and assess the direct impact of coastal proximity to redevelopment project success.

The 63 projects at 57 different locations represent over 17% of all projects that received funding from the six MDEQ brownfield redevelopment programs. The $63.3 million granted for these projects represents approximately 36% of the total funding allocated to the six programs. These case study projects would seem to have received a somewhat disproportionate amount of state funding, at least in terms of the number of projects. It would be dangerous to conclude from this, though, that greater state funding is important in insuring project redevelopment success. In fact, three of the projects that received the most MDEQ funding have as yet to show any significant redevelopment (Industrial Brownhoist project at $11.4 million, Bay City Waterfront project at $3 million, and Water Street project at $4.7 million).

In evaluating brownfield redevelopment success in Michigan, the case study projects were assessed across four impact areas. Specific metrics for each of these areas were developed through discussions among the research team, MDEQ staff, and Michigan Sea Grant staff, including:

- environmental site remediation prior to redevelopment,
- environmental impacts of the actual redevelopment project,
- economic and fiscal impacts, and
- social and community development impacts.

To evaluate environmental remediation impacts, the project team looked at the area cleaned, the type and amount of remediation work required, the remediation standard applied (residential, commercial, or industrial), and whether or not on-going "due care" activities are required. Much of this type of information is contained in the MDEQ project records and was also checked with field observations. Follow-up interviews are also conducted with local officials, project managers, and environmental consultants advising on the projects.

The evaluation of each project included the following to address the metrics in the project assessment instrument:

- archival review of MDEQ files located in Lansing and field offices, which included review of grant/loan application files, baseline assessment materials, Phase I and Phase II environmental assessment reports, and other project information;
- preliminary and follow-up field investigations to assess current project status and evaluate surrounding conditions;
- GIS analysis as appropriate to each individual site, and analysis of existing cartographic and photographic information of the project site and surrounding area;
- content analysis of local planning, development, and documents related to project, including newspaper and other media stories, public outreach materials, project advertising, and the like;
- key informant interviews as needed to complete the brownfield redevelopment project evaluation instrument and to provide additional insight into project issues.

Information for each of the 57 case study sites was compiled into a database and subjected to basic comparative analyses. The 41 successful projects were compared to eight projects currently undergoing redevelopment and to the eight projects that had seen little or no remediation and redevelopment activity. There were marked and significant differences between projects that had successfully been redeveloped and those that had not. This is hardly surprising, however, in that the assessment tool was intended to provide metrics for evaluation of successful (rather than unsuccessful) projects.

Assessing Brownfield Redevelopment Success

From a legislative standpoint, Michigan has thus far chosen to keep things rather simple in assessing brownfield redevelopment, concerning itself primarily with the amount of land that has been remediated, the amount of private investment that has been leveraged for both remediation and redevelopment activities, the number of new housing units constructed in the case of residential redevelopment, or the number of new jobs created in the case of commercial and industrial redevelopment projects. Although these measures are not trivial, they are clearly not enough. For instance, a brownfield property cleaned and redeveloped as a park or other public open space would certainly contribute to the amount of remediated land in a community, but there would likely be little direct private investment leveraged. The project would contribute few new jobs or housing units, and there would be little direct contribution to local fiscal resources. Yet, the social benefits could be tremendous, as would the indirect benefits for adjacent neighborhoods, and if properly planned, the project could spur other local development efforts.

The relative level of successfulness for a brownfield redevelopment project is not easy to identify. The literature on this topic is growing, but is still far from complete. The efforts to date may generally be divided into two broad categories. The first of these primarily consists of case studies of brownfield redevelopment projects, either individually or as a collection of projects within a given community or other geographic area. From these cases, discussions of relative success ensue, often leading to conclusions that identify key components or actions that made the project successful (Regional Analytics 2002; Wernstedt et al. 2004; Zavadskas and Antucheviciene 2006; Mohamed and Dancik 2007).

A second category of study proceeds to argue for a predetermined set of measures for success that can subsequently be applied to brownfield redevelopment projects. These studies may range from discussion of a single metric such as property value (Bacot and Odell 2006) to multiple metrics that often seem to evolve around the concept of a "triple bottom line" (examining environmental, economic, and social concerns) borrowed from discussions of sustainability (Lange and McNeil 2004; Pediaditi et al. 2006; Wedding and Crawford-Brown 2007; Paull 2008). It is this later discussion that most influenced the current research effort. However, this triple bottom line approach was somewhat modified.

In addition to examination of the environmental benefits realized from remediation activities, the environmental impacts of the redevelopment effort were examined in an effort to determine what sorts of environmental benefits are realized beyond any remediation activity. The redevelopment land-use type was evaluated for its potential for environmental pollution. Impacts in such areas as storm water management, groundwater protection, and buffering from environmentally sensitive areas were also examined. Project compatibility with adjacent developed and natural areas was reviewed. Each project was also checked for participation in established "green" programs such as the federal Energy Star program or LEED. This information was gathered through site evaluations and the study of redevelopment plans. Interviews with developers and local planning authorities were also useful in understanding how various redevelopment projects impact the environment after the brownfield remediation effort was completed.

From the standpoint of economic impacts there are employment gains, leveraged investment, and revitalized neighborhoods to be considered. Connections between the redevelopment project and larger regional, state, and global economies can also be significant. Often, though, it is rather difficult to assess the economic impact of a brownfield redevelopment project, since such effects are closely intertwined with other economic development programs and with other nearby redevelopment projects. In Michigan, brownfield redevelopment is frequently undertaken in concert with a variety of other economic development programs such as state renaissance zones, so-called Cool Cities (a state program inspired by the work of Richard Florida) projects, enterprise zones, and others. Further, brownfield redevelopment projects are also incorporated into general local revitalization efforts, representing but one significant piece of a much larger puzzle.

The fiscal aspect of brownfield redevelopment projects includes such concerns as the generation of new sources of local revenue derived from previously unproductive land. Brownfield redevelopment is also said to lower requirements for municipal investment in infrastructure to accommodate growth due to the reuse of existing infrastructure or because brownfield redevelopment projects tend to provide more compact forms of development than Greenfield projects at the urban periphery. As with other economic impacts, the picture here is rather cloudy. Most of the brownfield projects assessed have incorporated some form of tax credit or other incentive program, or are part of a brownfield redevelopment tax-increment financing district. This means that local communities often realize only small direct gains in local tax benefits from brownfield redevelopment, at least in the short-term. Further, many brownfield redevelopment projects are publicly held and are therefore exempt from local property taxes. Nevertheless, fiscal benefit is derived from the spillover effects of new jobs within a community, with accompanying new personal and business revenues, and from the potential for increase in the value of properties adjacent to the redevelopment site.

The community development impacts of brownfield redevelopment include providing space for government and social service activities, development of new affordable-housing units, creation of living-wage jobs, and new business generation. Also assessed was whether or not existing support structures such as public transportation are adequate to meet the needs of redevelopment projects. These measures are evaluated within the broader context of general community needs and desires as reflected in community master, housing, and parks and recreation plans. Community development leaders were also identified and interviewed about how they feel brownfield redevelopment projects impact services provided.

Similar evaluations of community plans and goals and interviews with community leaders were used to examine the social impacts of redevelopment projects. Public participation in the brownfield redevelopment process was assessed, as was the provision of public amenities such as parks and open space, waterfront access, and community center facilities, all of which are significant aspects of a variety of brownfield redevelopment projects in Michigan.

Analysis of the 63 projects indicates that the successful redevelopment projects have some broad characteristics in common. First, although some smaller projects such as the Tawas City and the Alpena Riverfront projects have impacts primarily within their immediate community, most successful projects recognize some importance for making broader regional, state, and even global connections. The economic success of projects often must depend on financing from outside the local area, and even from outside the state. Businesses such as the Whirlpool Corporation, Edgewater Automation (a developer of custom-automation equipment), and R&B Electronics, all benefiting from the state's brownfield redevelopment programs, are dependent on doing business in a global environment far outside the confines of Michigan. Even residential developers realize the impact created by homebuyers from outside the state. Many successful residential brownfield redevelopment projects, particularly in Michigan's coastal areas, include "second home" units and the associated recreational facilities such as marinas that are marketed to potential buyers in places such as the Chicago and Toronto metropolitan areas. Communities such as Traverse City, Marquette, and Ludington have taken advantage of "outside" capital, and the revitalization of Benton Harbor is absolutely dependent upon it.

A second feature common to a number of successful projects is a degree of regional planning cooperation, a remarkable feature within the political context of Michigan planning and development. As noted at the outset, Michigan is a strong home-rule state with a penchant for extending property rights as far as is practical. It is also a state that has divided its unincorporated areas into townships and given those townships opportunity for an extraordinary level of autonomy through a chartering process. All of this has led to notable levels of local competition when it comes to attracting development dollars, and a distinct lack of cooperation when it comes to urban planning (Michigan Land Use Leadership Council 2003).

Yet, brownfield redevelopment efforts do not seem to have quite the same localized character. A great number of Brownfield Redevelopment Authorities operate at the county level, rather than the local level, and in cases where county and municipal BRAs overlap, there seems to be a far greater level of cooperation than competition. Perhaps most remarkable is the Harbor Shores redevelopment project in Benton Harbor, where the Cities of Benton Harbor and St. Joseph, Benton Charter Township, and Berrien County have all entered into a cooperative agreement for the remediation and redevelopment of about 570 acres of brownfield sites that will bring tremendous economic and community development benefit to the entire region.

A third common element in successful brownfield redevelopment is the creation of a vision for redevelopment. It is clear that a basic remediation strategy to clean a site and make it ready for redevelopment is not enough. Even prior to the current economic recession, projects that did not have a definite and clear plan for redevelopment had a hard time attracting developers. A plan that is grounded in current conditions, but which looks forward five, 10, and even 20 years must be in place, and it must enjoy a significant level of community support.

The Mason Run project in Monroe, Michigan, clearly illustrates this. Located on a site once occupied by a large paper mill, the city saw a clear need to develop single-family housing at this location less than a mile from the downtown core. Further, the city identified the need for a significant amount of housing that would be suitable for households earning above the median income for the area—a need created by new development at the nearby industrial park and other remediated brownfield sites in the city. Working with its citizens, the city crafted a plan for redevelopment and located a developer that shared the vision of the plan. As a result, the city is realizing a new neighborhood of family homes that fits well within the context of existing homes, wetland areas, and recreational facilities. Currently about half complete and somewhat slowed by the current slump in the housing market, the developer and community are looking forward to completion of the approximately 300 homes over the next 10 years, which will be Monroe's newest neighborhood.

Related to this, the vision for redevelopment must be holistic and complete. It is not enough to begin a plan and intend to complete it at some indefinite future point in time. The Saginaw Riverfront in Bay City, Michigan, provides an illustration. A plan to consolidate several brownfield properties along the river near downtown was developed in the late-1990s, but it was incomplete in its consideration of the actual redevelopment for these properties. Where specific projects were identified, such as a new hotel and conference center, redevelopment has been very successful. In areas of the riverfront where actual projects still have not been identified, land sits cleared and ready, but redevelopment has not happened. There are dreams that some sort of tourism attraction might happen, but the overall vision remains unclear and the plan incomplete.

Ypsilanti's Water Street project provides another example of an incomplete and ungrounded vision. Over the past decade, the city has had at least two developers interested in the project, but the parties have not been able to agree on a vision and practical development scheme that is implementable. In the meantime, citizens of Ypsilanti have lost their enthusiasm for the project and city leaders are at a loss as to what might be done to redevelop the approximately 38-acre site.

Although it may seem obvious, it is imperative that the developer for each project, whatever it may be, not be over-extended in terms of financial backing and capacity to complete the project within a reasonable time-frame. Certainly problems associated with the current housing slump and general economic decline in the state have adversely impacted many brownfield redevelopment efforts. However, a number of successful projects have been started or are continuing through the depressed economic climate of the state. Work on these projects has indeed been slowed, particularly for residential projects. But the homes continue to move as developers have the resources to continue construction.

Strong project leadership is important, both for remediation activities and for redevelopment. For most of the case study projects, a clear project leader has been identified. That leader may be an individual or group, a non-profit entity or a city council and staff, so long as they are willing to take on the role of project "champion." It is inevitable that a brownfield redevelopment project encounter bumps along the road to successful completion. New areas of contamination are uncovered as redevelopment proceeds, remediation ends up costing far more than expected, the developer decides to add additional residential units on an area that was only remediated to commercial standards, commercial rentals and residential sales slow because of uncontrollable economic conditions—the list goes on. What is important is that someone, or group, or other entity is convinced of the environmental, social, and economic benefits of the project, to the point where they can steer it through all the unanticipated problems towards successful completion.

Finally, successful projects have all garnered and effectively utilized public support for brownfield redevelopment. Project leaders hold a clear understanding that community support can be one of the most important assets in promoting and implementing brownfield redevelopment efforts. Further, it is important that the public be involved from the outset of the project in order to counter the perception that decisions have already been made and deals cut. Even this, however, is no guarantee that community residents will not stand in opposition to key aspects of a redevelopment project.

The Harbor Shores project in Benton Harbor provides an illustration of this. A significant component of the redevelopment effort involves creation of a signature public golf course, part of which occupies what was once a little-utilized and run-down portion of a public beachfront park. The developers had addressed this issue early in their planning process and had elicited general public support for this part of the project, with agreements for beach access improvements, long-term contributions to park maintenance, and public access to an interconnected system of trails and bikeways throughout the redevelopment area. This early public involvement proved effective at quieting later opposition from a contingent of neighborhood activists opposed to the scope of the redevelopment project.

Michigan Brownfield Redevelopment Policy Concerns

Examination of the six MDEQ brownfield redevelopment programs clearly shows that state initiatives in this area have been instrumental in generating community reinvestment, as well as in aiding with cleanup of environmental contamination. These programs, and the policies that support them, have been successful in generating developer interest in brownfield properties through the overall reduction of private sector costs associated with brownfield sites. Further, such projects have had a positive impact on the surrounding areas, if for no other reason than the community has eliminated what is most often a nonproductive and blighted property (Hula 2002a). However, researchers working nearly a decade ago found that although brownfield redevelopment projects brought certain economic improvements to a community, no case could be made that a brownfield investment led to significant secondary development (Hula 2002b).

The evaluation here revealed a number of cases where an initial brownfield redevelopment project has been a catalyst for other redevelopment efforts in the surrounding area. The remediation of the former Ascot foundry site in St. Joseph and its redevelopment as the Edgewater commercial park has directly spurred investment in other commercial, as well as retail and residential projects, in the immediately adjacent area. Nearby, Benton Harbor's project has expanded to include a massive community development effort that has thus far led to numerous job-training programs, after-school programs, the development of approximately 400 units of affordable housing, and downtown revitalization.

Similar conclusions may be reached even for very small projects such as in Tawas City. There a small state investment in a riverfront improvement project, together with construction of a new city hall, has directly encouraged redevelopment of adjacent property, with plans to eventually reconstruct the small downtown area. Projects in Ludington and Marquette, St. Anne's Gate in Detroit, and others have all helped to spur investments in surrounding areas.

Other communities, however, have not been so successful. Although state-supported projects such as the Alpena Riverfront Area revitalization and the Doubletree Hotel and Conference Center in Bay City have been outstanding projects in and of themselves, they have not, thus far, directly stimulated significant levels of other new development in their communities. Part of the issue, as Hula (2002b) points out, is that such direct connections are notoriously difficult to demonstrate conclusively. More significant, though, is the amount of time that is involved. As indicated in any number of the case studies, successful projects require that communities be engaged in the redevelopment effort for the long haul. It can be a decade or more before the full benefits of brownfield revitalization come to fruition. Further, the general economic climate in Michigan over the past few years clearly clouds the picture and has undoubtedly delayed a wide variety of community reinvestment projects, not just brownfield redevelopment. The state, as well as local communities, must be prepared to wait out the apparent lack of progress in many as-yet-unsuccessful brownfield projects, since it is likely that many positive impacts will not be measurable in the near-term.

A focus on leveraging private sector investment makes it difficult for municipal leaders to develop and implement a broad community development plan when it comes to brownfield redevelopment. Too often there is a tendency to focus on the individual project with little attention paid to wider, collective concerns. However, it is clear that having a well-thought out and articulated development plan is an important factor in the success of brownfield redevelopment efforts that we have studied. Some of the case study communities, such as Ludington, Monroe, and Traverse City, have embedded their brownfield plans within their local comprehensive planning efforts and municipal zoning ordinances. Others such as Marquette and St. Joseph have opted to create brownfield plans and related implementation mechanisms coordinated with, but independent from, other local comprehensive planning and land-use regulation efforts.

The role of Brownfield Redevelopment Authorities is interesting in this regard. State statutes require that a development plan be in place for these entities to operate. However, the BRA is primarily a fiscal authority, and the plans it prepares are redevelopment financing plans, not more comprehensive community-development plans that consider a full range of social and economic concerns, as well as development impacts. Moreover, numerous developers and municipal officials indicated that brownfield property owners can be reluctant to have their properties listed in a BRA-financing plan out of fear of the stigma that the general public may attach to the site when it is recognized as a brownfield. The tendency is to list a site with a BRA only at the last minute when redevelopment work is about to commence and the financing framework must be in place.

What is missing in this is any sort of requirement that a community-redevelopment plan ensure that projects are developed in such a way as to maximize beneficial community impacts. The six MDEQ programs examined require that a developer be identified and the semblance of a redevelopment plan be in place, but these are parcel specific, and there is nothing to require that such plans be consistent with overall community preferences as reflected in more general community-development plans. Yet, the potential for such development coordination exists, with both the state and local BRAs working more closely together, and the state developing the capacity in such local entities to look beyond fiscal and other economic benefits. These sorts of state and local connections would likely improve the ability of MDEQ to follow up on the redevelopment aspects of its programs and aid in the promotion of the types of public-private partnerships that have become a hallmark of current efforts to redevelop brownfield sites (Hula 2002a, 2002b; Hamlin et al. 2005).

Developing local capacity to design and implement brownfield redevelopment plans that reflect local understandings and political culture is critical. For example, the private sector has been instrumental in championing brownfield redevelopment efforts in South West Michigan, while in South East Michigan civic leaders have tended to provide this boost. In Benton Harbor, it is the Whirlpool Corporation and its philanthropic foundation that have set the tone for that city's immense Harbor Shores undertaking, while in Monroe, city officials have been outspoken in their efforts at dealing with many brownfield issues. What the 63 case studies of this project have revealed is that there is no singular approach to creating public-private partnerships that works best at fostering successful brownfield redevelopment.

To be sure, both federal and state governments play legitimate and essential policy and fiscal oversight roles in local brownfield redevelopment efforts. They are also instrumental in the development of local capacity. Yet, both federal and state authorities must also take on the more decentralized roles. If environmental remediation dollars are to be maximized, the state must maintain a high degree of flexibility in allowing local communities to set desired brownfield redevelopment outcomes.

Finally, there must be some effort at controlling local competition through regional coordination. The state has allowed BRAs to function at the county level with local approval, which has been helpful in coordinating brownfield redevelopment efforts across municipalities. Further, adjoining jurisdictions have voluntarily entered into intergovernmental agreements to coordinate environmental remediation and redevelopment efforts. However, as illustrated in policy studies by the Michigan Land Use Institute (2005) and Good Jobs First (LeRoy et al. 2006) more must be done to coordinate state policy, particularly in the area of economic development. As these studies clearly indicate, there is a tendency for the state to invest disproportionately in suburban areas, often at the expense of older, urban jurisdictions where the preponderance of brownfield sites are located. Although it is not realistic to expect the state legislature to reverse such disproportional investment, there must be greater recognition that such expenditure patterns do exist, and that they tend to provide hidden subsidies that favor greenfield development over brownfield redevelopment.

How Successful, Michigan?

Within the environmental policy context, how successful has Michigan been at encouraging both remediation and redevelopment? From the standpoint of remediation, it can be argued that the state has been very successful. When MDEQ data for 365 projects representing just over $155 million in MDEQ funding support are examined, all but a very small handful of projects have been successful, at least insofar as remediation activities are concerned. These projects date between 1989 and 2008, which means that most have completed remediation work. Bear in mind, however, that MDEQ has no legislative mandate and only very limited capacity to follow up on the actual redevelopment of brownfield redevelopment projects. Further, what limited capacity does exist is often in association with other state agencies, such as the Michigan State Housing Authority or the Michigan Economic Development Council, that have their own particular interests and legislative mandates to pursue.

All indications from examination of the 63 brownfield redevelopment projects are that Michigan has been at least somewhat successful at redevelopment, despite the economic woes over the past half-decade. Certainly, a significant amount of redevelopment has occurred at approximately 500 of the 1,800 sites in which the state has invested funds for response and remediation activities (MDEQ 2008). Further, between half and two-thirds of those redevelopment projects have been in conjunction with the six MDEQ programs identified in this study. However, much work remains to assess brownfield redevelopment projects more fully to develop a methodology and database that will allow the state to track its investments in preparing brownfield sites for redevelopment, and to assist MDEQ in the evaluation of brownfield redevelopment project outcomes.

One area where Michigan's efforts do seem to be lacking relates to the environmental impacts of the redeveloped sites. Although evaluations show that the projects have done all that is "typical" in minimizing environmental impacts, developers of these projects, whether public or private, have not taken the extra step to provide leadership in creating "green," sustainable projects. Some of the projects evaluated use energy-efficient lighting, and residential projects offer Energy Star-rated appliance packages. None, however, are LEED certified. Further, very few use alternative storm water management systems such as bioswales and pervious surface treatments, rain gardens, and other low-impact strategies. Certainly, some of this is understandable. Both the Mason Run project in Monroe and the Edgewater project in St. Joseph have contaminated soils encapsulated on site. Porous pavement surfaces would not be acceptable, defeating environmental remediation efforts developed for the projects.

Perhaps no project is more telling in this regard than the Grand Landing project in Grand Haven. A multi-phased, mixed-use residential/commercial/retail project currently under development, it was to have included innovative on-site storm water management techniques. The developer's application for MDEQ support indicated the project would include permeable surfaces in the parking areas, bio-remediation, and on-site containment and cleaning of all stormwater runoff. However, local planning officials indicate that these techniques have not been used in the actual development so far. Rather, they have been eschewed in favor of expediency (planning commission evaluation of the innovative techniques was apparently taking more time that anticipated) and monetary savings due to the economic downturn. In all fairness, it should be noted that the project is a nice addition to the Grand Haven community, and takes advantage of its location near the center of the city, major transportation arteries, and, of course, its waterfront location along the Grand River and proximity to Michigan's coastal amenities.

Directions for Future Research

Issues related to long-term successfulness of brownfield redevelopment projects in Michigan came up in a number of interviews conducted with facility users. Obviously, people were comfortable enough to live in the homes, work at the jobs, shop at the stores, and use the recreational facilities resulting from redevelopment. Yet, some concerns were expressed about the ability of the state's flexible cleanup standards to protect the public health over the long term. Several individuals were surprised to learn that the state did not necessarily require that sites be totally cleaned and wondered aloud what long-term impacts there might be as a result of contaminated materials remaining on site.

On a related note, questions of the near- and long-term adequacy of technological solutions such as encapsulating contaminated material under streets arose. "What happens if the city crews come along in a year or two and need to dig up the street for some kind of repair? What's to keep kids from coming along and riding their bikes through that stuff when nobody's looking?"

Such questions also apply to the adequacy of institutional controls. Are restrictions applied through municipal zoning, subdivision, and other land-use regulations adequate to protect the public in the very long term? Institutional memory of the what, where, and why of such controls may tend to be set aside when it comes to some future proposal for a new economic development project that meets a pressing community need. And this is to say nothing of the pressures that the private market can exert to change municipal regulation. Clearly, such long-term issues need to be addressed as part of the definition of successful brownfield redevelopment.

A second concern for follow-up study relates to projects that are less than successful. Only a small handful of such projects were examined, primarily as a point of comparison related to the broad elements commonly held among successful redevelopment efforts. As mentioned, MDEQ does not have sufficient capacity to follow up on projects once the remediation work is complete. Thus, there is no accurate count of exactly how many projects have been started, but with little or no actual redevelopment work done.

Of the less than successful projects examined, some have not even begun assessment and remediation work as described in their applications for state funding support. Others have done some assessment work (including baseline assessment, Phase I, or Phase II activity), but little remediation has been completed. Still others have at least begun remediation activity. In all instances, none of the intended redevelopment work has actually been started, and there is no sign of any on-going activity on the sites. Clearly, the state's investment in such properties has had little return. A better understanding of the issues faced by these projects could provide valuable insight into essential elements of success and help local jurisdictions and developers avoid some of the pitfalls leading to project delays and failure.

On the matter of economic impacts, much work remains to be done. It is relatively easy to assess, for instance the number of jobs a redevelopment project creates: both temporary positions such as project construction, and permanent positions. But what of the impact that brownfield redevelopment has on surrounding property values? In this regard, obtaining access to current and historic property assessments can be difficult and time-consuming. This issue, however, is becoming less of a concern, as increasing numbers of jurisdictions digitize assessment data and make it available in various electronic formats, including over the web. There has also been promising work as to how such property value impacts can be assessed (Leigh and Coffin 2006). However, caution is in order here, because short-term efficiency may become the most feasible measure of success.

Finally, this research was conducted before the bursting of the housing bubble, making it relatively easy to identify impacts of remediation and redevelopment on adjacent property values (with the caveat that identifying and isolating variables that impact property values is never that easy). Moreover, many evaluative efforts treat property values in the aggregate compared against general federal and state policy shifts, rather than looking at specific properties against actual remediation and redevelopment efforts. Nevertheless, this research area does hold some promise in better understanding the full impacts of brownfield redevelopment over the long term.

References

Bacot, H. and C. Odell. 2006. "Establishing Indicators to Evaluate Brownfield Redevelopment." *Economic Development Quarterly* 20: 142–61.

Bartsch, C., E. Collaton and E. Pepper. 1996. *Coming Clean for Economic Development*. Washington, DC: Northeast-Midwest Institute.

Bartsch, C. and E. Collaton. 1997. *Brownfields: Cleaning and Reusing Contaminated Properties*. New York: Greenfield Press.

Center for Public Environmental Oversight and Urban Habitat Program. 1998. "Report on the Environmental Justice/Community Group Caucus at Brownfields '98." Report presented at Brownfields '98, Los Angeles.

Coastal Zone Management Act of 1972 (Public Law 92-583, 16 U.S.C. 1451-1456).

Copeland, C. 1997. *Superfund and the States: The State Role and Other Issues* (97-953 ENR). Washington, DC: Congressional Research Service.

Council for Urban Economic Development. 1999. *Brownfields Redevelopment: Performance Evaluation*. Washington, DC: US Environmental Protection Agency.

Davis, T. 2005. "Host of New Homes Dot the Landscape." *Ann Arbor News*, June 19. Available at http://www.mlive.com (accessed: June 30, 2005).

Davis, T.S. and K.D. Margolis. 1997. *Brownfields: A Comprehensive Guide to Redeveloping Contaminated Property*. Chicago: American Bar Association.

Dennison, M. 1998. *Brownfields Redevelopment*. Rockville, MD: Government Institutes.

Dyke, T.A. 2000. *Evaluating the Community Benefits of Brownfields Redevelopment*. Master of Community Planning Thesis, Massachusetts Institute of Technology.

Environmental Law Institute. 1991. *An Analysis of State Superfund Programs: 50-State Study, 1991 Update*. Washington, DC: U.S. Environmental Protection Agency, Office of Emergency and Remedial Response Hazardous Site Control Division.

Greenberg, M., K. Lowrie, H. Mayer, K.T. Miller and L. Solitare. 2001. "Brownfield Redevelopment as a Smart Growth Option in the United States." *The Environmentalist* 21: 129–43.

Hamlin, R., R. Hula, B. Cobarzan, C. Jackson-Elmoore and C. Leuca. 2008. *Brownfields: Making Programs Work for Michigan Communities*. Urban Policy Research Brief # 5. East Lansing, MI: Center for Community and Economic Development, Michigan State University. Available at www.ced.msu.edu/reports/Briefs%20-%2005f%20-%205%20-%20PQ.pdf (accessed: December 15, 2009).

Hula, R. 1999. *An Assessment of Brownfield Redevelopment Policies: The Michigan Experience*. Washington, DC: Price Waterhouse Coopers Endowment for the Business of Government.

——. 2002a. *The Michigan Brownfield Initiative and Private Market Redevelopment: An Assessment*. East Lansing: Michigan State University, Institute for Public Policy and Social Research.

——. 2002b. " There is Gold in Those Brownfields ... Maybe: Brownfield Reuse and Urban Economic Development in Michigan." In *Urban Policy Choices*, edited by D. Thorton and C. Weissert. East Lansing, MI: Michigan State University Press.

Hula, R. and R. Bromley. 2008. "Cleaning up the Mess: Redevelopment of Urban Brownfields." Conference paper, Midwest Political Science Association, Chicago, Illinois.

Hula, R. and A. Hemond. 2003. "Citizen Assessment of the Michigan Brownfield Initiative." Conference paper, Midwest Political Science Association National Conference, Chicago Illinois.

ICMA—International City/County Management Association. No date. *Getting to Smart Growth: 100 Policies for Implementation*. Washington, DC: Environmental Protection Agency. Available at http://www.epa.gov/smartgrowth (accessed: December 15, 2009).

Lang, M. 2001. *Brownfield Amendments of 2000*. East Lansing, MI: Victor Institute, Michigan State University.

Lange, D. and S. McNeil. 2004. "Clean It and They Will Come? Defining Successful Brownfield Development." *Journal of Urban Planning and Development* 130: 101–8.

Leigh, N.G. and S. Coffin. 2006. "Modeling the Relationship among Brownfields, Property Values, and Community Revitalization." *Housing Policy Debates* 16(2): 257–80.

LeRoy, G., A. Lack, K. Walter and P. Mattera. 2006. *The Geography of Incentives: Economic Development and Land Use in Michigan*. Washington, DC: Good Jobs First. Available at www.goodjobsfirst.org/pdf/michiganlanduse.pdf (accessed: December 15, 2009).

McCarthy, L. 2002. "The Brownfield Dual Land-Use Policy Challenge: Reducing Barriers to Private Redevelopment While Connecting Reuse to Broader Community Goals." *Land Use Policy* 19(4): 287–96.

Meyer, P.B. 2000. "Accounting for Stigma on Contaminated Lands: The Potential Contributions of Environmental Insurance Coverages." *Environmental Claims Journal* 12(3): 33–55.

——. 2007. "Brownfield Redevelopment for Community Regeneration: Preserving Neighborhoods and Avoiding Displacement." Conference paper, Association of Collegiate Schools of Planning, Milwaukee, WI.

Meyer, P.B. and Lyons, T.S. 2000. "Lessons from Private Sector Brownfield Redevelopers: Planning Public Support for Urban Regeneration." *Journal of the American Planning Association* 66(1): 46–57.

Michigan Department of Environmental Quality. 2006. *Consolidated Report on the Environmental Protection Bond Fund, Cleanup and Redevelopment Fund, Clean Michigan Initiative Bond Fund: Fiscal Year 2005*. Lansing, MI: Michigan DEQ. Available at http://www.deq.state.mi.us/documents/deq-rrd-FY05ConsolidatedReport.pdf (accessed: April 27, 2006).

——. 2008. *Consolidated Report on the Environmental Protection Bond Fund, Cleanup and Redevelopment Fund, and Clean Michigan Initiative Bond Fund: Fiscal Year 2008*. Lansing, MI: Department of Environmental Quality. Available at http://www.michigan.gov/deq (accessed: January 15, 2009).

Michigan Land Use Institute. 2005. *Follow the Money: Uncovering and Reforming Michigan's Sprawl Subsidies*. Beulah, MI: Michigan Land Use Institute. Available at http://mlui.org/growthmanagement/fullarticle.asp?fileid=16785 (accessed: Decmber 15, 2009).

Michigan Land Use Leadership Council. 2003. *Michigan's Land, Michigan's Future: Final Report of the Michigan Land Use Leadership Council*. Available at http://www.michiganlanduse.org/finalreport.htm (accessed: September 22, 2003).

Michigan Sea Grant. 2006. "Causes and Consequences of Environmental Change: Integrated Assessments." Ann Arbor, MI: Michigan Sea Grant College Program.

Mohamed, R. and B. Dancik. 2007. *Priming the Pump: Assessing the Investment Impact of Michigan's Site Assessment Fund*, Report # CS-2007-0. East Lansing, MI: Michigan State University, Land Policy Institute.

Mulcahy, J. 2005a. "Few are Following County's Blueprint," *Ann Arbor News*, June 19. Available at http://www.mlive.com (accessed: June 30, 2005).

——. 2005b. "Builders 'Racing to the Middle'," *Ann Arbor News*, June 20. Available at http://www.mlive.com (accessed: June 30, 2005).

——. 2005c. "Region Planning: More Talk than Action," *Ann Arbor News*, June 25. Available at http://www.mlive.com (accessed: June 20, 2005).

Patchin, P.J. 1994. "Contaminated Properties and the Sales Comparison Approach." *The Appraisal Journal* 62(3): 401–23.

Paull, E. 2008. *The Environmental and Economic Impacts of Brownfields Redevelopment*. Washington, DC: Northeast Midwest Institute.

Pediaditi, K., W. Wehrmeyer and J. Chenoweth. 2006. "Developing Sustainability Indicators for Brownfield Redevelopment Projects." *Engineering Sustainability* 159: 3–10.

Pendergrass, J. 1996. "Use of Institutional Controls as Part of a Superfund Remedy: Lessons from Other Programs." *Environmental Law Reporter* 26(3): 10109–23.

Pepper, E. 1997. *Lessons from the Field.* Washington, DC: Northeast-Midwest Institute.

Persky, J. and W. Wiewel. 1996. *Central City and Suburban Development: Who Pays and Who Benefits?* Chicago: University of Illinois Press.

Puentes, R. and M. Orfield. 2002. *Valuing America's First Suburbs: A Policy Agenda for Older Suburbs in the Midwest.* Washington, DC: Brookings Institute, Center for Urban and Metropolitan Policy.

Regional Analytics Inc. 2002. *A Preliminary Investigation into the Economic Impact of Brownfield Redevelopment Activities in Canada: Final Report.* Prepared for The National Round Table on the Environment and the Economy.

Rusk, D. 1993. *Cities without Suburbs.* Washington, DC: Woodrow Wilson Center Press. SEMCOG. 2003. *Land Use Change in Southeast Michigan: Causes and Consequences.* Detroit: SEMCOG. Available at http://www.semcog.org/cgi-bin/products/publications.cfm (accessed: December 15, 2009).

——. 2004. *Land Use Change in Southeast Michigan: Regional Summary, 1990–2000.* Detroit: SEMCOG. Available at http://www.semcog.org/cgi- bin/products/publications.cfm (accessed: December 15, 2009).

SEMCOG. Community Profiles. Available at http://semcog.org/Data/CommunityProfiles/index.htm (accessed: August 18, 2007).

Senate Report 107-002—*Brownfields Revitalization and Environmental Restoration Act of 2001.* Available at http://thomas.loc.gov/cgi-bin/cpquery/R?cp107:FLD010:@1(sr002 (accessed: December 15, 2009).

Siegel, L. 2001. *Community Advice: A Constructive Approach to Brownfields.* Mountain View, CA: Center for Public Environmental Oversight.

——. 2005. *The Do's and Don'ts of Community Involvement in Brownfields Revitalization.* Mountain View, CA: Center for Public Environmental Oversight.

——. 2008. *The Brownfields Assistance Project: Final Report.* Mountain View, CA: Center for Public Environmental Oversight.

Simons, R. 1998. *Turning Brownfields into Greenbacks: Developing and Financing Environmentally Contaminated Real Estate.* Washington, DC: Urban Land Institute.

U.S. Conference of Mayors. 2006. *Recycling America's Land: 2006 Brownfields Survey,* volume 6. Washington, DC: U.S. Conference of Mayors.

——. 2008. *Recycling America's Land: 2008 Brownfields Survey,* volume 7. Washington, DC: U.S. Conference of Mayors.

U.S. Environmental Protection Agency. 2010. "Brownfields and Land Revitalization." Available at http://epa.gov/brownfields (accessed: March 30, 2010).

U.S. General Accounting Office. 1997. *Superfund: State Voluntary Programs Provide Incentives to Encourage Cleanups* (GAO/RCE-97-66). Washington, DC: United States General Accounting Office.

Wedding, G.K. and D. Crawford-Brown. 2007. "Measuring Site-Level Success in Brownfield Redevelopments: A Focus on Sustainability and Green Building." *Journal of Environmental Management* 85: 483–95.

Wernstedt, K., L. Heberle, A. Alberini and P. Meyer. 2004. *The Brownfields Phenomenon: Much Ado about Something or the Timing of the Shrew.* Washington, DC: Resources for the Future.

Yount, K.R. 2003. "What Are Brownfields? Finding a Conceptual Definition." *Environmental Practice* 5(1): 25–33.

Zavadskas, E.K. and J. Antucheviciene. 2006. "Development of an Indicator Model and Ranking of Sustainable Revitalization Alternatives of Derelict Property: A Lithuanian Case Study." *Sustainable Development* 14: 287–99.

Appendix 13.1   Michigan brownfield redevelopment

DNRE Project Case Studies

Project Name	Project Location		DNRE Program	DNRE Award Amount	Year Awarded	DNRE Funded Activities	Prior Property Usage	Redevelopment Project Type	Current Status
	County	Municipality							
Diamond REO	Ingham	Lansing	Site Reclamation Grant	$2,000,000	1992	Due Care Implementation, Other Environmental Response Action	Industrial	Commercial and light industrial	Substantially complete
Star Watch Case	Mason	Ludington	Site Reclamation Grant	$855,700	1993	Due Care Implementation, Other Environmental Response Action	Industrial	Residential and recreational	Substantially complete
Mulberry Place	Wayne	Wyandotte	Site Reclamation Grant	$182,000	1993	Due Care Implementation, Demolition, Interim Response Action	Industrial and mixed	Residential	Substantially complete
Lakeshore Blvd Ext.	Marquette	Marquette	Site Assessment Grant	$148,630	1994	Due Care Planning	Industrial	Recreational	Substantially complete
SSM Industrial Park	Chippewa	Sault St Marie	Site Assessment Grant	$180,000	1994	Due Care Planning	Industrial	Industrial	Substantially complete

Appendix 13.1 Continued

300 Marquette/ WCRR	Mackinac	Saint Ignace	Site Assessment Grant	$62,753	1994	Due Care Planning	Industrial	Recreational and commercial	Current
			Site Reclamation Grant	$27,183	1994				
Manistique Industrial Park	Schoolcraft	Manistique	Site Assessment Grant	$630,000	1994	Due Care Planning	Industrial	Industrial	Current
Baseball Stadium	Ingham	Lansing	Site Reclamation Grant	$922,813	1995	Due Care Implementation, Other Environmental Response Action	Industrial and commercial	Recreational and commercial	Substantially complete
Cheboygan Industrial Park	Cheboygan	Cheboygan	Site Assessment Grant	$23,500	1996	Due Care Planning	Industrial	Not redeveloped	No activity
Old Car Ferry Dock	Schoolcraft	Manistique	Site Assessment Grant	$70,000	1996	Due Care Planning	Industrial	Commercial, residential, and recreational	Substantially complete
MagneTek	Shiawassee	Owosso	Site Reclamation Grant	$1,558,000	1996	Due Care Implementation, Other Environmental Response Action	Industrial	Industrial	Substantially complete

Appendix 13.1 Continued

Monroe Steel Casting Plant	Monroe	Monroe	Site Reclamation Grant	$1,090,000	1996	Due Care Implementation, Other Environmental Response Action	Industrial	Residential	Substantially complete
Traverse City Ironworks/ River's Edge	Grand Traverse	Traverse City	Site Reclamation Grant	$1,582,975	1997	Site Assessment, Due Care Implementation, Other Environmental Response Action	Industrial	Residential	Substantially complete
H. Brown Company Superfund Site	Kent	Walker	Site Reclamation Grant	$289,300	1998	Due Care Implementation, Other Environmental Response Action	Industrial	Light industrial	Substantially complete
Alpena Riverfront Area	Alpena	Alpena	Site Assessment Grant	$48,000	1998	Due Care Planning	Mixed	Mixed and recreational	Substantially complete
St. Anne's Gate	Wayne	Detroit	Site Reclamation Grant	$808,000	1998	Due Care Implementation, Other Environmental Response Action	Mixed residential commercial	Residential	Substantially complete

Appendix 13.1 Continued

Former Marx Manufacturing Facility	Wayne	Taylor	Site Reclamation Grant	$200,000	1998	Due Care Implementation, Other Environmental Response Action	Industrial	Commercial	Substantially complete
Harbour View Center	Grand Traverse	Traverse City	Revitalization Revolving Loan Fund	$269,535	1998	Due Care Implementation, Demolition, Interim Response Action	Industrial	Mixed residential and commercial	Substantially complete
Atlantic Automotive (North of Main Industrial Area)	Berrien	Benton Harbor	Site Reclamation Grant	$2,000,000	1998	Due Care Implementation, Other Environmental Response Action	Industrial	Industrial	Current
American Brownhoist (also see Bay City Waterfront)	Bay	Bay City	Site Reclamation Grant	$1,416,000	1998	Preliminay investigation; funds returned to MDEQ	Industrial	Not redeveloped	Cancelled (grant funds returned)
Park Street Redevelopment	Grand Traverse	Traverse City	Site Reclamation Grant	$661,800	1999	Site investigation; remove USTs and contaminated soil	Industrial	Mixed residential, commercial, and recreational	Substantially complete

Appendix 13.1 Continued

							Military base		
Grosse Ile Airport and Commerce Park	Wayne	Grosse Ile Twp	Waterfront Redevelopment Grant	$749,000	2000	Infrastructure; due care implementation: other environmental response action		Commercial and airport	Substantially complete
			Site Reclamation Grant	$220,000	1999				
			Revitalization Revolving Loan Fund	$850,000	1999				
American Foundry Property	Washtenaw	Milan	Site Reclamation Grant	$384,410	1999	Site investigation, soil excavation and removal; concrete crushing	Industrial	Residential	Substantially complete
Bay City Waterfront	Bay	Bay City	Waterfront Redevelopment Grant	$3,063,000	1999	Property Acquisition, Infrastructure, Business Relocation	Industrial	Not redeveloped	No activity
Race Street Waterfront Redevelopment	Ingham	Lansing	Waterfront Redevelopment Grant	$667,500	1999	Public infrastructure and facility improvements	Industrial	Recreational	Substantially complete

Appendix 13.1   Continued

Project	County	City	Grant	Amount	Year	Activity	Current Use	Proposed Use	Status
South Rail Yard	Marquette	Marquette	Waterfront Redevelopment Grant	$1,100,000	1999	Property Acquisition	Industrial	Mixed residential, commercial, and recreational	Current
Mason Run	Monroe	Monroe City	Revitalization Revolving Loan Fund	$1,000,000	1999	Due Care Implementation, Other Environmental Response Action	Industrial	Residential and recreational	Current
			Site Reclamation Grant	$1,000,000	1999				
			Brownfield Redevelopment Loan	$800,000	2004				
Edgewater	Berrien	St. Joseph	Waterfront Redevelopment Grant	$776,675	1999	Infrastructure; due care implementation; other environmental response action	Industrial	Commercial and industrial	Substantially complete
			Brownfield Redevelopment Grant	$651,000	2001				
Ludington Waterfront	Mason	Ludington	Waterfront Redevelopment Grant	$180,450	2000	Infrastructure, Demolition	Mixed industrial and commercial	Mixed commercial, residential, and recreational	Substantially complete

Appendix 13.1 Continued

Muskegon Waterfront	Muskegon	Muskegon	Waterfront Redevelopment Grant	$2,311,418	2000	Infrastructure, Demolition, Other Environmental Reponse Action	Mixed industrial and commercial	Mixed recreational and commercial	Substantially complete
Karl Schmidt Unisia Property	Van Buren	South Haven	Brownfield Redevelopment Grant	$843,500	2000	Due Care Implementation, Other Environmental Response Action, Demolition	Industrial	Not redeveloped	No activity
Former Grand Rapids Die Cast	Kent	Walker	Revitalization Revolving Loan Fund	$295,000	2006	Due Care Implementation, Demolition, Interim Response Action	Industrial	Industrial	Substantially complete
Grand Haven Waterfront	Ottawa	Grand Haven	Waterfront Redevelopment Grant	$880,000	2000	Property Acquisition	Mixed industrial and commercial	Mixed recreational and commercial	Substantially complete

Appendix 13.1 Continued

Water Street	Washtenaw	Ypsilanti	Waterfront Redevelopment Grant	$3,728,000	2000	Property Acquisition, Assessment, Due Care Implementation	Mixed industrial, commercial, and residential	Not redeveloped	No activity
			Brownfield Redevelopment Loan	$500,000	2004				
			Waterfront Redevelopment Grant	$500,000	2004				
River Rouge Oxbow Restoration	Wayne	Dearborn	Waterfront Redevelopment Grant	$2,000,000	2000	Infrastructure, Restoration, Other Environmental Response Action	Waterway	Recreational waterway	Substantially complete
Former Grand Rapids Die Cast	Kent	Walker	Revitalization Revolving Loan Fund	$295,000	2000	Due Care Implementation, Other Environmental Response Action	Industrial	Industrial	Substantially complete
			Site Reclamation Grant	$405,000	2000				
Leo Joe Smith Trust Site-Walmart	Wayne	Taylor	Site Reclamation Grant	$342,000	2000	Due Care Implementation, Other Environmental Response Action	Commercial	Commercial	Substantially complete

Appendix 13.1 Continued

	Bay	Bay City	Brownfield Redevelopment Grant	$1,000,000	2001	Due Care Implementation, Other Environmental Response Action	Mixed commercial industrial	Commercial and recreational	Substantially complete
Bay City Conference Center			Site Reclamation Grant	$477,750	2001				
Alpena Third and River (Riverview Condos)	Alpena	Alpena	Site Reclamation Grant	$61,000	2001	Due Care Implementation	Industrial and commercial	Not redeveloped	No activity
Former Fletcher Paper	Alpena	Alpena	Site Assessment Grant	$264,500	2001	Due Care Planning	Industrial	Commercial and recreational	Current
Muskegon Lakeshore Smart Zone	Muskegon	Muskegon	Brownfield Redevelopment Grant	$1,000,000	2002	Due Care Implementation	Mixed	Mixed	Substantially complete
Grand Traverse Commons— Phases I and II	Grand Traverse	Traverse City	Brownfield Redevelopment Grant	$1,000,000	2003	Demolition, Due Care Implementation	State hospital	Residential, commercial, and recreational	Substantially complete
			Brownfield Redevelopment Grant	$1,000,000	2005				

Appendix 13.1 Continued

Borden Building	Isabella	Mt. Pleasant	Brownfield Redevelopment Grant	$184,690	2006	Demolition, Property Aquisition, Infrastructure	Industrial	Municipal offices	Substantially complete
			Waterfront Redevelopment Grant	$409,400	2004				
Waterfall Park	Bay	Bay City	Waterfront Redevelopment Grant	$150,000	2004	Site acquisition and cleanup	Public right of way	Recreational	Substantially complete
River Rasin Battlefield	Monroe	Monroe	Brownfield Redevelopment Grant	$1,000,000	2004	Demolition, Due Care Implementation	Industrial	Recreational	Current
Baker Oil	Calhoun	Springfield	Brownfield Redevelopment Grant	$700,000	2005	Other Environmental Response Action,	Industrial	Commercial	Substantially complete
			Brownfield Redevelopment Loan	$300,000	2005	Due Care Implementation, Demolition			
Will Branscumb Property	Berrien	Benton Harbor	Brownfield Redevelopment Grant	$130,000	2005	Due Care Implementation, Demolition	Industrial	Not redeveloped	No activity
Hastings Library	Barry	Hastings	Brownfield Redevelopment Grant	$213,000	2005	Due Care Implementation	Municipal	Public library	Substantially complete

Appendix 13.1 Continued

									Current
Grand Landing	Ottawa	Grand Haven	Brownfield Redevelopment Grant	$1,000,000	2006	Due Care Implementation	Industrial and commercial	Mixed residential and commercial	
			Brownfield Redevelopment Loan	$1,000,000	2006				
Elberta Waterfront Park	Benzie	Elberta	Site Reclamation Grant	$446,400	2003	Due Care Implementation, Other	Industrial	Recreational	Substantially complete
			Brownfield Redevelopment Grant	$735,387	2006	Environmental Response Action			
			Brownfield Redevelopment Loan	$250,000	2006				
R&B Electronics	Chippewa	Sault Saint Marie	Brownfield Redevelopment Grant	$155,000	2006	Phase I/Phase II Investigation, Other Environmental Response Action	Industrial	Industrial	Substantially complete
Habitat for Humanity	Ingham	Lansing	Brownfield Redevelopment Grant	$249,750	2006	Phase I/Phase II Investigation, Due Care Implementation	Industrial	Residential	Substantially complete

Appendix 13.1   Continued

Traverse City Place	Grand Traverse	Traverse City	Brownfield Redevelopment Grant	$1,000,000	2006	Due Care Implementation, Other Environmental Response Action	Industrial and commercial	Not redeveloped	No activity
			Brownfield Redevelopment Loan	$1,000,000	2006				
Harbor Shores	Berrien	Benton Harbor St. Joseph	Brownfield Redevelopment Grant	$1,000,000	2007	Demolition, Due Care Implementation	Industrial and commercial	Residential, commercial, and recreational	Current
			Brownfield Redevelopment Loan	$1,000,000	2007				
Tawas City Downtown	Iosco	Tawas City	Waterfront Redevelopment Grant	$60,000	2008	Property acquisition, infrastructure, due care implementation	Commercial	Recreational	Substantially complete

Index